MMPI-2
Assessing Personality
and Psychopathology

MMPI-2
Assessing Personality and Psychopathology
Second Edition

John R. Graham
Professor of Psychology
Kent State University

New York   Oxford
OXFORD UNIVERSITY PRESS
1993

Oxford University Press

Oxford   New York   Toronto
Delhi   Bombay   Calcutta   Madras   Karachi
Kuala Lumpur   Singapore   Hong Kong   Tokyo
Nairobi   Dar es Salaam   Cape Town
Melbourne   Auckland   Madrid

and associated companies in
Berlin   Ibadan

Copyright © 1990, 1993 by Oxford University Press, Inc.

Published by Oxford University Press, Inc.
198 Madison Avenue, New York, New York 10016-4314

Oxford is a registered trademark of Oxford University Press

Library of Congress Cataloging-in-Publication Data
Graham, John R. (John Robert), 1940–
MMPI-2 : assessing personality and psychopathology /
John R. Graham. — 2nd ed.
p. cm.   Includes bibliographical references and indexes.
ISBN 0-19-507922-1
1. Minnesota Multiphasic Personality Inventory. I. Title.
II. Title: MMPI-two.
[DNLM: 1. MMPI.   WM 145 G7393m 1993]
RC473.M5G73   1993
155.2′83—dc20
DNLM/DLC
for Library of Congress   93-9161

9 8 7 6

Printed in the United States of America
on acid-free paper

To My Best Friend, Mary Ann

Foreword

A great deal has happened with the MMPI over the past few years. The MMPI revision, begun more than a decade ago, culminated in an instrument for adults that eliminated the problems that had overtaken the original inventory as it approached its fiftieth year of use. Since its publication in 1989, the revised form of the MMPI, the MMPI-2, has been subjected to a considerable amount of empirical research and broad clinical application. Of course, changes in an instrument as widely used as the MMPI also became the subject of some initial controversy as well. Some psychologists had considerable professional loyalty to the original MMPI and found it difficult to shift into gear with the new instrument, instead hanging onto the out-of-date form. The early questions raised about the revised MMPI have now cleared and even the revision's staunchest critics have acknowledged its broadened acceptance (Webb, Levitt, & Rojdev, 1993). The broad acceptance of the MMPI-2 over the past three years is due, in no small part, to the great efforts and teachings of Jack Graham. The first edition of this text, published in 1990, gave psychologists an important resource in their transition to the MMPI-2. It was the first major textbook on the use of the MMPI-2 by a psychologist who had been instrumental in its development.

Graham played a very central role in the development of MMPI-2 and MMPI-A. He served as one of the original MMPI restandardization committee (along with James Butcher, Grant Dahlstrom, and Auke Tellegen) and contributed to the development of the revised forms in numerous ways. His ideas were central to the formulation of the original revision plan for the MMPI, he contributed items to both the adult and adolescent experimental booklets, he collected a substantial amount of the normal response data on which the new norms were developed, he organized and conducted several clinical validation studies, and he has been prominent in communicating the results of the restandardization study through extensive writings and professional education programs through the MMPI-2 Workshop series.

Graham's 1990 book was everything that assessment-oriented psychologists expected: It provided basic information about the original scales as they were revised for MMPI-2 and contained a clear, comprehensive introduction to the new scales that were designed to provide the clinician with an expanded assessment tool. Graham's unique approach to clinical assessment was carried over into his MMPI-2 text.

Since 1990, a considerable amount of new information has become available on the MMPI-2. A number of new scales have been developed, substantial new validation studies have become available, and interpretative procedures for the MMPI-2 have been clarified. In addition, a new form of the MMPI for adolescents, MMPI-A (Butcher, Williams, Graham, Archer, Tellegen, Ben-Porath, & Kaemmer, 1992), was published in 1992—for the first time in the more than a half century of MMPI applications. The MMPI-A provided an appropriate form of the instrument for young people between the ages of 14 and 18 and expanded the range of assessment information for younger subjects.

Graham's latest revision of his important textbook on the MMPI was devoted to updating and incorporating the latest information on MMPI-2 and in providing a general introduction to the MMPI-A. He has clearly maintained the interpretive tradition established by the previous versions of his textbook—solid description of empirical research, clear explanations, willingness to take important stands on controversial points, and presentation of knowledge-based, clinically oriented, and practical interpretative strategies for the MMPI-2.

Graham's earlier editions of his textbook have always been extremely accurate as well as comprehensive in scope and easy to read. The present book offers another excellent addition to this classic line. As with past editions, I welcome the occasion to use this important textbook and to recommend it to others whom I teach in my MMPI-2 training programs.

Department of Psychology   James N. Butcher
University of Minnesota

Preface to the Second Edition

When the first edition of *MMPI-2: Assessing Personality and Psychopathology* was published in 1990, the revised version of the MMPI—the MMPI-2—had just become available and was not yet widely used by clinicians. Except for information included in the MMPI-2 manual, no data had been published concerning the psychometric characteristics of the revised instrument. Some information included in the first edition was taken (by permission) from the MMPI-2 manual. Other information was known to this author because of his involvement in the development of the MMPI-2.

Now, three years after the publication of the MMPI-2, the instrument is being widely used and has been evaluated very positively by reviewers (Archer, 1992b; Nichols, 1992). Although ample evidence shows that the original instrument and the revised version have a great deal in common, there are also important differences between them. This book updates the first edition by considering all that we have learned about the MMPI-2 since it was published in 1989. In addition, readers are introduced to the adolescent version of the test, the MMPI-A. As was the case with the first edition, this book is appropriate for use as a textbook in personality assessment courses and as a reference guide for professionals who use the MMPI-2 and the MMPI-A in research and clinical work.

Chapter 1 describes the rationale underlying the original MMPI and its development. It also discusses the revision of the MMPI and summarizes similarities and differences between the original MMPI and the MMPI-2. Chapter 2 discusses MMPI-2 materials and procedures for administering, scoring, and coding the resulting profile of scores. Chapter 3 is devoted to the validity scales. The validity scales are considered singly and in combination, and strategies are discussed for detecting persons who approach the MMPI-2 in an invalid manner. Chapter 4 discusses each of the ten standard clinical scales and suggests interpretive inferences for high scores on each scale. The interpretation of two- and three-point code types and other profile configurations is covered in Chapter 5. Chapter 6 presents several different approaches to content interpretation. Content-homogeneous subscales, including those developed by Harris and Lingoes, are considered, and a set of content scales, developed specifically for the MMPI-2, is presented. The use of critical items also is discussed. Chapter 7 covers some important supplementary scales, including two alcohol abuse scales and a marital distress scale developed since the MMPI-2 was published. Chapter 8 examines psychometric characteristics of the MMPI-2, including reliability and validity. Data concerning the comparability of the MMPI and the MMPI-2 are presented and discussed.

Chapter 9 includes an expanded consideration of the use of the MMPI-2 with special groups (older adults, ethnic minorities, medical patients, correctional subjects, and subjects in nonclinical settings). Chapter 10 presents a detailed interpretive strategy developed by the author and illustrates the strategy with several cases. Chapter 11 focuses on computerized administration, scoring, and interpretation and considers major professional and ethical issues associated with computerized use of the MMPI-2. A sample computerized interpretation is compared with a clinician-generated interpretation of the same data. Chapter 12 discusses the development of the adolescent version of the test (MMPI-A), presents a strategy for interpreting MMPI-A results, and illustrates the strategy with a case.

A guiding principle throughout the preparation of this book was that material should be presented in a way that will be most directly useful to students learning about the MMPI-2 and to practitioners using the instrument clinically. Thus, no attempt was made to include exhaustive technical information about the MMPI or the MMPI-2. However, enough information is included to permit the reader to evaluate the appropriateness of the MMPI-2 for various kinds of subjects and assessment tasks. Other sources, such as the test manuals (Hathaway and McKinley, 1983; Butcher, Dahlstrom, Graham, Tellegen, & Kaemmer, 1989) and the two-volume *MMPI Handbook* (Dahlstrom, Welsh, & Dahlstrom, 1972, 1975) are available for those who require more information about the original MMPI or the MMPI-2.

Just as the MMPI came to be used widely in the United States and around the world, the MMPI-2 is likely to be even more popular as a personality assessment instrument. If used appropriately, the MMPI-2 and the MMPI-A are tools that can make assessment tasks more efficient and more fruitful. I hope that this book will contribute to more effective use of the MMPI-2 and the MMPI-A in assessing personality and psychopathology.

Kent, Ohio   J. R. G.
November 1992

Preface to the First Edition

In the five decades after work on the MMPI was initiated at the University of Minnesota by Starke Hathaway and J. Charnley McKinley, the instrument came to be widely used in the United States and around the world. It was administered routinely to clients in hospitals, clinics, and private practices and to nonclinical subjects in situations such as employment screening and marital counseling. The first and second editions of *The MMPI: A Practical Guide* (Graham, 1977, 1987) provided students and practicing clinicians with information needed to learn about the MMPI and to interpret the instrument in clinical practice.

Recently, the MMPI was updated and restandardized (Butcher, Dahlstrom, Graham, Tellegen, & Kaemmer, 1989). The revision involved a modernization of the content and language of test items, elimination of objectionable items, collection of nationally representative normative data, and development of some new scales. Although all of these changes have made the revised instrument, MMPI-2, a better tool for assessing personality and psychopathology, it has rendered existing reference works, such as *The MMPI: A Practical Guide*, inadequate. The present book is appropriate for use as a textbook in personality assessment courses and as a reference guide for professionals who use the MMPI-2 in research and clinical work.

Chapter 1 describes the rationale underlying the MMPI and presents historical information about scale development and standardization of the original instrument. The revision of the MMPI is discussed, and similarities and differences between the original MMPI and the MMPI-2 are summarized. Chapter 2 describes MMPI-2 test materials and procedures for administering and scoring the instrument and for coding the resulting profile of scores. Chapter 3 is devoted to consideration of the validity scales. The standard validity scales are considered singly and in combination, and several new validity scales are presented. Each of the ten standard clinical scales is discussed in Chapter 4, and suggestions are made for interpreting scores at various T-score levels. The interpretation of two- and three-point code types and other profile configurations is covered in Chapter 5. Chapter 6 presents several different approaches to content interpretation. The concept of critical items is introduced. Content-homogeneous subscales, including those developed by Harris and Lingoes, are considered, and a new set of content scales, developed especially for the revised item pool of the MMPI-2, is presented. Chapter 7 covers some frequently used supplementary scales and introduces several new scales for the MMPI-2. In Chapter 8 psychometric characteristics of the MMPI-2 (e.g., reliability and validity) are examined, and the instrument's use with special populations (adolescents, medical patients, ethnic groups, correctional subjects, and nonclinical subjects) is discussed. Chapter 9 presents the author's general strategy for interpreting the MMPI-2, and the strategy is illustrated with several cases. Finally, Chapter 10 focuses on computerized administration, scoring, and interpretation of the MMPI-2, and it considers major professional and ethical issues involved in computerized use of the instrument. A sample computerized interpretation is presented and compared with a clinician-generated interpretation of the same data.

A guiding principle throughout the preparation of this book was that material should be presented in a way that will be most directly useful in learning about the MMPI-2 and using the instrument clinically. Thus, no attempt was made to include exhaustive technical information and research data about the MMPI or the MMPI-2. However, enough information is included to permit the reader to evaluate the appropriateness of the MMPI-2 for various kinds of subjects and assessment tasks. Other sources, such as the test manuals (Hathaway & McKinley, 1983; Butcher, Dahlstrom, Graham, Tellegen, & Kaemmer, 1989) and the two-volume *MMPI Handbook* (Dahlstrom, Welsh, & Dahlstrom, 1972, 1975) are readily available for those who require more information about the original MMPI or the MMPI-2.

Just as the MMPI came to be used widely in the United States and around the world, I am convinced that the MMPI-2 will be even more popular as a personality assessment instrument. If used appropriately, the MMPI-2 is a tool that can make assessment tasks more efficient and more fruitful. It is hoped that this book will help clinicians better understand the use of the MMPI-2 in clinical assessment.

Kent, Ohio   J. R. G.
September 1989

Acknowledgments

I wish to thank those who supported the completion of this book. I am grateful to the authors and publishers who granted permission to reproduce their works. Beverly Kaemmer at the University of Minnesota Press was especially helpful in this regard. National Computer Systems generously provided the computerized interpretive report included in Chapter 11. Carolyn Williams supplied the adolescent case materials used in Chapter 12 and made helpful suggestions concerning their interpretation. Yossi Ben-Porath, Jim Butcher, and Carolyn Williams read sections of the manuscript and made helpful comments. Auke Tellegen provided helpful consultation concerning several of the prototypes for invalid responding discussed in Chapter 3. The Department of Psychology at Kent State University supplied clerical support. Jaime Carr, Melissa Franks, and Annette Rees contributed significantly to the manuscript through their clerical and editorial efforts. Mary Ann Stephens offered intellectual and emotional support during the entire time that the manuscript was being completed. Although I gratefully acknowledge all of the support noted above, I readily accept final responsibility for the finished product.

Contents

MMPI-2
Assessing Personality
and Psychopathology

1

Development of the MMPI and the MMPI-2

DEVELOPMENT OF THE MMPI

Original Purpose

The MMPI was first published in 1943. The test authors, Starke Hathaway, Ph.D., and J. Charnley McKinley, M.D., who were working in the University of Minnesota Hospitals, expected that the MMPI would be useful for routine diagnostic assessments. During the 1930s and 1940s a primary function of psychologists and psychiatrists was to assign appropriate psychodiagnostic labels to individual cases. An individual interview or mental status examination and individual psychological testing usually were used with each patient. It was possible that a group-administered paper-and-pencil personality inventory would provide a more efficient and reliable way of arriving at appropriate psychodiagnostic labels.

Rationale

Hathaway and McKinley used the empirical keying approach in the construction of the various MMPI scales. This approach, which requires one to determine empirically items that differentiate between groups of subjects, is a common technique today but represented a significant innovation at the time of the MMPI's construction. Most prior personality inventories had been constructed according to a logical keying approach. With this approach, test items were selected or generated rationally according to face validity, and responses were keyed according to the subjective judgment of the test author concerning what kinds of responses were likely to be indicative of the attributes being measured. Both clinical experience and research data seriously questioned the adequacy of this logical keying approach. Increasingly, it became apparent that subjects could falsify or distort their responses to items in order to present themselves in any way they chose. Further, empirical studies indicated that the subjectively keyed responses often were not consistent with differences actually observed between groups of subjects. In the newly introduced empirical keying procedure, responses to individual test items were treated as unknowns, and empirical item analysis was utilized to identify test items that differentiated between criterion groups. Such an approach to item responses overcame many of the difficulties associated with the earlier, subjective approaches.

Clinical Scale Development

The first step in the construction of the basic MMPI scales was to collect a large pool of potential inventory items.[1] Hathaway and McKinley selected a wide variety of personality-type statements from such sources as psychological and psychiatric case histories and reports, textbooks, and earlier published scales of personal and social attitudes. From an initial pool of about a thousand statements the test authors selected 504 that they judged to be reasonably independent of each other.

The next step was to select appropriate criterion groups. One criterion group, referred to as the Minnesota normals, consisted primarily of relatives and visitors of patients in the University of Minnesota Hospitals. This group was augmented by several other groups of normal subjects, including a group of recent high school graduates who were attending precollege conferences at the University of Minnesota, a group of Work Progress Administration workers, and some medical patients at the University of Minnesota Hospitals. The second major group of subjects, referred to as clinical subjects, was made up of psychiatric patients at the University of Minnesota Hospitals. This second group included patients representing all of the major psychiatric categories being used clinically at the time of the test construction. Clinical subjects were divided into subgroups of discrete diagnostic samples according to their clinically determined diagnostic labels. Whenever there was any doubt about a patient's clinical diagnosis or when more than one diagnosis was given, the patient was not included in this clinical reference group. The different subgroups of clinical subjects formed were hypochondriasis, depression, hysteria, psychopathic deviate, paranoia, psychasthenia, schizophrenia, and hypomania.

The next step in scale construction was to administer the original 504 test items to the Minnesota normals and to the patients in each of the clinical groups. An item analysis was conducted separately for each of the clinical groups in order to identify the items in the pool of 504 that differentiated significantly between the specific clinical group, other clinical groups, and a group of normal subjects. Individual MMPI items that were identified by this procedure were included in the resulting MMPI scale for that clinical group.

[1]Information concerning clinical and validity scale development is abstracted from a series of articles by Hathaway (1956, 1965); Hathaway and McKinley (1940, 1942); McKinley and Hathaway (1940, 1944); McKinley, Hathaway, and Meehl (1948); and Meehl and Hathaway (1946).

In an attempt to cross-validate the clinical scales, each scale (e.g., depression scale) was administered to new groups of normal subjects, clinical subjects with that particular clinical diagnosis, and subjects with other clinical diagnoses. If significant differences were found among scores for the normal group, the specific clinical group, and the group of other clinical subjects, the clinical scale was considered to have been adequately cross-validated and thus was ready for use in the differential diagnosis of new patients whose diagnostic features were unknown.

At a somewhat later time, two additional clinical scales were constructed. First, the Masculinity-Femininity (Mf) scale originally was intended to distinguish between homosexual men and heterosexual men. Because of difficulties in identifying adequate numbers of items that differentiated between these two groups, Hathaway and McKinley subsequently broadened their approach in the construction of the Mf scale. In addition to the all too few items that did discriminate between homosexual and heterosexual men, other items were identified that were differentially endorsed by normal male and female subjects. Also, a number of items from the Terman and Miles I scale (1936) were added to the original item pool and included in the Mf scale. Second, the Social Introversion (Si) scale was developed by Drake (1946) and came to be included as one of the basic MMPI scales. Drake selected items for the Si scale by contrasting item response frequencies for a group of college women who participated in many extracurricular activities and a group who participated in few or no extracurricular activities. Subsequently, the scale's use was extended to men as well as women.

Validity Scale Development

Hathaway and McKinley also developed four scales, hereafter referred to as the validity scales, whose purpose was to detect deviant test-taking attitudes. The Cannot Say scale, or category, represented by a question mark, is simply the total number of items in the MMPI either omitted or responded to as both true and false by the individual taking the test. Obviously, the omission of large numbers of items, which tends to lower the scores on the clinical scales, calls into question the interpretability of the whole resulting profile of scores.

The L scale, originally called the Lie scale of the MMPI, was designed to detect rather unsophisticated and naive attempts on the part of test subjects to present themselves in an overly favorable light. The L-scale items were rationally derived and cover everyday situations in order to assess the strength of a person's unwillingness to admit even very minor weaknesses in character or personality. An example of an L-scale item is "I do not read every editorial in the newspaper every day." Most people would be quite willing to admit that they do not read every editorial every day, but persons determined to present themselves in a favorable light might not be willing to admit to such a perceived shortcoming.

The F scale of the MMPI was designed to detect individuals whose approach to the test-taking task is different from that intended by the test authors. F-scale items were selected by examining the endorsement frequency of the Minnesota normal group and identifying the items endorsed in a particular direction by fewer than 10 percent of the normals. Obviously, because few normal people endorse an item in that direction, a person who does endorse the item in that direction is exhibiting a deviant response. A large number of such deviant responses calls into question the extent to which a test subject complied with the test instructions when completing the MMPI.

The K scale of the MMPI was constructed by Meehl and Hathaway (1946) to identify clinical defensiveness. It was noted that some clearly abnormal subjects who took the MMPI obtained scores on the clinical scales that were not as elevated as would be expected given their clinical status. Items in the K scale were selected empirically by comparing the responses of a group of patients who were known to be clinically deviant but who produced normal scores on the clinical scales of the MMPI with a group of people producing normal clinical scale scores and for whom there was no indication of psychopathology. A high K-scale score was intended to indicate defensiveness and call into question the person's responses to all of the other items.

The K scale also was used later to develop a correction factor for some of the clinical scales. Meehl and Hathaway reasoned that, if the effect of a defensive test-taking attitude, as reflected by a high K score, is to lower scores on the clinical scales, perhaps one might be able to determine the extent to which the scores on the clinical scales should be raised to reflect more accurately a person's behavior. By comparing the efficiency of each clinical scale with various portions of the K scale added as a correction factor, Meehl and Hathaway determined the appropriate weights of the K-scale score for each clinical scale to correct for the defensiveness indicated by the K-scale score. Certain clinical scales were not K-corrected at all because the simple raw score on those clinical scales seemed to result in the most accurate prediction about a person's clinical condition. Other scales have proportions of K—ranging from .2 to 1.0—added to elevate the clinical scales appropriately.

Modified Approach to MMPI Utilization

After a decade of clinical use and additional validity studies, it became apparent that the MMPI was not adequately successful for its original purpose, namely, the valid psychodiagnosis of a new patient. Although patients in any particular clinical category (e.g., depression) were likely to obtain high scores on the corresponding clinical scale, they also often obtained high scores on other clinical scales. Also, many normal subjects obtained high scores on one or more of the clinical scales. Clearly, the clinical scales were not pure measures of the symptom syndromes suggested by the scale names.

A number of different reasons have been suggested for the MMPI's failure to fulfill completely its original purpose. From further research it became apparent that many of the clinical scales of the MMPI are highly intercorrelated, making it unlikely that only a single scale would be elevated for an individual. These intercorrelations are due, to a large extent, to item overlap between scales. Also, the unreliability of the specific psychiatric diagnoses of subjects used in the development of the MMPI scales contributes to their failure to differentiate among clinical groups.

Although the limited success of the MMPI scales in differentiating among clinical groups might have been bothersome in the 1940s, this limitation is not particularly critical today. Currently, practicing clinicians place less emphasis on diagnostic labels per se. Accumulating evidence suggests that psychiatric nosology is not as useful as medical diagnosis. Information in a psychiatric chart that a patient's diagnosis is schizophrenia, for example, does not tell us much about the etiology of the disorder for that individual or about recommended therapeutic procedures.

For this reason, the MMPI came to be used in a way quite different from that in which it originally had been intended. It was assumed that the clinical scales were measuring something other than error variance because reliable differences in scores were found among individuals known to differ in other important ways. The modified approach to the MMPI treated each of its scales as an unknown and, through clinical experience and empirical research, the correlates of each scale were identified (indeed, more than ten thousand studies have been published about the MMPI). According to this approach, when a person obtained a score on a particular scale, the clinician attributed to that person the characteristics and behaviors that through previous research and experience had been identified for other individuals with similar scores on that scale. To lessen the likelihood that excess meaning would be attributed because of the clinical scale names, the following scale numbers were assigned to the original scales, and today they replace the clinical labels:

Present Scale Number	Original Scale Name
1	Hypochondriasis
2	Depression
3	Hysteria
4	Psychopathic Deviate
5	Masculinity-Femininity
6	Paranoia
7	Psychasthenia
8	Schizophrenia
9	Hypomania
0	Social Introversion

Thus, for example, when discussing a patient among themselves, MMPI users would refer to him or her as a "four-nine" or a "one-two-three," descriptive phrases in shorthand that communicate to the listener the particular behavior descriptions associated with the "4–9" or "1–2–3" syndrome.

In addition to identifying empirical correlates of high scores on each of the above numbered scales, it also is possible to identify empirical correlates for low scores and for various combinations of scores on the scales (e.g., highest scale in the profile, two highest scales in the profile). Some investigators have developed very complex rules for classifying individual profiles and have identified behavioral correlates of profiles that meet the criteria (Gilberstadt & Duker, 1965; Marks, Seeman, & Haller, 1974). Thus, even though the MMPI was not particularly successful in terms of its original purpose (differential diagnosis of clinical groups believed in the 1930s to be discrete psychiatric types), it has proved possible, subsequently, to use the test to generate descriptions of and inferences about individuals (normal subjects and patients) on the basis of their own profiles. It is this behavioral description approach to the utilization of the test in everyday practice that has led to its great popularity among practicing clinicians.

DEVELOPMENT OF THE MMPI-2

Reasons for the Revision

The original MMPI was a widely used instrument. Several national surveys revealed that it was the most frequently used personality test in the United States (Harrison, Kaufman, Hickman, & Kaufman, 1988; Lubin, Larsen, & Matarazzo, 1984). It was commonly applied in inpatient (Lubin, Larsen, Matarazzo, & Seever, 1985) and outpatient (Lubin et al., 1985; Piotrowski & Keller, 1989) psychiatric and medical (Piotrowski & Lubin, 1990) settings. Counseling psychologists and community counselors used the test extensively (Bubenzer, Zimpfer, & Mahrle, 1990; Watkins, Campbell, & McGregor, 1988). Even clinicians who might seem unlikely to value the MMPI, reported it to be an important assessment instrument. The use of the MMPI by members of the Society for Personality Assessment, a group typically associated with projective techniques, was second only to the Rorschach for personality assessment (Piotrowski, Shery, & Keller, 1985). Members of the American Association for Behavior Therapy indicated that it is important for professional clinicians to be skilled in the use of the MMPI (Piotrowski & Keller, 1984). Although the MMPI was developed for use with adults, it was the most widely used objective assessment measure for adolescent clients (Archer, Maruish, Imhof, & Piotrowski, 1991).

In spite of its widespread use, critics expressed concern about certain aspects of the instrument. Until the publication of the MMPI-2 in 1989, the

MMPI had not been revised since its publication in 1943. There were serious concerns about the adequacy of the original standardization sample. That sample consisted of 724 persons who were visiting friends or relatives at the University of Minnesota Hospitals. The sample was one of convenience, and little effort had been made to ensure that it was representative of the U.S. population. Standardization subjects came primarily from the geographic area around Minneapolis, Minnesota. All were white, and the typical person was about 35 years of age, married, residing in a small town or rural area, working in a skilled or semiskilled trade (or married to a man of this occupational level), and having about eight years of formal education (Dahlstrom, Welsh, & Dahlstrom, 1972). Hathaway and Briggs (1957) later refined this sample by eliminating persons with incomplete records or faulty background information. The refined sample was the one typically used for converting raw scores on supplementary MMPI scales to T scores. In addition to concerns that the original standardization sample was not representative of the general population, there were concerns that the average American citizen had changed since the normative data had been collected in the late 1930s.

There also were concerns about the item content of the original MMPI. Some of the language and references in the items had become archaic or obsolete. For example, not many contemporary subjects could respond meaningfully to the item involving "drop the handkerchief" because the game had not been popular among children for many years. Likewise, references to sleeping powders and streetcars were largely inappropriate for contemporary subjects.

Some of the items of the original MMPI included sexist language that was not in accord with contemporary standards concerning the use of such language in psychological tests. Certain items, such as those dealing with Christian religious beliefs, were judged inappropriate for many contemporary test subjects. Many test subjects found items dealing with sexual behavior and bowel and bladder functions to be irrelevant to personality assessment and therefore objectionable.

Because the original MMPI items had never been subjected to careful editorial review, a number of them included poor grammar and inappropriate punctuation. Some of the idioms were troublesome for test subjects with limited formal education.

Finally, there was concern that the original MMPI item pool was not broad enough to permit assessment of certain characteristics judged important by many test users. For example, few items concerned suicide attempts, use of drugs other than alcohol, and treatment-related behaviors. Although many supplementary scales were developed using the original MMPI item pool, the success of these scales often was limited by the inadequacy of the item pool.

MMPI researchers and users had considered the need for revision and restandardization for quite some time. In 1970 the entire MMPI Symposium was devoted to the topic of revision (Butcher, 1972). However, the enormity of the task and the unavailability of funds delayed revision plans for over a decade. In 1982 the University of Minnesota Press appointed a restandardization committee, consisting of James N. Butcher, W. Grant Dahlstrom, and John R. Graham, to consider the need for and feasibility of a revision of the MMPI. Based on the recommendations of the committee, a decision was made to revise the MMPI. Funds to support the revision were provided by the University of Minnesota Press. The test distributor, National Computer Systems, provided support in the form of test materials, forms, and scanning and scoring of data.

Goals of the Restandardization Project

From the start of the restandardization project, it was determined that every effort would be made to maintain continuity between the original MMPI and its revision. This would assure that the considerable research base that had accumulated since the test's publication would still be relevant to the new version.

A primary goal of the project was to collect a contemporary normative sample that would be more representative of the general population than had been true of Hathaway's original sample. Additionally, efforts would be made to improve the MMPI item pool by rewriting some of the items, deleting others that had been judged to be objectionable, and generating new items that would expand the content dimensions of the item pool.

Major revisions of the existing validity and clinical scales were not anticipated as part of the restandardization project, although it was hoped that the project would produce data that later could lead to improvements in the basic scales. Also, it was hoped that items that were added to the item pool would be useful in generating new scales.

Preparing the Experimental Booklet

Preparing the experimental booklet (Form AX) involved several simultaneous processes. To maintain continuity between the original and revised forms of the MMPI, a decision was made to include all 550 unique items in Form AX. The second occurrences of the 16 repeated items, originally included to facilitate early machine scoring, were deleted. They no longer served any useful purpose and also disturbed many test subjects who assumed incorrectly that they were included to determine if the subject was responding consistently.

Of the 550 items, 82 were rewritten for Form AX, and 15 were reworded to eliminate reference to a specific gender. For example, "Any man who is willing to work hard has a good chance of succeeding" was changed to "Anyone who is

---

[2]Although not involved in the early stages of the restandardization project, Auke Tellegen later was appointed to the restandardization committee.

willing to work hard has a good chance of succeeding." In other items, idiomatic or obsolete expressions were replaced with more contemporary wordings. For example, "irritable" was substituted for "cross," "bad behavior" for "cutting up," and "often" for "commonly." References that had become dated were replaced. For example, "sleeping powders" was changed to "sleeping pills," and "bath" was changed to "bath or shower." A number of changes were directed at eliminating subcultural bias. For example, "I go to church almost every week" was changed to "I attend religious services almost every week."

Most of the item changes were slight, and all of them were made with the objective of preserving original meaning while using more acceptable, contemporary language. Data were collected and analyzed to ensure that changes did not significantly affect endorsement patterns (Ben-Porath & Butcher, 1989a).

A second major change in the item pool involved adding new items. The committee reviewed the content dimensions in the original MMPI item pool, and it sought recommendations from experts in personality measurement and clinical assessment concerning content dimensions that should be added to the pool. The committee generated 154 items that were added to the item pool, bringing the Form AX booklet length to 704 items. Items were added in such content areas as drug abuse, suicide potential, Type A behavior patterns, marital adjustment, work attitudes, and treatment amenability.

Normative Data Collection

Procedures were developed with the goal of obtaining a large normative group that was broadly representative of the United States population (Butcher, Dahlstrom, Graham, Tellegen, & Kaemmer, 1989). Census data from 1980 were used to guide subject solicitation. To ensure geographic representativeness seven testing sites (Minnesota, Ohio, North Carolina, Washington, Pennsylvania, Virginia, and California) were selected. Potential subjects were selected in a particular region primarily from community or telephone directories. Letters were sent to prospective subjects explaining the nature of the project and asking them to participate. After an initial trial period, it was decided that individual subjects would be paid fifteen dollars for their participation, and couples who participated together would be paid forty dollars. Subjects were tested in groups in convenient locations in their communities. To ensure representativeness of the sample, subjects from special groups were added. These included military personnel and Native Americans. In addition to completing Form AX of the MMPI, all subjects completed a biographical-information form and a life-events form. Couples completed two additional forms describing the nature and length of their relationships and rating each other on 110 characteristics, using a revised form of the Katz Adjustment Scales (Katz & Lyerly, 1963).

Using these procedures, approximately twenty-nine hundred subjects were tested. After eliminating subjects because of test invalidity or incompleteness of other forms, a final sample of 2,600 community subjects (1138 men and 1462 women) was constituted. Of this number, 841 couples were included in the sample. To collect test-retest data, 111 female subjects and 82 male subjects were retested about a week after the initial testing.

Racial composition of the sample was as follows: Caucasian, 81 percent; African-American, 12 percent; Hispanic, 3 percent; Native-American, 3 percent; and Asian-American, 1 percent. Subjects ranged in age from 18 to 85 years (M = 41.04; SD = 15.29) and in formal education from 3 to 20+ years (M = 14.72; SD = 2.60). Most men (61.6%) and women (61.2%) in the sample were married. Approximately 32 percent of the men and 21 percent of the women had professional or managerial positions, and approximately 12 percent of the men and 5 percent of the women were laborers. The median family income was $30–35,000 for men and $25–30,000 for women. Approximately 3 percent of male subjects and 6 percent of female subjects in the normative sample indicated that they were involved in treatment for mental health problems at the time of their participation in the study.

Clearly, the normative sample for the MMPI-2 is more representative of the general population than was Hathaway's original sample. Although higher educational levels seem to be overrepresented, these reflect the types of persons who are likely to take the test. Also, Butcher (1990b) has shown that there is only a negligible relationship between educational level of the MMPI-2 normative subjects and scores on the MMPI-2 validity and clinical scales.

Adolescent Data Collection

Concurrently with the adult data collection, a large normative sample of adolescent subjects also was assembled. The adolescents were solicited from school rosters in most of the same cities where adult data were being collected. A separate experimental booklet (Form TX) was used, and the adolescent subjects also completed biographical-information and life-events forms. Form TX included the 550 unique items in the original MMPI (some of them in rewritten form) as well as a number of the items that had been added to Form AX. In addition, new items were added to Form TX to cover content dimensions relevant to adolescents but not included in the original MMPI item pool.

A separate form of the test, MMPI-A, has been published for use with adolescent subjects (Butcher et al., 1992). Chapter 12 of this book contains additional information concerning the development and use of the MMPI-A.

Additional Data Collection

To provide data necessary for making decisions, such as which items from the Form AX booklet would be included in the final revised booklet, data were

Table 1.1. Item Changes and Deletions for the Basic Validity and Clinical Scales

Scale	Number of Items			Types of Changes			
	Deleted	Remaining	Changed	A	B	C	D
L	0	15	2	1	1	0	0
F	4	60	12	1	5	6	0
K	0	30	1	0	1	0	0
Hs	1	32	5	0	1	3	1
D	3	57	2	1	1	0	0
Hy	0	60	9	0	4	2	3
Pd	0	50	4	0	2	1	1
Mf	4	56	6	1	2	1	2
Pa	0	40	2	1	0	0	1
Pt	0	48	2	0	0	1	1
Sc	0	78	13	0	1	7	5
Ma	0	46	7	4	2	1	0
Si	1	69	6	0	3	2	1
Not on any scale	—	—	16	3	7	3	3

Note: A = elimination of possibly sexist wording; B = modernization of idioms and usage; C = grammatical clarification; D = simplification.

Source: Butcher, J.N., Dahlstrom, W.G., Graham, J.R., Tellegen, A., & Kaemmer, B. (1989). *Minnesota Multiphasic Personality Inventory-2 (MMPI-2): Manual for administration and scoring.* Minneapolis: University of Minnesota Press. Copyright © 1989 by University of Minnesota Press. Reproduced by permission.

collected from a variety of additional subject groups. These included psychiatric patients, alcoholics, chronic pain patients, marital counseling clients, college students, and job applicants.

Development of the Final Booklet

The final version of the revised MMPI (MMPI-2) includes 567 items from the Form AX booklet. Several criteria were employed in deciding which items were to be included in the final booklet. First, all items entering into the standard validity and clinical scales were provisionally included, as were items needed to score supplementary scales judged to be important. Certain items were maintained because they would be included in new scales developed from the item pool.

From this provisional item pool, a number of items were deleted because they were judged on the basis of previous research (Butcher & Tellegen, 1966) to be objectionable. These items dealt with religious attitudes and practices, sexual preferences, and bowel and bladder functions. Table 1.1 indicates for each validity and clinical scale the number of items deleted, remaining, and changed. Not many items were changed in most scales, and even fewer were deleted from these basic scales.

In summary, the MMPI-2 is similar in most ways to the original MMPI. The MMPI-2 booklet includes the items necessary for scoring the standard validity and clinical scales. Although not all of the supplementary scales that could be scored from the original MMPI can be scored from MMPI-2, many of them can. Much of the research concerning interpretation of the original MMPI still applies directly to the MMPI-2. Improvements in the MMPI-2 include a more contemporary and representative standardization sample, updated and improved items, deletion of objectionable items, and some new scales. The following chapters will include further discussion of similarities and differences between the original MMPI and the MMPI-2.

2

Administration and Scoring

QUALIFICATIONS OF TEST USERS

The MMPI-2 is easily administered and scored by hand or by using computerized procedures. Although these procedures can be completed by a clerk or secretary, the MMPI-2 is a sophisticated psychological test. Its use is restricted to qualified professionals who have adequate training in test theory, personality structure and dynamics, and psychopathology and psychodiagnosis. Additionally, users of MMPI-2 should have detailed knowledge of the inventory itself. Familiarity with all material included in the MMPI-2 manual is essential. Additionally, users should be familiar with MMPI-2 interpretive procedures presented in books such as this one.

WHO CAN TAKE THE MMPI-2?

The MMPI-2 is intended for use with subjects who are 18 years of age or older. The MMPI-A (Butcher et al., 1992) should be used with subjects who are younger than 18. Because both the MMPI-A and MMPI-2 were normed on 18-year-old subjects, either test can be used with subjects of this age. The clinician should decide for each individual case whether to use the MMPI-A or MMPI-2 with 18-year-olds. Ordinarily, one would probably select the MMPI-A for 18-year-olds who are still in high school and the MMPI-2 for 18-year-olds who are in college, working, or otherwise living a more independent lifestyle. As long as visual disabilities or other physical problems do not interfere, there is no upper age limit to who can take the test.

   The clinical condition of potential examinees is an important consideration in deciding who can take the MMPI-2. Completion of the test is a lengthy and tedious task for many subjects. Persons who are very anxious or agitated often find the task almost unbearable. Frequently it is possible to break the testing session into several shorter periods for such individuals. Also, persons who are confused may not be able to understand or to follow the standard instructions. Such persons sometimes can complete the test if the items are presented by means of a standardized audio tape.

ADMINISTERING THE MMPI-2

For most persons the test can be administered either individually or in groups, using the forms of the test and answer sheets most convenient for the examiner. For persons of average or above-average intelligence, without complicating factors, the testing time typically is between 1 and 1½ hours. For less intelligent individuals, or those with other complicating factors, the testing time may exceed 2 hours. Although it might at times seem more convenient to have the test subject take the MMPI-2 home to complete, this procedure is unacceptable. The test should always be completed in a professional setting with adequate supervision. This increases the likelihood that the test will be taken seriously and that the results will be valid and useful.

Before the MMPI-2 is administered, the examiner should establish rapport with test subjects. The best way to ensure the subject's cooperation is to explain why the MMPI-2 is being administered, who will have access to the results, and why it is in the best interest of the test subject to cooperate with the testing.

The test should be administered in a quiet, comfortable place. The examiner or a proctor should be readily available to monitor the test taking and to answer questions that may arise. Care should be taken to make sure that the test subject carefully reads the test instructions and understands them. Questions that arise during the course of testing should be handled promptly and confidentially. Most questions can be handled by referring the test subject back to the standardized test instructions.

National Computer Systems (NCS) offers computer software that permits administration of the MMPI-2 by means of a personal computer. Test subjects follow directions on the computer monitor, where test items are displayed, and record their responses using several identified keys on the computer keyboard. Although the results of research studies concerning the equivalence of computerized and conventional administrations are not completely consistent, differences associated with type of administration typically have been small and probably of little practical consequence. This conclusion is consistent with early studies indicating that the MMPI is a very robust instrument and that administration using the booklet, card forms, or tape recordings of the original MMPI yielded comparable results (Cottle, 1950; MacDonald, 1952; Wiener, 1947). Thus, it probably can be assumed that computer administration will yield results comparable to use of the standard test booklet and answer sheets. An important consideration in deciding whether or not to use computer administration is the computer time required. A single administration of MMPI-2 can tie up a personal computer for several hours. The issue of equivalence is considered in more detail in Chapter 11 of this book.

TESTING MATERIALS

Unlike the original MMPI, the MMPI-2 exists in only one booklet form. An optional hardcover version of the booklet is available for use when subjects may not have a table or desk surface on which to take the test. However, the order of the items in both the softcover and hardcover versions is identical. Items are arranged so that those required for scoring the standard validity and clinical scales appear first in the booklets. If only the standard scales are required, test subjects need only complete the first 370 items. However, if less than the entire test is completed, many of the supplementary scales available for MMPI-2 cannot be scored. Usually, if test subjects can complete 370 items, they can complete 567.

For subjects who may have difficulty completing the standard form of the test, several alternatives are available. A standardized tape-recorded version of the items is available from NCS. This version is useful for semiliterate persons and for persons with disabilities that make completion of the standard form difficult or impossible. A standardized Spanish version of the test is also available from NCS.

Almost since the MMPI was first published, there have been efforts to develop short or abbreviated forms of the test. However, except when administering the first 370 items in the booklet when only the standard validity and clinical scales are needed, the use of shortened forms of MMPI-2 is not acceptable. Research with the original MMPI clearly indicates that short forms are not adequate substitutes for the standard instrument (Butcher, Kendall, & Hoffman, 1980; Dahlstrom, 1980).

Several different answer sheets are available for the MMPI-2. The one to be used depends on how the examiner plans to score the test. If the test is to be hand-scored, one kind of answer sheet should be used. If it is to be computer-scored, a different kind of answer sheet is indicated. Before administering the test, examiners should determine how scoring will be accomplished. Reference to the test manual or to a NCS test catalog will provide specific information about which answer sheets to use. It should be emphasized that none of the answer sheets designed for use with the original MMPI can be used with MMPI-2.

SCORING THE MMPI-2

Once the subject's responses to the MMPI-2 items have been recorded on an answer sheet, scoring can be accomplished by computer or by hand. As mentioned earlier, special answer sheets must be used if computer scoring is to be done. Computer scoring can be completed in several different ways. NCS distributes computer software that permits users to score the standard validity and clinical scales, as well as numerous supplementary scales, using their own personal computers. If the test has been computer-administered, test responses are stored in the computer's memory, and scoring programs can be applied directly to them. Another option, which is especially attractive for high-volume users, is to attach a scanner to the personal computer. Answer sheets are scanned quickly, and the responses are subjected to scoring programs. A final option for computer scoring is to forward subjects' responses to NCS in Minneapolis, where they are scored and returned to users. For those who do not require immediate availability of results, answer sheets can be mailed to NCS, where they are scored and returned by mail. From most locations in the United States the entire process can be accomplished in a week. For those who require faster turnaround time, a teleprocessing service is offered by NCS. A personal computer keyboard and modem are used to access the scoring services of NCS via telephone lines. The 567 items can be entered by a clerical person in approximately 10 minutes, and the resulting scores are returned immediately and printed out on the user's personal computer printer. All of the computerized scoring options offered by NCS are paid for on a per use basis. Persons interested in available computer services should contact NCS for details.

Many persons, particularly those who are not high-volume users, may prefer to hand-score the MMPI-2, using hand-scoring templates available from NCS. Scoring keys are available for the standard validity and clinical scales and for numerous supplementary scales. The scoring templates are quite easy to use. Each template is placed over an answer sheet that is designed especially for hand-scoring purposes. The number of blackened spaces is counted and represents the raw score for the scale in question. Care should be taken when scoring scale 5 (Masculinity-Femininity) to use the scoring key appropriate for the test subject's gender. For some scales only one scoring key is available, and for others two keys (front and back) are required. Obviously, when two keys are used, the raw score is the total number of blackened spaces on the front and back of the answer sheet. Raw scores for the standard validity and clinical scales are recorded in spaces provided on the answer sheet itself. For supplementary scales, raw scores are recorded in spaces provided on separate profile sheets. Although hand-scoring the MMPI-2 is a simple clerical task, it should be completed with great care, as counting and recording errors are rather common.

CONSTRUCTING THE PROFILE

Profile sheets are available from NCS for the standard validity scales and for numerous supplementary scales. The standard profile sheet to be used routinely for MMPI-2 is one that applies a K-correction to some of the raw scores of the clinical scales. NCS also offers profile sheets that permit construction of non-K-corrected profiles. It is recommended that K-corrected scores be used for most purposes. Almost all of the information available concerning MMPI and MMPI-2 interpretation is based on K-corrected scores, and it cannot be applied directly to noncorrected scores and profiles of scores. The test manual (Butcher et al., 1989) discusses circumstances when the noncorrected scores may be preferable. Generally speaking, K-corrected scores may lead to overestimates of deviance from nonclinical subjects, particularly those who are well educated. In such cases, the noncorrected scores may more accurately reflect the subjects' adjustment level compared with the normative sample. However, one should not use the noncorrected scores to generate statements concerning personality characteristics and symptoms of test subjects.

A first step in constructing a profile of the standard validity and clinical scales is to transfer the raw scores from the answer sheet to appropriate blanks at the bottom of the profile sheet, making sure that the profile is the appropriate one for the person's gender. At this time it also is important to be certain that identifying data (name, age, date, education, etc.) are recorded on the profile sheet.

At this point a K-correction is added to the raw scores for scales 1 (Hs), 4 (Pd), 7 (Pt), and 9 (Ma) (formerly, the Hypochondriasis, Psychopathic Deviate, Psychasthenia, and Hypomania scales, respectively). The proportion of a person's K-scale raw score that is to be added to each of these scales is indicated on the profile form. On these scales the total score (the original raw score plus the K-correction) is calculated and recorded in the appropriate blank on the profile sheet.

For each scale the examiner should refer to the numbers in the column above the scale label. The number in the column corresponding to the raw score (K-corrected if appropriate) on the scale is marked by the examiner either with a small x or a small dot. Raw scores on the Cannot Say (?) scale are recorded on the profile sheet but are not plotted as part of the profile. Care should be taken when plotting the scale 5 (Mf) scores. For men higher raw scores yield higher T scores, whereas for women higher raw scores yield lower T scores. After a dot or x has been entered in the column above each scale label, the MMPI-2 profile for the person examined is completed by connecting the plotted dots or x's with each other. Traditionally, the three validity scales are joined to each other but are not connected with the ten clinical scale scores. Similar procedures are used to construct profiles for supplementary scales.

Because T scores are printed at each side of the profile sheet, by plotting the scores in the manner described here, the raw scores for each scale can be converted visually to T scores. A T score has a mean of 50 and a standard deviation of 10. For scales L, F, K, 5, and 0 the T scores used are linear, whereas for scales 1, 2, 3, 4, 6, 7, 8, and 9 special uniform T scores are used. More information concerning characteristics of uniform T scores and differences between linear and uniform scores can be found in Chapter 8. The T-score conversions provided on the profile sheet are based on the responses of the MMPI-2 normative standardization sample. Thus, a T score of 50 for any particular scale indicates that a person's score is equal to the average or mean score for the normative subjects of the test subject's same gender. Scores greater than 50 indicate scores higher than the average for the normative sample, and scores below 50 indicate scores lower than the average for the normative sample.

CODING THE PROFILE

Although it is possible to derive useful information by interpreting an examinee's T score on a single scale in isolation, much of the information relevant to interpretation of MMPI-2 protocols is *configural* in nature. Thus, in addition to interpreting individual scores, it is necessary to consider the pattern of the scores in relation to each other. To facilitate profile interpretation, coding is a procedure for recording most of the essential information about a profile in a concise form and for reducing the possible number of different profiles to a manageable size. Coding conveys information about the scores on scales relative to one another and also indicates an absolute range within which scores fall. It also permits easy grouping of similar profiles, using all or only part of the code.

Two major coding systems were utilized with the original MMPI: Hathaway's (1947) original system and a more complete system developed by Welsh (1948). Welsh's system is the only one used in recent years, and a slight modification of it is the one recommended for use (Butcher & Williams, 1992).

Modified Welsh Code

Step 1

Use the number instead of the name for each scale.

Hs — 1	Pa — 6
D — 2	Pt — 7
Hy — 3	Sc — 8
Pd — 4	Ma — 9
Mf — 5	Si — 0

Step 2

Record the ten numbers of the clinical scales in order of T scores, from the highest on the left to the lowest on the right.

Step 3

To the right of and separated from the clinical scales, record the three validity scales (L,F,K) in the order of the T scores, with the highest on the left and

Table 2.1. Example of Modified Welsh Code

Scale Name	Scale No.	Raw Score	T score
Cannot Say	—	5	—
L	—	6	61
F	—	12	73
K	—	11	41
Hypochondriasis	1	19	66
Depression	2	39	91
Hysteria	3	23	54
Psychopathic Deviate	4	18	40
Masculinity-Femininity	5	33	64
Paranoia	6	5	34
Psychasthenia	7	43	85
Schizophrenia	8	46	84
Hypomania	9	10	31
Social Introversion	0	10	31

Modified Welsh Code: 2*78″ ' 1+5–3/4:690   F'+L–/K

the lowest on the right. Do not include the ? scale. The set of clinical scales and the set of validity scales are coded separately. No supplementary scales are included in the coding.

Step 4

When adjacent scales are within one T-score point, they are underlined. When adjacent scales have the same T score, they are placed in the ordinal sequence found on the profile sheet and underline.

Step 5

To indicate scale elevations, appropriate symbols are inserted after scale numbers as follows:

120 and above	″
110–19	'
100–109	**
90–99	*
80–89	″
70–79	'
65–69	+
60–64	-
50–59	/
40–49	:
30–39	#
29 and less	to the right of #

If a ten-point T-score range does not contain any scale, the appropriate symbol for that elevation must be included. It is not necessary to include a symbol to the left of the scale with the highest score or to the right of the scale with the lowest score.

Step 6

Repeat steps 4 and 5 for the validity scales.

As a practice exercise, the reader might wish to cover the code at the bottom of Table 2.1 and code the T scores into the Welsh code, using the instructions given above.

3

The Validity Scales

For the MMPI-2 to yield maximally accurate and useful information, it is necessary that the test subject approach the test-taking task in the manner indicated in the instructions. Subjects are instructed to read each item, consider its content, and give a direct and, as far as possible, honest response to the item, using the true-false response format provided. To the extent that extreme deviations from these procedures occur, the resulting protocol should be considered invalid and should not be interpreted further. Less extreme deviations in test-taking attitude should be taken into account when the resulting scores are interpreted.

Although Hathaway and McKinley hoped that the empirical keying procedure used in developing the MMPI would make such distortions less likely than in earlier face-valid inventories, they recognized the importance of assessing test-taking attitudes. The four validity indicators developed specifically to assess test-taking attitude with the original MMPI have been maintained with the MMPI-2. In addition, three new validity indicators were developed specifically for MMPI-2.

In addition to providing important information about test-taking attitudes, the original validity scales of the MMPI also can be used as sources of inference about extratest behaviors. Both aspects of the four original validity indicators will be considered in this chapter. The items included in each of the original validity scales and the keyed response for each item are presented in Appendix A of this book.[1] T-score transformations for raw scores on the L, F, and K scales can be found in Appendix B. Appendix J reports information about items included in the three new validity scales and keyed responses for the items. T-score transformations for raw scores on the three new validity scales can be found in Appendix K.

---

[1]Item numbers in Appendix A and elsewhere in this book correspond to those in the MMPI-2 booklet. Appendix J of the MMPI-2 manual (Butcher et al., 1989) includes a table for converting the MMPI-2 item numbers to group-form item numbers of the original MMPI.

CANNOT SAY (?) SCORE

The Cannot Say (?) score is simply the number of omitted items (including items answered both true and false). There are many reasons why people omit items on the MMPI-2. Occasionally, items are omitted because of carelessness or confusion. Omitted items also can reflect an attempt to avoid admitting undesirable things about oneself without directly lying. Indecisive people, who cannot decide between the two response alternatives, may leave many items unanswered. Some items are omitted because of a lack of information or experience necessary for a meaningful response.

Regardless of the reasons for omitting items, a large number of such items can lead to lowered scores on other scales. Therefore, the validity of a protocol with many omitted items should be questioned. The MMPI-2 manual suggests that protocols with thirty or more omitted items must be considered highly suspect, if not completely invalid. This criterion seems to be too liberal. This author's own practice is to interpret with great caution protocols with more than ten omitted items and not to interpret at all those with more than thirty omitted items. As indicated in Chapter 2, however, the best procedure is to ensure that few or no items are omitted. If encouraged before beginning the MMPI-2 to answer all items, most people usually will omit only a very few. Also, if the examiner scans answer sheets at the time the test is completed and encourages individuals to try to answer previously omitted items, most people will complete all or almost all of the items.

L SCALE

As indicated in Chapter 1, the L scale originally was constructed to detect a deliberate and rather unsophisticated attempt on the part of subjects to present themselves in a favorable light (Meehl & Hathaway, 1946). All of the fifteen rationally derived items in the original L scale have been maintained in MMPI-2. The items deal with minor flaws and weaknesses to which most people are willing to admit (e.g., "I do not read every editorial in the newspaper every day" ; "I do not like everyone I know.") However, individuals who deliberately are trying to present themselves in a very favorable way are not willing to admit even such minor shortcomings. Such people produce high L-scale scores.

Although most L-scale items are not answered in the scored direction (false) by most people, many normal individuals endorse several of the items in the scored direction. The average number of L items endorsed in the scored direction by subjects in the MMPI-2 normative sample was approximately three. Although it was reported that better-educated, brighter, and more sophisticated people from higher social classes scored lower on the L scale of the original MMPI (Graham, 1987), preliminary analyses suggest that relationships between T scores on the MMPI-2 L scale and these demo-graphic characteristics are trivial and need not be considered when interpret-ing L-scale scores (Butcher, 1990b).

High Scores on the L Scale

When a test subject has a high (T > 65) score on the L scale, one should entertain the possibility that the person is not being honest and frank in answering items on the inventory. The result of such a test-taking attitude is that the individual's scores on most or all of the clinical scales have been low-ered artificially in the direction of appearing better adjusted psychologically. T scores on the L scale between 55 and 65 are suggestive of defensiveness, but the resulting protocol can be interpreted if the defensiveness is taken into account. (See the later section in this chapter concerning interpretation of defensive profiles). However, T scores above 65 suggest such extreme denial and/or defensiveness that the protocol should not be interpreted.

Because all of the L-scale items are keyed in the false direction, a person who answers false to all or most of the MMPI-2 items will produce a very ele-vated score on the L scale. The True Response Inconsistency (TRIN) scale, which is described later in this chapter, can help to determine if a false response bias accounts for an elevated L-scale score. TRIN-scale T scores greater than 80 (in the false direction) indicate a false response bias that invalidates the protocol.

In addition to suggesting a defensive test-taking attitude, high L-scale scores (T = 55–65) tend to be associated with other important characteristics and behaviors. High scorers on the L scale tend to be overly conventional and socially conforming. They are rigid and moralistic, and they overevaluate their own worth. They utilize repression and denial excessively, and they appear to have little or no insight into their own motivations. Also, they have little awareness of the consequences to other people of their behavior. In cer-tain rare cases, particularly when the F and K scales are also elevated, an extremely high L-scale score may be suggestive of a full-blown clinical confu-sion that may be either organic or functional in nature.

Low Scores on the L Scale

Below-average scores (T < 50) on the L scale usually indicate that the person responded frankly to the items and was self-confident enough to be able to admit to minor faults and shortcomings. Low scorers have been described as perceptive, socially responsive, self-reliant, and independent. They also appear to be strong, natural, and relaxed, and they function effectively in leadership roles. They are able to communicate their ideas effectively, although at times they may impress others as somewhat cynical and sarcastic.

Sometimes below-average L-scale scores suggest a deviant test-taking attitude in which subjects are being overly self-critical and may be exaggerating problems and negative characteristics. This interpretation of low L-scale scores is most appropriate when the K-scale score also is quite low and the F-scale score is very high. Additional information concerning exaggeration is provided later in this chapter.

An all-true response set also produces very low T scores on the L scale. The TRIN scale, which is described later in this chapter, can be helpful in determining if a low L-scale score is due to a true response bias. T scores greater than 80 on the TRIN scale (in the true direction) indicate a true response bias that invalidates the protocol.

Summary of Descriptors for the L Scale[2]

High L-scale scores (T > 55) are indicative of persons who:

1. are trying to create a favorable impression of themselves by not being honest in responding to the items
2. may be defensive, denying, and repressing
3. may be confused
4. manifest little or no insight into their own motivations
5. show little awareness of consequences to other people of their behavior
6. overevaluate their own worth
7. tend to be conventional and socially conforming
8. are unoriginal in thinking and inflexible in problem solving
9. are rigid and moralistic
10. have poor tolerance for stress and pressure

Low L-scale (T < 50) scores are indicative of persons who:

1. probably responded frankly to the items
2. are confident enough about themselves to be able to admit to minor faults and shortcomings
3. in some cases may be exaggerating negative characteristics
4. are perceptive and socially reliant
5. are seen as strong, natural, and relaxed
6. are self-reliant and independent
7. function effectively in leadership roles
8. communicate ideas effectively
9. may be described by others as cynical and sarcastic

---

[2]The reader should recognize that the descriptors listed in this and subsequent summaries are modal ones and that all descriptors will not apply necessarily to all individuals with a given score or configuration of scores. The descriptors should be viewed as hypotheses to be validated by reference to other test and nontest data.

F SCALE

The F scale originally was developed to detect deviant or atypical ways of responding to test items (Meehl & Hathaway, 1946). The sixty-four items in the original F scale were those answered in the scored direction by fewer than 10 percent of adult normal subjects. Several of the F-scale items were deleted from the MMPI-2 because of objectionable content, leaving the F scale with sixty items in the revised instrument.

A factor analysis of the original F scale (Comrey, 1958a) identified nineteen content dimensions, assessing such diverse characteristics as paranoid thinking, antisocial attitudes or behavior, hostility, and poor physical health. A person can obtain a high F-scale score by endorsing items in some, but not necessarily all, of these nineteen content areas. In general, and because the scales of the MMPI-2 are intercorrelated, high scores on the F scale usually are associated with elevated scores on the clinical scales, especially on scales 6 and 8. In the MMPI-2 normative sample, scores on the F scale are related to age and ethnicity, with adolescents, African Americans, Native Americans, and Hispanics scoring approximately five T-score points higher on the F scale than other groups.

As used by the practicing clinician, the F scale serves three important functions. First, it is an index of test-taking attitude and is useful in detecting deviant response sets. Second, if one can rule out profile invalidity, the F scale is a reliable indicator of degree of psychopathology, with higher scores suggesting greater psychopathology. Finally, scores on the F scale can be used to generate inferences about other extratest characteristics and behaviors.

High Scores on the F Scale

When T scores on the F scale are greater than 100, the possibility of an invalidating response set should be considered. Persons who obtain scores at this level may have responded in a random way to the MMPI-2 items. The Variable Response Inconsistency (VRIN) scale, which is described later in this chapter, is helpful in detecting random responding. If the F-scale T score is greater than 100 *and* the VRIN T score is greater than 80, random responding is likely. Additional information concerning random responding is presented later in this chapter.

Persons who obtain T scores greater than 100 on the F scale may have used a true response bias in responding to the items. If this is the case, one would expect T scores on the True Response Inconsistency scale (TRIN), which is described later in this chapter, to be greater than 80 (in the true direction). One must also consider the possibility that F-scale scores in this range have resulted from test subjects' attempts to "fake bad" when taking the MMPI-2. Procedures for detecting this response set also are discussed later in this chapter.

Among hospitalized psychiatric patients, F-scale T scores greater than 100 are suggestive of serious psychopathology. Persons with such scores may be manifesting delusions and visual and/or auditory hallucinations. They may be withdrawn, have reduced speech, and have short attention spans. They typically show poor judgment and may lack knowledge of reasons for hospitalization. Psychotic diagnoses often are assigned to such persons. However, there may also be extratest signs of organicity.

T scores between 80 and 99 on the F scale suggest the possibility of malingering or exaggeration of symptoms as a plea for help. Persons with scores at this level may have been quite resistant to the testing procedure. However, scores at this level may also be indicative of clearly psychotic symptoms and behaviors.

T scores between 65 and 79 on the F scale sometimes are associated with very deviant social, political, or religious convictions. However, persons with scores at this level may manifest clinically severe neurotic or psychotic disorders. When persons who are relatively free of serious psychopathology obtain F-scale scores at this level, they often are described as moody, restless, dissatisfied, changeable, unstable, curious, complex, opinionated, and opportunistic.

Mild elevations (T = 50–65) may indicate endorsement of items in a particular problem area such as work, health, or family relationships. Persons with scores at this level typically function adequately in most aspects of their life situations.

Low Scores on the F Scale

T scores below 50 on the F scale indicate that items were answered as most normal persons answer them. Persons with scores at this level are likely to be socially conforming and free of disabling psychopathology. One should also consider the possibility that persons who score at this level may have "faked good" in responding to the MMPI-2 items. More information about this response set is presented later in this chapter.

Summary of Descriptors for the F Scale

T scores equal to or greater than 100 are indicative of persons who:

1. may have responded randomly to MMPI-2 items
2. may have responded true to all of the MMPI-2 items or false to all of the MMPI-2 items
3. may have been "faking bad" when taking the MMPI-2
4. if hospitalized psychiatric patients, may manifest:

    a. delusions
    b. visual and/or auditory hallucinations
    c. reduced speech
    d. withdrawal
    e. poor judgment
    f. short attention span
    g. lack of knowledge of reasons for hospitalization
    h. psychotic diagnosis
    i. some extratest signs of organicity

T scores in a range of 80–99 are indicative of persons who:

1. may be malingering
2. may be exaggerating symptoms and problems as a plea for help
3. may be quite resistant to the testing procedure
4. may be clearly psychotic

T scores in a range of 65 to 79 are indicative of persons who:

1. may have very deviant social, political, or religious convictions
2. may manifest clinically severe neurotic or psychotic disorders
3. if relatively free of psychopathology, are described as:
    a. moody
    b. restless
    c. dissatisfied
    d. changeable, unstable
    e. curious and complex
    f. opinionated
    g. opportunistic

T scores in a range of 50 to 65 are indicative of persons who:

1. have endorsed items relevant to a particular problem area
2. typically function adequately in most aspects of their life situations

Low Scores on the F Scale

T scores that are below 50 are indicative of persons who:

1. answered items as most normal people do
2. are likely to be free of disabling psychopathology
3. are socially conforming
4. may have "faked good" in responding to the MMPI-2 items

K SCALE

When early experience with the MMPI indicated that the L scale was quite insensitive to several kinds of test distortion, the K scale was developed as a more subtle and more effective index of attempts by examinees to deny psychopathology and to present themselves in a favorable light or, conversely, to exaggerate psychopathology and to try to appear in a very unfavorable light (Meehl & Hathaway, 1946; McKinley et al., 1948). High scores on the K scale thus were thought to be associated with a defensive approach to the test, whereas low scores were thought to be indicative of unusual frankness and self-critical attitudes. In addition to identifying these deviations in test-taking attitudes, a statistical procedure also was developed for correcting scores on some of the clinical scales. (See discussion of the K-correction in Chapter 1.)

The original K scale included thirty items that were empirically identified by contrasting item responses of abnormal individuals who produced normal profiles with item responses of a group of normal subjects. The MMPI-2 version of the K scale includes all thirty of these original items. The items in the K scale cover several different content areas in which a person can deny problems (e.g., hostility, suspiciousness, family dissention, lack of self-confidence, excessive worry). The K-scale items tend to be much more subtle than items in the L scale; therefore, it is less likely that a defensive person will recognize the purpose of the items and will be able to avoid detection.

Subsequent research and experience have indicated that the K scale is much more complex than was originally believed. Scores on the K scale of the original MMPI were related to educational level (Graham, 1987). Better-educated persons tended to score higher on the scale. Thus, it was recommended that educational levels of test subjects be taken into account in interpreting scores on the K scale. Data available at this time suggest that the relationship between educational level and K-scale scores of the MMPI-2 is minimal and that the same interpretations can be made of K-scale scores regardless of persons' educational levels (Butcher, 1990b; Butcher, Graham, Dahlstrom, & Bowman, 1990).

Although above-average scores on the K scale typically indicate defensiveness, moderate elevations sometimes reflect ego strength and psychological resources. There is no definite way to determine when elevated K-scale scores indicate clinical defensiveness and when they indicate more positive characteristics. However, if elevated K-scale scores are found for persons who do not seem to be disturbed psychologically and who appear to be functioning reasonably well, the possibility that the K-scale score is reflecting positive characteristics rather than defensiveness should be considered.

Because most of the data concerning interpretation of MMPI scores were based on K-corrected scores, the K correction was maintained with the MMPI-2. However, there has not been much research to support the routine use of the K-correction. The results of research studies concerning the K-correction have been mixed at best (Dahlstrom et al., 1972). Several studies have indicated that the K-correction did not lead to more accurate predictions for test subjects (Clopton, Shanks, & Preng, 1987; McCrae et al., 1989; Silver & Sines, 1962; Wooten, 1984). The use of the K-correction was especially problematic in normal samples (Hsu, 1986). To date there have been no published studies of the efficacy of the K-correction for the MMPI-2.

It is this author's recommendation that K-corrected scores be used routinely for the MMPI-2. This procedure permits test interpreters to draw on the MMPI data that were based on K-corrected scores. However, there are some circumstances when noncorrected scores can be useful. First, as was the practice when the original MMPI was utilized with adolescents, the current adolescent version of the test (MMPI-A) does not use K-corrected scores. Also, when dealing with test subjects who can be assumed to be psychiatrically normal (e.g., job applicants), noncorrected scores may give a better indication of subjects' standing on scales in relation to the normative sample. National Computer Systems distributes profile sheets on which both corrected and noncorrected scores can be plotted. However, caution should be taken not to use the noncorrected scores to generate inferences using MMPI (or MMPI-2) data that utilized K-corrected scores. In other words, in some circumstances noncorrected scores may give a better indication of subjects' overall adjustment level, but the noncorrected scores should not be used to generate more detailed and specific inferences about test subjects. Clearly, additional research is needed to determine the extent to which corrected and noncorrected scores of the MMPI-2 lead to accurate predictions about test subjects.

High Scores on the K Scale

T scores above 55 on the K scale indicate that test subjects may have approached the test-taking task more defensively than the average person. The higher the score, the more likely it is that the subject was being clinically defensive. T scores between 56 and 65 indicate defensiveness that needs to be taken into account in interpreting other scores on the test. More information concerning interpretation of defensive profiles is presented later in the chapter.

T scores greater than 65 on the K scale strongly suggest a response set that invalidates the profile. The score that best identifies a fake-good response set probably should be determined in each setting where the MMPI-2 is used. In a study by Graham, Watts, and Timbrook (1991), T-score cutoffs of 50 or greater for men and 55 or greater for women yielded the best discrimination between valid and fake-good protocols.

High scores on the K scale also can be indicative of a false response set. The TRIN scale can be helpful in determining if an elevated K scale score is due to such a response set. A high K-scale score (T > 65) *and* a T score on the

TRIN scale greater than 80 (in the false direction) strongly suggest that false responses were given to items without regard for their content.

In addition to providing information about test-taking attitudes, scores on the K scale are associated with other characteristics. High K-scale scorers may be trying to maintain an appearance of adequacy, control, and effectiveness. High scorers tend to be shy and inhibited, and they are hesitant about becoming emotionally involved with other people. In addition, they are intolerant and unaccepting of unconventional beliefs and behavior in other people. They lack self-insight and self-understanding. Delinquency is unlikely among people with high scores on the K scale. When high K-scale scores are accompanied by marked elevations on the clinical scales, it is likely that the person is quite seriously disturbed psychologically but has little or no awareness of such problems. When moderately high scores are found for persons who do not seem to be disturbed psychologically and who appear to be functioning reasonably well, they may be reflecting ego strength and other positive characteristics.

Average Scores on the K Scale

T scores in an average range (40–55) on the K scale suggest a healthy balance between positive self-evaluation and self-criticism. Such people tend to be well adjusted psychologically and to manifest few signs of emotional disturbance. They are independent, self-reliant, and capable of dealing with problems in their daily lives. They tend to have high intellectual abilities and wide interests, and to be ingenious, enterprising, versatile, and resourceful. They think clearly and approach problems in a reasonable and systematic way. In social situations, they mix well with other people, are enthusiastic and verbally fluent, and tend to take an ascendant role.

Low Scores on the K Scale

Low T scores on the K scale (< 40) may be indicative of a true response set or of a deliberate attempt to present oneself in an unfavorable light. (See the discussion of profile invalidity below for details about these two response sets). The TRIN scale can be helpful in determining if a low K-scale score indicates a true response set. A low K-scale score (T < 40) *and* a T score on the TRIN scale greater than 80 (in the true direction) strongly indicate the likelihood of a true response set.

Low scores also may indicate that subjects are exaggerating problems as a plea for help or that they are experiencing confusion that may be either organic or functional in nature. Low scorers tend to be very critical of themselves and of others and to be quite self-dissatisfied. They may be ineffective in dealing with problems in their daily lives, and they tend to have little insight into their own motives and behavior. They are socially conforming and tend to be overly compliant with authority. They are inhibited, retiring, and shallow, and they have a slow personal tempo. They tend to be rather awkward socially and to be blunt and harsh in social interactions. Their outlook toward life is characterized as cynical, skeptical, caustic, and disbelieving, and they tend to be quite suspicious about the motivations of other people.

Summary of Descriptors on the K Scale

Very high scores (T > 65) on the K scale are indicative of persons who:

1. may have responded false to most of the MMPI-2 items
2. may have tried to "fake good" in responding to the MMPI-2 items

Moderately high scores (T = 56–65) on the K scale are indicative of persons who:

1. may have approached the test-taking task in a defensive manner
2. may be trying to give an appearance of adequacy, control, and effectiveness
3. are shy and inhibited
4. are hesitant about becoming emotionally involved with people
5. are intolerant, unaccepting of unconventional attitudes and beliefs in other people
6. lack self-insight and self-understanding
7. are not likely to display overt delinquent behavior
8. if clinical scales also are elevated, may be seriously disturbed psychologically but have little awareness of this
9. if not seriously disturbed psychologically, may have above-average ego strength and other positive characteristics

Average scores on the K scale (T = 40–55) are indicative of persons who:

1. maintained a healthy balance between positive self-evaluation and self-criticism in responding to the MMPI-2 items
2. are psychologically well adjusted
3. show few overt signs of emotional disturbance
4. are independent and self-reliant
5. are capable of dealing with problems in daily life
6. exhibit wide interests
7. are ingenious, enterprising, versatile, and resourceful
8. think clearly and approach problems in reasonable and systematic ways
9. mix well socially
10. are enthusiastic and verbally fluent
11. take an ascendant role in relationships

Low K-scale scores (T < 40) are indicative of persons who:

1. may have responded true to most of the MMPI-2 items
2. may have tried to "fake bad" when responding to the MMPI-2 items
3. may be exaggerating problems as a plea for help
4. may exhibit acute psychotic or organic confusion
5. are critical of self and others and are self-dissatisfied
6. are ineffective in dealing with the problems of daily life
7. show little insight into their own motives and behavior
8. are socially conforming
9. are overly compliant with authority
10. have a slow personal tempo
11. are inhibited, retiring, and shallow
12. are socially awkward
13. are blunt and harsh in social situations
14. are cynical, skeptical, caustic, and disbelieving
15. are suspicious about the motivations of other people

BACK-PAGE INFREQUENCY (Fb) SCALE

The Back-Page Infrequency (Fb) scale originally was developed for the experimental booklet used in the normative data collection for the MMPI-2 (Butcher et al., 1989). The procedures used to develop Fb were similar to those used in the development of the standard F scale. Since the items in the standard F scale appeared early in the experimental booklet, that scale did not offer evidence of the validity of responses to items appearing later in the 704-item booklet. The original Fb scale included 64 items that appeared later in the experimental booklet to which fewer than 10 percent of normal subjects responded in the scored direction. The version of the Fb scale included in MMPI-2 has 40 of the original 64 items.

In a protocol for which the standard F scale score is indicative of a valid approach, an elevated Fb score could indicate that the subject responded to items in the second half of the test booklet in an invalid manner. In this situation, one could interpret the standard scales that are based on items that occur early in the booklet, but supplementary and content scales that are based on items that occur later in the booklet should not be interpreted. Of course, if the standard F-scale score is indicative of invalidity, the protocol should not be interpreted at all.

Because this is a relatively new scale, few research data are available concerning optimal cutoff scores for identifying invalid records. Until additional research data become available, the best practice probably is to utilize the same T-score cutoffs for the Fb scale as for the F scale. Because of the differing number of items in the F and Fb scales, one should never compare raw scores on these two scales.

As with the F scale, persons who respond randomly to MMPI-2 items throughout the test will have very elevated Fb-scale scores and elevated scores (T > 80) on the VRIN scale. Subjects who respond true to most of the MMPI-2 items or who "fake bad" in responding to many of the items also will produce very elevated scores on the Fb-scale. For a true response bias, the elevated Fb-scale score will be accompanied by a TRIN-scale T score greater than 80 (in the true direction).

VARIABLE RESPONSE INCONSISTENCY (VRIN) SCALE

The Variable Response Inconsistency (VRIN) scale was developed for the MMPI-2 as an additional validity indicator (Butcher et al., 1989). It provides an indication of subjects' tendencies to respond inconsistently to MMPI-2 items. The VRIN scale consists of sixty-seven pairs of items with either similar or opposite content. Each time a subject answers items in a pair inconsistently, one raw-score point is added to the score on the VRIN scale. For some item pairs two true responses result in a point being scored for the scale; for other item pairs two false responses result in a point being added; and for still other item pairs a true response and a false response result in a point being added. This scale is very complicated to score, and it is recommended that it be scored by computer. If hand scoring is done, considerable care should be exercised to avoid errors.

The MMPI-2 manual (Butcher et al., 1989) indicates that interpretation of the VRIN scale requires caution until more empirical data are available. However, the manual suggests that a raw score equal to or greater than 13 (T > 80) indicates inconsistent responding that probably invalidates the resulting protocol. A random response set produces a T score on the VRIN scale of 96 for men and 98 for women. In an all-true or an all-false response set the VRIN-scale T score will be near 50. Subjects who deliberately are "faking bad" on the MMPI-2, as well as those who honestly are admitting to serious psychopathology, produce about average T scores on the VRIN scale.

In summary, The VRIN scale was developed to identify subjects who respond to the MMPI-2 items inconsistently and whose resulting protocols therefore should not be interpreted. Such inconsistent responding typically results when subjects do not read the content of the items and respond instead in a random or near-random way to the items.

The VRIN scale will be most useful when it is used along with the F scale. A high F-scale score and a high VRIN-scale score would support the notion that the subject has responded to the MMPI-2 items in a random manner. However, a high F-scale score and a low or moderate VRIN-scale score would be suggestive of a protocol that did not result from random responding or confusion. Instead, one would suspect that the protocol came either from a severely disturbed person who responded validly to the items or from a person who approached the items with the intention of appearing more disturbed than really was the case. Because an all-true or all-false response set would be likely to produce a high F-scale score and an average VRIN-scale score, those sets must also be considered. The TRIN scale, which is described next, is helpful in determining if a protocol is the product of a true or false response set.

TRUE RESPONSE INCONSISTENCY (TRIN) SCALE

The True Response Inconsistency (TRIN) scale was developed for the MMPI-2 to identify subjects who respond inconsistently to items by giving true responses to items indiscriminantly (acquiescence) or by giving false responses to items indiscriminantly (nonacquiescence) (Butcher et al., 1989). In either case, the resulting profile may be invalid and uninterpretable.

The TRIN scale consists of twenty-three pairs of items that are opposite in content. Two true responses to some item pairs or two false responses to other item pairs would indicate inconsistent responding. The TRIN raw score is obtained by subtracting the number of pairs of items to which subjects responded inconsistently with two false responses from the number of pairs of items to which subjects responded inconsistently with two true responses, and then adding a constant value of 9 to the difference. TRIN scale raw scores can range from 0 to 23. Higher TRIN-scale raw scores indicate a tendency to give true responses indiscriminately, and lower TRIN-scale raw scores indicate a tendency to give false responses indiscriminately. When TRIN-scale raw scores are converted to T scores, raw scores above 9 and below 9 are converted to T scores greater than 50, with the likelihood of a true or false response set indicated by the letters T or F following the T scores. As with the VRIN scale, the TRIN scale involves complex scoring that is best done by computer. If hand scoring is done, considerable care should be exercised.

Because only limited data are available concerning the TRIN scale, interpretation should be done cautiously until additional data are available (Butcher et al., 1989). The MMPI-2 manual suggests that, as rough guidelines, TRIN-scale raw scores of 13 or more (T > 80 in the direction of true) or of 5 or less (T > 80 in the direction of false) may be suggestive of indiscriminate responding that might invalidate the protocol.

PROFILE INVALIDITY

Some MMPI-2 users consider any protocol invalid and uninterpretable that has more than thirty omitted items or a T score greater than 65 on one or more of the standard validity scales. Although this practice is a very conservative one that is not likely to result in labeling as valid profiles that are in fact invalid, it represents an oversimplified view of profile validity and causes many valid profiles to be discarded. For example, the MMPI-2 manual (Butcher et al., 1989) indicates that F-scale scores in a T-score range of 71–90 can be indicative of psychosis. Gynther, Altman, and Warbin (1973) demonstrated that profiles with F-scale T scores equal to or greater than 100 on the original MMPI can have reliable extratest personality and behavioral correlates (e.g., disorientation, hallucinations, delusions, short attention span). Thus, a more sophisticated approach to profile validity is indicated.

Some subjects approach the MMPI-2 with such deviant test-taking attitudes that the resulting protocols are simply not interpretable. For example, subjects who respond in a random manner to the MMPI-2 items or who approach the test with a deliberate and extreme attempt to feign psychopathology will produce a protocol that should not be interpreted. Other subjects do not follow the test instructions exactly, but their deviant responding is less extreme. For example, clients who are seeking psychological or psychiatric treatment for the first time may tend to exaggerate symptoms and problems to some extent as a plea for help. These tendencies must be taken into account when the resulting protocol is interpreted, but they do not necessarily make the protocol uninterpretable.

Deviant Response Sets and Styles

To produce a valid MMPI-2 protocol a person must read and consider the content of each item and respond to the item as true or false. Occasionally, individuals respond in a stylistic way (e.g., false to each item) without reference to item content. Such behavior usually occurs among people who lack adequate reading skills, who are too confused to follow directions, or who have a very negativistic attitude toward the assessment procedure. Sometimes subjects are highly motivated to appear more or less well adjusted on the MMPI-2 than is actually the case for them, and they respond to item content in terms of the impressions they want to present of themselves rather than in terms of actual self-perceptions.

In ideal circumstances the test examiner should be aware of such response tendencies. Efforts should be made to assure that test subjects follow the standard instructions for completing the MMPI-2. If cooperation cannot be elicited, the test should not be administered. However, particularly in situations where large numbers of people are tested at once, some persons complete the MMPI-2 without following standard instructions. It is important for the MMPI-2 user to know how to detect the resulting invalid protocols. If profile invalidity is suspected, it may be helpful to consider what is known about the test subject's behavior from observation. Lack of congruence between the profile and observed behavior could be accounted for by the adoption of a response set or style.

If a particular MMPI-2 protocol is deemed invalid, the examiner may be able to discuss the situation with the test subject and readminister the test. Often a second testing yields a valid and interpretable protocol. If retesting is not possible or if it does not yield a valid protocol, no interpretation should be attempted. Further, it should be understood that the only thing that an invalid protocol tells us about a test subject is that, because the test items were not responded to in a valid manner, the resulting scores do not present an accurate picture of what the person really is like. For example, it is tempting to conclude that a person who presents a fake-good protocol, one in which even an average number of symptoms and problems are denied, is really a very maladjusted person who is trying to conceal that maladjustment. Such a conclusion is not justified. The person could just as well be a well-adjusted person who, because of circumstances, felt the need to present himself or herself as even better adjusted than he or she really is. For example, such motivation often is present when parents complete the MMPI-2 as part of a child custody evaluation.

Random Responding

One deviant response set involves a random or near-random response to the test items. A person may respond in a clearly random manner or may use an idiosyncratic response pattern such as marking every block of eight items as true, true, false, false, true, true, false, false, or every block of six items as true, false, true, false, true, false, and repeating this pattern with each such subsequent block. Because the responses are made without regard to item content, the resulting protocol must be considered invalid. The profile configurations resulting from a completely random response set are shown in Figures 3.1A and 3.1B. In the random response profile, the F-scale T score is very elevated (usually greater than 100), the K scale is at or near a T score of 50, and the L scale is moderately elevated (T = 60–70). The Fb scale also is quite elevated, usually at about the same T-score level as the F scale. The clinical scales are characterized by generally elevated scores, usually with the highest score on scale 8 and the second-highest score on scale 6. Scales 5 and 0 are likely to be below 70.

Similar patterns of scores on the standard validity and clinical scales also are obtained when subjects approach the test with a fake-bad or a true response set. The VRIN scale is very helpful in determining when the very deviant scores should be attributed to a random response set rather than to the fake-bad or true sets. The random response set results in a very high VRIN-scale score (Wetter, Baer, Berry, Smith, & Larsen, 1992; Berry, Baer, & Harris, 1991). Although a completely random response pattern yields a VRIN-scale T score of 96 for men and 98 for women, any T score greater than 80 on the VRIN scale suggests random responding to enough items to invalidate the protocol. Obviously, a protocol resulting from random responding should not be interpreted.

It should be understood that the random profiles in Figures 3.1A and 3.1B and the other invalid profiles presented in this section are modal profiles that would result if all items in the MMPI-2 were answered in the invalid manner. In practice, subjects may begin the MMPI-2 in a valid manner and then change to an invalid approach later in the test. Thus, many invalid profiles will approximate the modal ones presented here, but they will not match them exactly.

All-True Responding

If all of the MMPI-2 items are answered in the true direction, the resulting profiles look like the ones presented in Figures 3.2A and 3.2B. The salient features of the profile are an extremely elevated F-scale score (usually well above a T score of 100), L- and K-scale T scores well below 50, and extreme elevations on the clinical scales on the right side of the profile, usually with the highest scores on scales 6 and 8. The Fb scale will also be quite elevated, usually at about the same level as the F scale. Other forms of deviant responding (e.g., fake-bad, all-true) also produce elevated scores on the traditional validity and some of the clinical scales. The TRIN-scale score is very helpful in detecting a true response set. Although true responses to all items in the MMPI-2 will result in a TRIN-scale score of 118 for men and 120 for women, any T score greater than 80 on the TRIN scale indicates indiscriminant true responding that invalidates the protocol. Obviously, a profile resulting from all-true responding should under no circumstances be interpreted.

All-False Responding

The person who responds false to all of the MMPI-2 items will produce a profile like the ones shown in Figures 3.3A and 3.3B. Note the simultaneous elevations on scales L, F, and K and the more elevated scores on the clinical scales on the left side of the profile. The T scores on the Fb and VRIN scales will be near 50 in the all-false response set. The TRIN scale will be especially helpful in detecting a false response set. A completely false response set will yield a TRIN-scale T score of 114 for men and 118 for women, but any T score greater than 80 on the TRIN scale indicates indiscriminant false responding that invalidates the profile. Obviously, a profile resulting from all-false responding should not be interpreted.

Negative Self-Presentation

Faking Bad. Test subjects may be motivated to present an unrealistically negative impression when completing the MMPI-2. An extreme case of negative self-presentation would be when a person deliberately responds to test items in a manner that is thought to communicate that the test subject is very psychologically disturbed when in fact that is not the case. This response set often is referred to as "faking bad" or as malingering.

Figure 3.1A. K-corrected male profile indicative of random responding. Reproduced by permission of University of Minnesota Press.

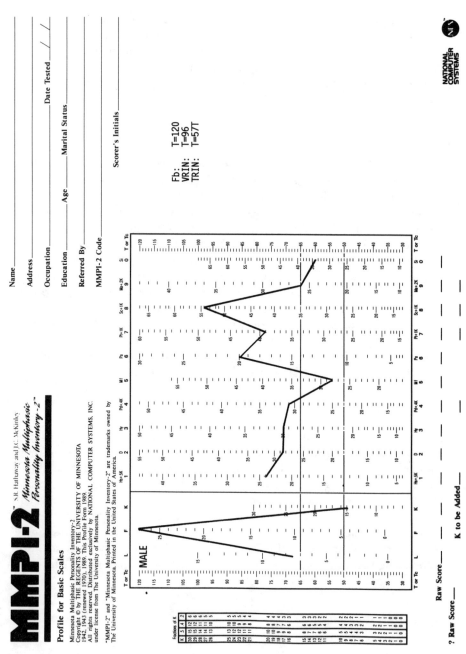

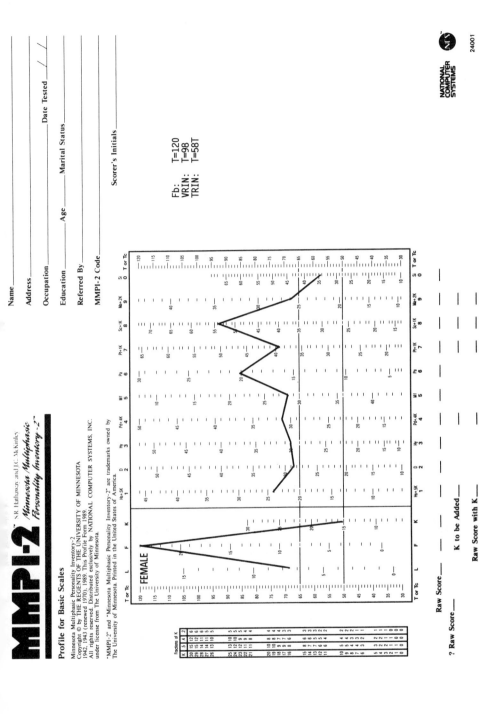

Figure 3.1B. K-corrected female profile indicative of random responding. Reproduced by permission of University of Minnesota Press.

41

Figure 3.2A. K-corrected male profile indicative of all-true responding. Reproduced by permission of University of Minnesota Press.

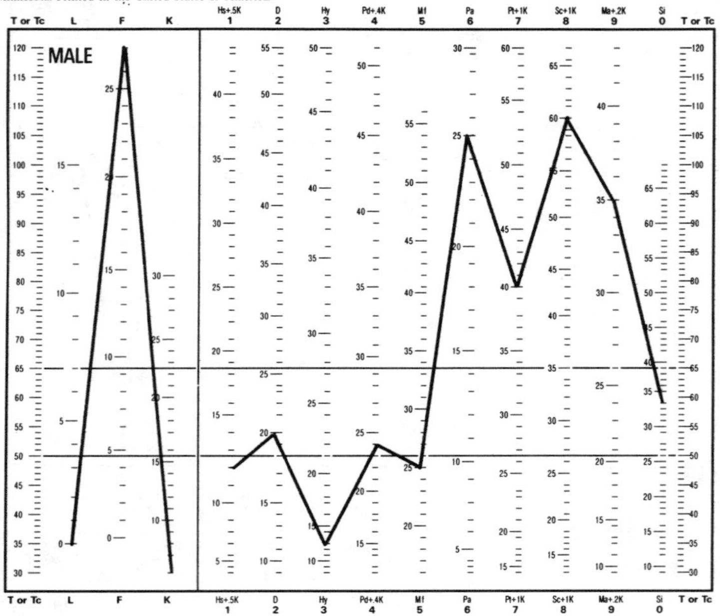

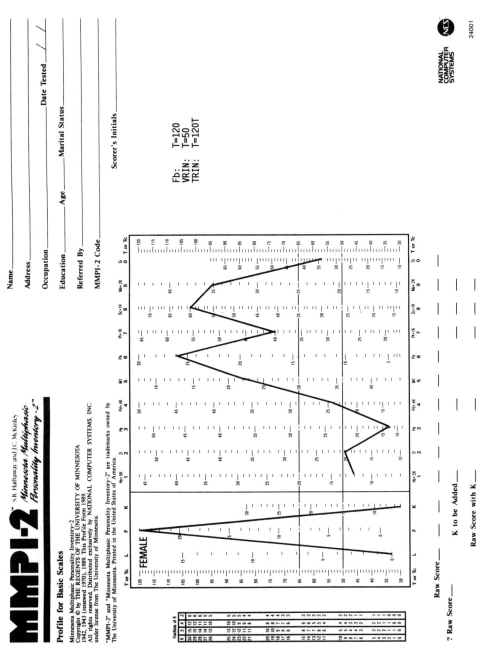

Figure 3.2B. K-corrected female profile indicative of all-true responding. Reproduced by permission of University of Minnesota Press.

Figure 3.3A. K-corrected male profile indicative of all-false responding. Reproduced by permission of University of Minnesota Press.

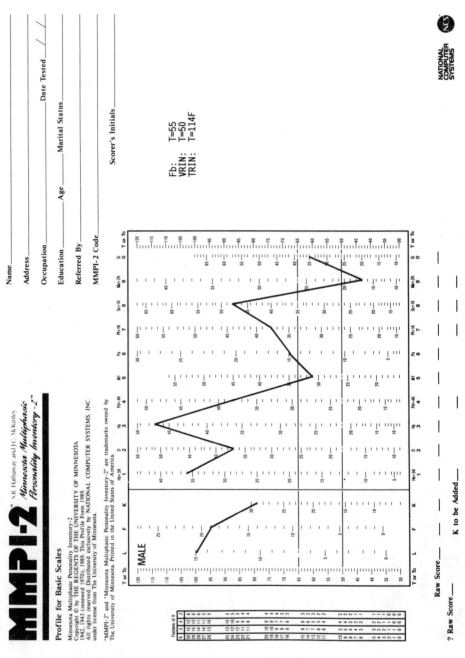

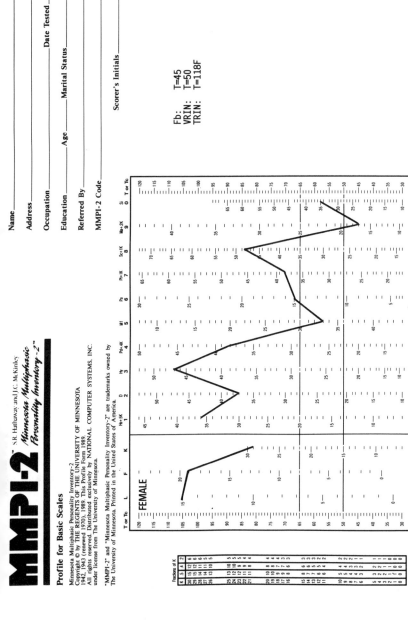

Figure 3.3B. K-corrected female profile indicative of all-false responding. Reproduced by permission of University of Minnesota Press.

45

As the data in Figures 3.4A and 3.4B indicate, when normal persons attempt to simulate serious psychopathology, they tend to overendorse deviant items, producing scores much higher than those of seriously disturbed patients. The profiles in Figures 3.4A and 3.4B are based on data from a study by Graham, Watts, and Timbrook (1991) in which groups of male and female college students took the MMPI-2 with standard instructions and again with instructions to present themselves as if they had serious psychological or emotional problems. The student data were compared with data from a sample of hospitalized psychiatric patients.

The fake-bad profile is characterized by a very elevated F-scale T score (usually well above 100). Likewise, the Fb scale is elevated, usually at about the same level as the F scale. Because the subject who is faking bad is responding consistently, although not honestly, to the content of the items, both the TRIN scale and VRIN scale scores are not significantly elevated. Scores on the clinical scales are very elevated, with scales 6 and 8 typically being the most elevated. Scales 5 and 0 typically are the least elevated clinical scales in the fake-bad profile.

Gough (1950) found that people who were trying to create the impression of severe psychopathology scored considerably higher on the F scale than on the K scale. He suggested that the difference between the F scale *raw score* and the K scale *raw score* can serve as a useful index for detecting fake-bad profiles. Gough (1950) and Meehl (1951) indicated that when such an index number is positive and is greater than 9, a profile should be considered as a fake-bad profile. Carson (1969) suggested that a cutoff score of +11 yields more accurate identification of fake-bad profiles. Although a single cutoff score cannot be established for all settings, whenever the F-scale raw score is greater than the K-scale raw score the possibility of faking bad should be considered, and as the difference becomes greater, the likelihood of a fake-bad profile becomes greater.

At first glance the fake-bad profile looks similar to the profile one might expect to obtain from a person who is actually very psychologically disturbed. However, there are some important differences. First, the F-scale score is usually higher for the fake-bad profile. The usual range of F-scale T scores for a person who has been diagnosed as psychotic is 71–90, whereas in the fake-bad profile the F-scale T score is well above 100. In addition, in a fake-bad profile the clinical scales tend to be more extremely elevated than in a valid profile from a disturbed person.

Considerable research was conducted to ascertain the extent to which the fake-bad response set could be detected by the validity scales of the original MMPI. Schretlen (1988) reviewed fifteen studies and concluded that the profiles of persons who are faking abnormality on the MMPI can be accurately distinguished from nonpathological profiles. He added that it is more difficult to differentiate MMPI profiles that have been faked from those produced by persons with genuine mental disorders. Berry et al. (1991) used meta-analytic procedures to compare results of twenty-eight studies that examined the ability of the MMPI validity scales to detect malingering. They concluded that the

ψ= Double - M

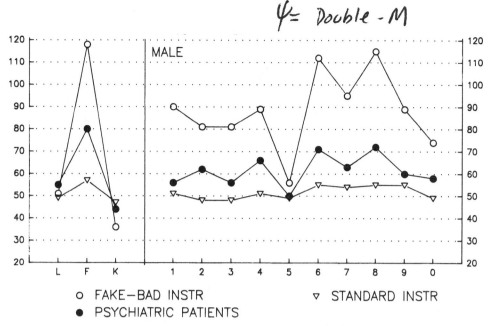

○ FAKE—BAD INSTR   ▽ STANDARD INSTR
● PSYCHIATRIC PATIENTS

Figure 3.4A. K-corrected profiles of male students with standard and fake-bad instructions and of male psychiatric patients. (*Source:* Graham, J.R., Watts, D., Timbrook, R.E. (1991). Detecting fake-good and fake-bad MMPI-2 profiles. *Journal of Personality Assessment, 57,* 264–77. Copyright © 1991 by Lawrence Erlbaum Associates, Inc. Reproduced by permission.)

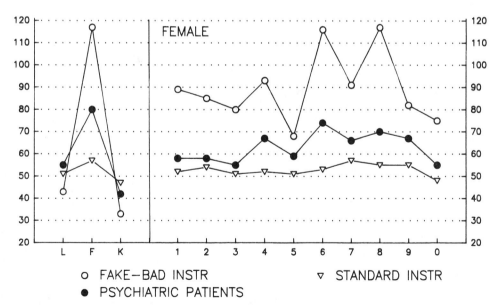

○ FAKE—BAD INSTR   ▽ STANDARD INSTR
● PSYCHIATRIC PATIENTS

Figure 3.4B. K-corrected profiles of female students with standard and fake-bad instructions and of female psychiatric patients. (*Source:* Graham, J.R., Watts, D., Timbrook, R.E. (1991). Detecting fake-good and fake-bad MMPI-2 profiles. *Journal of Personality Assessment, 57,* 264–77. Copyright © 1991 by Lawrence Erlbaum Associates, Inc. Reproduced by permission.)

MMPI validity scales are quite successful in detecting malingered versus honest profiles. Although discrimination between normals answering honestly and normals faking bad, or malingering, was greater than between normals faking bad, or malingering, and actual psychiatric patients, the latter discrimination was quite respectable. The F scale seemed to produce the best discrimination, but the F minus K index also was effective. Gough's Dissimulation Scale (Gough, 1950) also was an effective discriminator, but it is not routinely scored on the MMPI-2.

To date there has been only one published study of the effectiveness of MMPI-2 scales in detecting faking bad or malingering (Graham, Watts, & Timbrook, 1991). That study concluded that the standard validity scales of the MMPI-2 are effective in discriminating between profiles of normals taking the MMPI-2 with standard instructions and with instructions to fake bad and psychiatric patients taking the test with standard instructions. The F scale was the most effective indicator of faking, with an optimal cutoff score (raw score > 18) correctly classifying 97 percent of faked profiles for men and 95 percent for women. The classification rates for men and women with standard instructions were 97 percent and 100 percent, respectively. Large proportions of the psychiatric patients (90% of the men and 95% of the women) also were correctly classified by the F scale, but a much higher cutoff score (>27 for men and >29 for women) was indicated than for the normal subjects. Using the same cutoff score with psychiatric patients led to a very large number of patients being classified as faking bad. Graham et al. noted that the MMPI-2 cutoff scores that yielded optimal classification were similar to those previously reported for the original MMPI.

Exaggeration. Sometimes individuals who really have psychological symptoms and problems exaggerate them in responding to the MMPI-2 items. This is a rather common occurrence among persons who are trying to communicate to others that they desperately need professional help. In these circumstances the resulting profile of scores will depend on what actual symptoms and problems are being exaggerated. Thus, it is not possible to identify a prototype for this response set. The major clue that one has that such a response set might be operating is that scores on the F scale and the clinical scales seem to be much higher than would be expected given the person's history and observations made of the person during the interview and/or testing. Exaggeration does not completely invalidate a profile, but interpretations must be modified to take into account that the obtained scores represent an overreporting of symptoms and problems.

Positive Self-Presentation

Faking Good. Sometimes persons completing the MMPI-2 are motivated to deny problems and to appear better-off psychologically than is in fact the case. This response set is relatively common when persons who complete the

MMPI-2 as part of a job application process or child custody evaluation. In its most blatant form this tendency is referred to as "faking good."

In the fake-good response set, the L and K scales are likely to be elevated significantly, and the T score on the F scale may be well below 50. Research with the original MMPI indicated that the standard validity scales could identify persons who fake good but not as accurately as those who fake bad (Exner, McDowell, Pabst, Stackman, & Kirk, 1963; Gough, 1950; Grayson & Olinger, 1957; Grow, McVaugh, & Eno, 1980; Hunt, 1948; Lanyon, 1967; McAnulty, Rappaport, & McAnulty, 1985; Otto, Lang, Megargee, & Rosenblatt, 1988; Rapaport, 1958; Rice, Arnold, & Tate, 1983; Walters, 1988). Gough (1950) proposed that fake-good profiles should have K-scale scores higher than F-scale scores and that the raw-score difference between these two scales could be used as an index of faking good. Subsequent research with the MMPI indicated that the proposed relationship between the K and F scales typically is present when persons fake good, but the difference in raw scores between the two scales has not proven to be an effective index for discriminating valid from fake-good protocols (Cofer, Chance, & Judson, 1949; Gough, 1950; Grow et al., 1980; Hunt, 1948; McAnulty et al., 1985).

Baer, Wetter, and Berry (1992) presented the results of a meta-analysis of data from 25 MMPI studies in which subjects responding honestly were compared to subjects underreporting psychopathology. The analysis confirmed that the L and K scales were effective in discriminating between valid and fake-good protocols and that detection of faking good is less accurate than detection of faking bad. Although several supplementary scales showed some promise in detecting underreporting of psychopathology, Baer et al. recommended that, until additional research is available about these supplementary scales, clinicians consider the L and K scales when making judgments about underreporting of psychopathology on the MMPI-2.

To date there has only been one published study concerning the detection of faking good with the MMPI-2. Graham, Watts, and Timbrook (1991) administered the MMPI-2 to male and female college students with standard instructions and with instructions to present a very positive impression of themselves as if they were being evaluated for a job they really wanted. The mean profiles obtained under the two instructional sets are presented in Figures 3.5A and 3.5B. Consistent with prior research with the MMPI, subjects in the fake-good condition had somewhat lower scores on most of the clinical scales. The validity scale pattern previously reported for the MMPI also was present. Subjects who were instructed to fake good had T scores well above 50 on the L and K scales and below 50 on the F scale.

Graham, Watts, and Timbrook (1991) found that it was more difficult to detect fake-good than fake-bad profiles. Although L plus K, and K minus F raw-score indexes were relatively effective in detecting the fake-good set, the L-scale raw score worked as well as, and in some cases better than, any other measure. It was concluded that optimal raw-score cutoffs depended on whether it was more important to identify the fake-good profiles or the honest

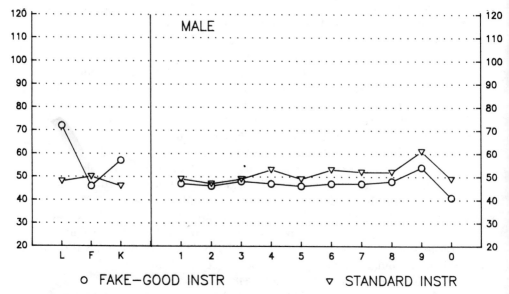

Figure 3.5A. K-corrected profiles of male students with standard and fake-good instructions. (*Source:* Graham, J.R., Watts, D., Timbrook, R.E. (1991). Detecting fake-good and fake-bad MMPI-2 profiles. *Journal of Personality Assessment, 57,* 264–77. Copyright © 1991 by Lawrence Erlbaum Associates, Inc. Reproduced by permission.)

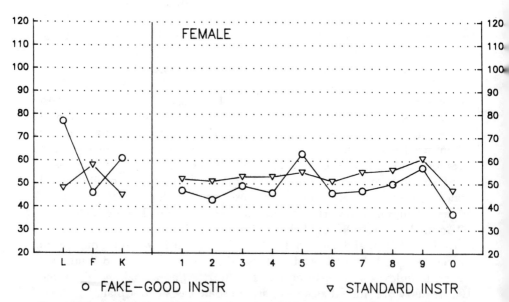

Figure 3.5B. K-corrected profiles of female students with standard and fake-good instructions. (*Source:* Graham, J.R., Watts, D., Timbrook, R.E. (1991). Detecting fake-good and fake-bad MMPI-2 profiles. *Journal of Personality Assessment, 57,* 264–77. Copyright © 1991 by Lawrence Erlbaum Associates, Inc. Reproduced by permission.)

ones. In their samples, for example, an L-scale raw score of 8 correctly classified 93 percent of the honest profiles but only 67 percent of the fake-good profiles. To classify correctly 96 percent of the fake-good male profiles, an L-scale raw-score cutoff of 4 was necessary. However, this lower L-scale score correctly classified only 70 percent of the valid profiles. Similar results were reported for the female subjects.

Defensive Profiles. Sometimes persons are motivated to present unrealistically favorable impressions but do not do so as blatantly as in the fake-good response sets. For example, persons taking the MMPI-2 as part of employment screening or child custody evaluations may want to emphasize positive characteristics and minimize negative ones. The resulting profile may underestimate problems and symptoms but it is not necessarily uninterpretable.

In a defensive profile the L- and K-scale scores typically are more elevated than the F-scale score. However, scores on these two scales will not be as elevated as in the fake-good response set. Figure 3.6 presents an example of a defensive profile. This pattern of scores was produced by a father in a contested child custody case. Note the pattern on the validity scales, with the L- and K-scale scores being considerably more elevated than the F-scale score. It is interesting to note that this man produced moderately elevated scores on scales 4 and 9 even though he was generally defensive in responding to the test items. Perhaps he did not view the kinds of characteristics represented in the items in these scales to be particularly negative.

Sometimes in a defensive profile either the L- or K-scale score is elevated but not both. This is because the two scales seem to be measuring somewhat different aspects of test-taking attitudes. Persons who have elevated scores on the K scale are generally denying symptoms and problems. Persons who have elevated scores on the L scale are trying to present a picture of themselves as honest, moral, and conforming. Although these two aspects of test-taking attitude often occur simultaneously, in some circumstances they do not.

Butcher (1985) has suggested procedures for interpreting profiles for which the validity scale configuration is suggestive of defensiveness. If there are any clinical scales elevated above a T score of 65, they should be interpreted using the standard correlates of the scales. It should be recognized that these scale elevations may reflect more significant problems because they are obtained when test subjects are trying to present the most favorable view of themselves. Because in a defensive profile the person is attempting to present an overly favorable view of his or her functioning, clinical scale T scores in a 60 to 65 range should be considered significant. If all of the clinical scale T scores in a defensive profile are below 60, the profile is not providing much useful information about the subject. One cannot tell if such a profile is indicative of a well-adjusted person who is motivated to appear even more well adjusted or of a poorly adjusted person who is trying to appear to be well adjusted.

In summary, although detecting the fake-good response set is possible with the MMPI-2, accuracy generally is not as great as in detecting the fake-bad

Figure 3.6. K-corrected male profile produced by defensive subject. Reproduced by permission of University of Minnesota Press.

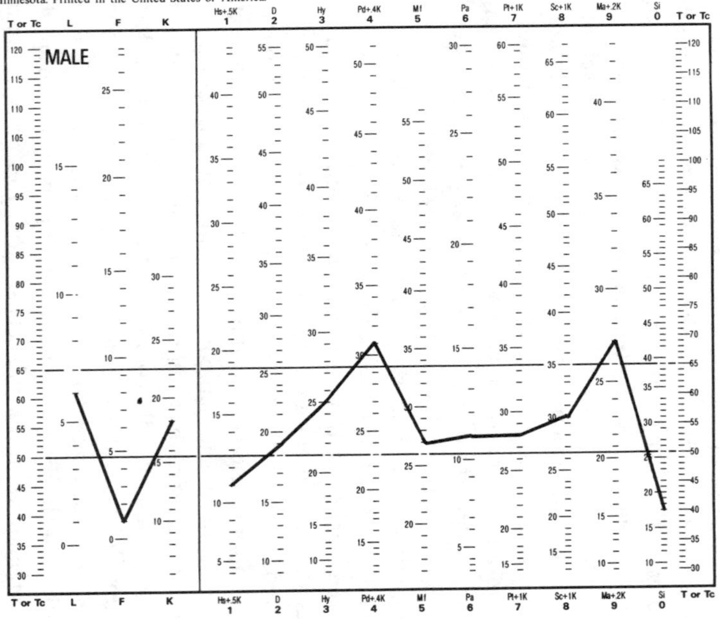

52

response set. The L-scale raw score appears to be the most effective way to detect this response set. Optimal L-scale cutoff scores depend on which kind of classification is more important (i.e., detecting honest protocols versus fake-good protocols). As is the case with other response sets, one should not interpret protocols resulting from a fake-good response set. One's inferences in such cases should be limited to the conclusion that the resulting pattern of scores on the clinical scales does not represent accurately what the person is really like. One should not infer that the person is covering up serious psychological problems. It is equally likely that the person is of average adjustment and, because of circumstances (e.g., child custody proceedings) is simply overstating positive characteristics and understating negative ones. It is possible to interpret defensive protocols if appropriate adjustments are made to take the defensiveness into account.

OTHER INDICATORS OF TEST INVALIDITY

Some authors (e.g., Greene, 1980; 1989) have suggested additional ways of detecting invalid MMPI profiles. Gough (1954) developed the Dissimulation (Ds) scale to identify persons who are simulating or exaggerating psychopathology. Persons who were instructed to dissimulate scored higher on this scale. Although the Ds scale was relatively effective in identifying malingering, it did not work as well as the F scale (Berry et al., 1991), so it is not routinely scored in the MMPI-2. The test-retest (TR) index (Buechley & Ball, 1952) was the total number of the sixteen repeated items in the MMPI that a subject answered inconsistently. Because the sixteen repeated items were deleted from the MMPI-2, the TR index is not relevant to the revised instrument. The Carelessness scale (Greene, 1978) consisted of twelve pairs of items that were judged to be psychologically opposite in content. This scale was thought to identify subjects who were careless in responding to the MMPI items. The VRIN scale of the MMPI-2 is a more sophisticated approach to measuring response inconsistency and should be used instead of the Carelessness scale.

Lees-Haley, English, and Glenn (1991) developed an MMPI-2 scale for detecting faking among personal-injury claimants. Although the effort to develop such a scale should be applauded, significant problems with the research belie the study's conclusions concerning how well the scale works.

Greene (1980, 1991) and others have suggested that the relative endorsement of subtle and obvious items of the MMPI and MMPI-2 can be helpful in identifying fake-good or fake-bad response sets. As discussed in Chapter 7, more accurate identification of fake-good and fake-bad response sets can be accomplished by using the standard validity scales of the MMPI-2 than by using the subtle-obvious subscales. Therefore, the use of the subtle-obvious subscales is not recommended.

4

The Clinical Scales

A primary goal of this chapter is to discuss the nature of each MMPI-2 clinical scale in an attempt to elucidate the dimensions being assessed by the scale. In addition, descriptive material is presented for high scorers on each scale. This material is based on an examination of previously reported data for the clinical scales of the original MMPI and consideration of new data concerning extratest correlates of the clinical scales of the MMPI-2. Empirical validity studies for the original MMPI are summarized in Chapter 8. Likewise, that chapter describes research studies conducted with the MMPI-2 to determine extratest correlates of the clinical scales.

As noted earlier, the clinical scales of the MMPI-2 are basically the same as those found in the original MMPI. A few items were deleted from some of the scales because they had become dated or because they were judged to have objectionable content, usually having to do with religious beliefs or bowel or bladder function. Some of the items in the clinical scales were modified slightly to modernize them, to eliminate sexist references, or to improve readability. The items included in each clinical scale and the keyed response for each item are presented in Appendix A.

The definition of a high score on the clinical scales has varied considerably in the literature and from one clinical scale to another. Some researchers have considered MMPI T scores above 70 as high scores. Others have defined high scores in terms of the upper quartile in a distribution. Still others have presented descriptors for several T-score levels on each scale. Another approach has been to identify the highest scale in the profile (high point) irrespective of its T-score value.

Low scores also have been defined in different ways in the literature, sometimes as T scores below 40 and other times as scores in the lowest quartile of a distribution. This latter approach has led to scores well above the mean being considered as low scores. Compared with high scores, limited information is available in the literature concerning the meaning of low scores.

Original MMPI data have supported the notion that low scores on a particular scale indicate the absence of problems and symptoms characteristic of high scorers on that scale. Other data have suggested that low scores on some scales are associated with problems and negative characteristics. Still other data have been interpreted as indicating that both high and low scores on certain scales indicate similar problems and negative characteristics.

Several studies with the MMPI-2 have attempted to clarify the meaning of low scores on the clinical scales. Keiller and Graham (1993) examined extratest characteristics of high-, medium-, and low-scoring nonclinical subjects on the eight clinical scales. They concluded that low scores convey important information but not as much as high scores. Low scores were associated with fewer than the average number of symptoms and problems and above-average adjustment. In this nonclinical sample, low scores on the clinical scales were not held to indicate problems and negative characteristics.

Timbrook and Graham (1992) examined ratings of symptoms for psychiatric inpatients who had high, average, or low scores on each of the clinical scales of the MMPI-2. As in the Keiller and Graham (1993) study of nonclinical subjects, the high scores of the patients conveyed much more information than the low scores. As expected, high scorers on the clinical scales were rated as having more severe symptoms than low scorers, and the symptoms were generally consistent with previously reported data for the MMPI clinical scales. However, for several scales, ratings of symptoms differed for low and average scorers. For some scales the lower scorers were rated as having less severe symptoms than high and average scorers. For example, low-scoring women on scale 7 were rated as less depressed than both high and average scorers on that scale. For other scales, the low scorers were rated as having more severe symptoms than average and high scorers. For example, low-scoring men on scale 2 were rated as more uncooperative than average scorers and high scorers on that scale. Low-scoring women on scale 9 were rated higher on unusual thought content than average scorers and high scorers on that scale. Some of the findings of the Timbrook and Graham study probably emerged because of the intercorrelations of the clinical scales. For example, the men who had low scores on scale 2 could have had high scores on scale 4. This would account for the uncooperativeness associated with low scores on scale 2. We would not necessarily expect to find that men who score low on scale 2 but who do not score high on scale 4 would also be rated as more uncooperative.

Based on the empirical data concerning the meaning of low scores on the MMPI and the MMPI-2, this author recommends a very conservative approach to the interpretation of low scores on the MMPI-2 clinical scales. In nonclinical settings (e.g., personnel selection) low scores in a valid protocol should be interpreted as indicating more positive adjustment than high or average scores. However, if the validity scales indicate that the test was completed in a defensive manner, low scores should not be interpreted at all. In clinical settings, it is recommended that low scores on the clinical scales not be interpreted. The exceptions are scales 5 and 0, for which limited inferences can be made about low scorers. (See later sections.) Any inferences that could be made about a person based on low scores could probably also be made (and with greater confidence) based on that person's high scores. Clearly, much more research is needed concerning the meaning of low scores in a variety of settings. Perhaps in the future we will have enough research data to support making specific inferences on the basis of low scores, but for now such a practice is not recommended.

The approach used in this chapter will be to suggest interpretive inferences about persons who have high scores on each of the clinical scales. In general, T scores greater than 65 are considered to be high scores, although inferences about persons with scores at several different T-score levels are presented for some scales. It should be understood that the T-score levels used have been established somewhat arbitrarily and that clinical judgment will be necessary in deciding which inferences should be applied to scores at or near the cutoff scores for the levels. It should also be understood that not every inference presented will apply to every person who has a T score at that level. In general, greater confidence should be placed in inferences based on extreme scores. All inferences should be treated as hypotheses to be considered in the context of other information available about the test subject.

SCALE 1 (HYPOCHONDRIASIS)

Scale 1 originally was developed to identify patients who manifested a pattern of symptoms associated with the label of hypochondriasis. The syndrome was characterized in clinical terms by preoccupation with the body and concomitant fears of illness and disease. Although such fears usually are not delusional in nature, they tend to be persistent. An item deletion because of objectionable content reduced scale 1 from 33 items in the original MMPI to 32 items in the MMPI-2.

Of all of the clinical scales, scale 1 seems to be the most clearly homogeneous and unidimensional. All of the items deal with somatic concerns or with general physical competence. Factor analysis (Comrey, 1957b) has indicated that much of the variance in scale 1 is accounted for by a single factor, one that is characterized by the denial of good health and the admission of a variety of somatic symptoms. Patients with bona fide physical problems typically show somewhat elevated T scores on scale 1 (approximately 60). Elderly subjects tend to produce scale 1 scores that are slightly more elevated than those of the general adult standardization subjects, probably reflecting the declining health typically associated with aging.

Interpretation of High Scores on Scale 1

For persons with extremely high scores on scale 1 (T > 80), dramatic and sometimes bizarre somatic concerns should be suspected. If scale 3 also is elevated, the possibility of a conversion disorder should be considered. If scale 8 is very elevated along with scale 1, somatic delusions may be present.

Persons with more moderate elevations on scale 1 (T = 60–80) tend to have generally vague, nonspecific complaints. When specific symptoms are elicited, they tend to be epigastric in nature. Chronic weakness, lack of energy, and sleep disturbance also tend to be characteristic of high scorers. As stated above, medical patients with bona fide physical problems generally obtain T scores of about 60 on this scale. When medical patients produce T scores much above 60, one should suspect a strong psychological component to the illness. Moderately high scores on scale 1 tend to be associated with diagnoses such as somatoform disorders, somatoform pain disorders, anxiety disorders, and depressive disorders. Acting-out behavior is rare among high scale 1 scorers.

High scale 1 scorers (T > 60) in both psychiatric and nonpsychiatric samples tend to be characterized by a rather distinctive set of personality attributes. They are likely to be selfish, self-centered, and narcissistic. Their outlook toward life tends to be pessimistic, defeatist, and cynical. They are generally dissatisfied and unhappy and are likely to make those around them miserable. They complain a great deal and communicate in a whiny manner. They are demanding of others and are very critical of what others do, although they are likely to express hostility in rather indirect ways. High scorers on scale 1 often are described as dull, unenthusiastic, unambitious, and lacking ease in oral expression. High scorers generally do not exhibit much manifest anxiety, and in general they do not show signs of major incapacity. Rather, they appear to be functioning at a reduced level of efficiency. Problems are much more likely to be long-standing in nature than situational or transient.

Extremely high and moderately high scorers typically see themselves as physically ill and are seeking medical explanations and treatment for their symptoms. They tend to lack insight concerning the causes of their somatic symptoms, and they resist psychological interpretations. These tendencies, coupled with their generally cynical outlook, suggest that they are not good candidates for psychotherapy or counseling. They tend to be very critical of their therapists and to terminate therapy if they perceive the therapist as suggesting psychological reasons for their symptoms or as not giving them enough support and attention.

Summary of Descriptors for Scale 1

High scores on scale 1 indicate persons who:

1. have excessive bodily concern
2. may have conversion disorders or somatic delusions (if T > 80)
3. describe somatic complaints that generally are vague but if specific are likely to be epigastric in nature
4. complain of chronic weakness, lack of energy, and sleep disturbance
5. if medical patients, may have a strong psychological component to their illnesses

6. are likely to be diagnosed as somatoform, somatoform pain, depressive, or anxiety disorders
7. are not likely to act out in psychopathic ways
8. seem selfish, self-centered, and narcissistic
9. have pessimistic, defeatist, and cynical outlook toward life
10. are unhappy and dissatisfied
11. make others miserable
12. complain
13. communicate in a whiny manner
14. are demanding and critical of others
15. express hostility indirectly
16. are described as dull, unenthusiastic, and unambitious
17. lack ease in oral expression
18. generally do not exhibit much manifest anxiety
19. seem to have functioned at a reduced level of efficiency for long periods of time
20. see themselves as medically ill and seek medical treatment
21. lack insight and resist psychological interpretations
22. are not very good candidates for psychotherapy or counseling
23. become critical of therapists
24. terminate therapy prematurely when therapists suggest psychological reasons for symptoms or are perceived as not giving enough attention and support

SCALE 2 (DEPRESSION)

Scale 2 was developed originally to assess symptomatic depression. The primary characteristics of symptomatic depression are poor morale, lack of hope in the future, and a general dissatisfaction with one's life situation. Of the 60 items originally composing scale 2, 57 were retained in the MMPI-2. Many of the items in the scale deal with various aspects of depression such as denial of happiness and personal worth, psychomotor retardation and withdrawal, and lack of interest in one's surroundings. Other items in the scale cover a variety of other symptoms and behaviors, including somatic complaints, worry or tension, denial of hostile impulses, and difficulty in controlling one's own thought processes. Scale 2 seems to be an excellent index of examinees' discomfort and dissatisfaction with their life situations. Whereas very elevated scores on this scale suggest clinical depression, more moderate scores tend to be indicative of a general attitude or lifestyle characterized by poor morale and lack of involvement. Scale 2 scores are related to age, with elderly subjects scoring approximately five to ten T-score points higher than the mean for the total MMPI-2 normative sample. Subjects who recently have been hospitalized or incarcerated tend to show moderate elevations on scale 2 that reflect dissatisfaction with current circumstances rather than clinical depression.

Interpretation of High Scores on Scale 2

High scorers on this scale (particularly if the T scores exceed 70) often display depressive symptoms. They may report feeling depressed, blue, unhappy, or dysphoric. They tend to be pessimistic about the future in general and more specifically about the likelihood of overcoming their problems and making a better adjustment. They talk about committing suicide. Self-depreciation and guilt feelings are common. Behavioral manifestations may include lack of energy, refusal to speak, crying, and psychomotor retardation. Patients with such high scores often receive depressive diagnoses. Other symptoms of high scorers include physical complaints, bad dreams, weakness, fatigue or loss of energy, agitation, tension, and fearfulness. They also are described as irritable, high-strung, and prone to worry and fretting. They may have a sense of dread that something bad is about to happen to them.

High scorers also show a marked lack of self-confidence. They report feelings of uselessness and inability to function in a variety of situations. They act helpless and give up easily when faced with stress. They see themselves as having failed to achieve adequately in school and at their jobs.

A lifestyle characterized by withdrawal and lack of intimate involvement with other people is common. High scorers tend to be described as introverted, shy, retiring, timid, seclusive, and secretive. Also, they tend to be aloof and to maintain psychological distance from other people. They may feel that others do not care about them, and their feelings are easily hurt. They often have a severely restricted range of interests and may withdraw from activities in which they previously participated. They are extremely cautious and conventional in their activities, and they are not very creative in problem solving.

High scorers may have great difficulty in making even simple decisions, and they may feel overwhelmed when faced with major life decisions such as vocation or marriage. They tend to be very overcontrolled and to deny their own impulses. They are likely to avoid unpleasantness and will make concessions in order to avoid confrontations.

Because high scale 2 scores are suggestive of great personal distress, they may indicate a good prognosis for psychotherapy or counseling. There is some evidence, however, that high scorers may tend to terminate treatment prematurely when the immediate crisis passes.

Summary of Descriptors for Scale 2

High scores on scale 2 indicate persons who:

1. display depressive symptoms (particularly if T scores exceed 70)
2. feel blue, unhappy, or dysphoric
3. are quite pessimistic about the future

4. may talk about committing suicide
5. have feelings of self-depreciation and guilt
6. may cry, refuse to speak, and show psychomotor retardation
7. often are given depressive diagnoses
8. report bad dreams, physical complaints, weakness, fatigue, and loss of energy
9. are agitated and tense
10. are described as irritable, high-strung, and prone to worry and fretting
11. have a sense of dread that something bad is going to happen to them
12. lack self-confidence
13. feel useless and unable to function
14. act helpless and give up easily
15. feel like failures in school or at work
16. have lifestyles characterized by withdrawal and lack of involvement with other people
17. are introverted, shy, retiring, timid, seclusive, and secretive
18. are aloof and maintain psychological distance from other people
19. may feel that others do not care about them
20. have their feelings easily hurt
21. have restricted range of interests
22. withdraw from activities in which they previously participated
23. are very cautious and conventional and are not creative in problem solving
24. have difficulty making decisions
25. feel overwhelmed when faced with major life decisions
26. are overcontrolled and deny own impulses
27. avoid unpleasantness and make concessions to avoid confrontations
28. because of personal distress, are likely to be motivated for psychotherapy or counseling
29. may terminate therapy prematurely when immediate crisis passes

SCALE 3 (HYSTERIA)

This scale was developed to identify patients who were having hysterical reactions to stress situations. The hysterical syndrome is characterized by involuntary psychogenic loss or disorder of function.

All of the 60 items in the original version of scale 3 were retained in the MMPI-2. Some of the items deal with a general denial of physical health and a variety of rather specific somatic complaints, including heart or chest pain, nausea and vomiting, fitful sleep, and headaches. Another group of items involves a general denial of psychological or emotional problems and of discomfort in social situations. Although these two clusters of items are reasonably independent in normal subjects, persons using hysterical defenses seem to score high on both clusters.

Scale 3 scores are related to intellectual ability, with brighter persons scoring higher on the scale. In addition, high raw scores are much more common among women than among men in both normal and psychiatric populations.

It is important to take into account the level of scores on scale 3. Whereas marked elevations (T > 80) suggest a pathological condition characterized by classical hysterical symptomatology, moderate levels are associated with characteristics that are consistent with hysterical disorders but that do not include the classical hysterical symptoms. As with scale 1, patients with bona fide medical problems for whom there is no indication of psychological components to the conditions tend to obtain T scores of about 60 on this scale.

Interpretation of High Scores on Scale 3

Marked elevations on scale 3 (T > 80) may be suggestive of persons who react to stress and avoid responsibility by developing physical symptoms. Their symptoms usually do not fit the pattern of any known organic disorder. They may include, in some combination, headaches, stomach discomfort, chest pains, weakness, and tachycardia. Nevertheless, such persons may be symptom-free most of the time, but under stress the symptoms may appear suddenly; they are likely to disappear just as suddenly when the stress subsides.

Except for the physical symptoms, high scorers tend to be relatively free of other symptoms. Although they sometimes describe themselves as prone to worry, lacking energy, feeling worn-out, and having sleep disturbances, they are not likely to report severe anxiety, tension, or depression. Hallucinations, delusions, and suspiciousness are rare. The most frequent diagnoses for high scale 3 scorers among psychiatric patients are conversion disorder and psychogenic pain disorder.

A salient feature of the day-to-day functioning of high scorers is a marked lack of insight concerning the possible underlying causes of their symptoms. In addition, they show little insight concerning their own motives and feelings.

High scorers are often described as extremely immature psychologically and at times even childish or infantile. They are quite self-centered, narcissistic, and egocentric, and they expect a great deal of attention and affection from others. They often use indirect and devious means to get the attention and affection they crave. When others do not respond appropriately, they may become hostile and resentful, but these feelings are likely to be denied and not expressed openly or directly.

High scorers on scale 3 tend to be emotionally involved, friendly, talkative, enthusiastic, and alert. Although their needs for affection and attention drive them into social interactions, their interpersonal relationships tend to be rather superficial and immature. They are interested in people primarily because of what they can get from them rather than because of a sincere interest in them.

Occasionally high scorers will act out in a sexual or aggressive manner with little apparent attention to or understanding of what they are doing. When confronted with the realities of their behavior, they may act surprised and feel resentful and persecuted.

Because of their needs for acceptance and affection, high scorers may initially be quite enthusiastic about counseling and psychotherapy. However, they seem to view themselves as having medical problems and want to be treated medically. They are slow to gain insight into the underlying causes of their behavior and are quite resistant to psychological interpretations. If therapists insist on examining the psychological causes of symptoms, premature termination of therapy is likely. High scorers may be willing to talk about problems in their lives as long as they are not perceived as causing their symptoms, and they often respond quite well to direct advice and suggestion.

When high scorers become involved in therapy, they often discuss worry about failure in school or work, marital unhappiness, lack of acceptance by their social groups, and problems with authority figures. Histories often include a rejecting father to whom females reacted with somatic complaints and to whom males reacted with rebellion and overt hostility.

Summary of Descriptors for Scale 3

High scores on scale 3 indicate persons who:

1. react to stress and avoid responsibility through development of physical symptoms
2. may report headaches, stomach discomfort, chest pains, weakness, or tachycardia
3. have symptoms that may appear and disappear suddenly
4. do not report severe emotional turmoil
5. rarely report hallucinations, delusions, or suspiciousness
6. if psychiatric patients, receive diagnoses of conversion disorder or psychogenic pain disorder
7. lack insight concerning the causes of their symptoms
8. lack insight about their own motives and feelings
9. are psychologically immature and childish or infantile
10. are self-centered, narcissistic, and egocentric
11. expect a great deal of attention and affection from others
12. use indirect and devious means to get attention and affection
13. do not express resentment and hostility openly
14. tend to be emotionally involved, friendly, talkative, and alert
15. have superficial and immature interpersonal relationships
16. are interested in what other people can do for them
17. occasionally act out in a sexual or aggressive manner with little apparent insight into their actions
18. initially are enthusiastic about treatment
19. view themselves as having medical problems and want medical treatment

20. are resistant to psychological interpretations
21. are likely to terminate treatment if therapists insist on examining the psychological causes of symptoms
22. may be willing to talk about their problems as long as they are not perceived as causing their somatic symptoms
23. often respond well to direct advice and suggestion
24. when involved in therapy, discuss failure at work or school, marital unhappiness, lack of acceptance, and problems with authority figures
25. have histories of rejecting fathers

SCALE 4 (PSYCHOPATHIC DEVIATE)

Scale 4 was developed to identify patients diagnosed as psychopathic personality, asocial or amoral type. Whereas subjects included in the original criterion group were characterized in their everyday behavior by such delinquent acts as lying, stealing, sexual promiscuity, excessive drinking, and the like, no major criminal types were included. All fifty of the items in the original scale were maintained in the MMPI-2. The items cover a wide array of topics, including absence of satisfaction in life, family problems, delinquency, sexual problems, and difficulties with authorities. Interestingly, the keyed responses include both admissions of social maladjustment and assertions of social poise and confidence.

Scores on scale 4 tend to be related to age, with younger persons scoring slightly higher than older persons. In the MMPI-2 normative samples, Caucasian and Asian-American subjects scored somewhat lower on scale 4 (5–10 T-score points) than African-American, Native-American, and Hispanic subjects.

One way of conceptualizing what scale 4 assesses is to think of it as a measure of rebelliousness, with higher scores indicating rebellion and lower scores indicating an acceptance of authority and the status quo. The highest scorers on the scale rebel by acting out in antisocial and criminal ways; moderately high scorers may be rebellious but may express the rebellion in more socially acceptable ways; and low scorers are apt to be overly conventional and accepting of authority.

Interpretation of High Scores on Scale 4

Extremely high scores (T > 75) on scale 4 tend to be associated with difficulty in incorporating the values and standards of society. Such high scorers are likely to engage in a variety of asocial, antisocial, and even criminal behaviors. These behaviors may include lying, cheating, stealing, sexual acting out, and excessive use of alcohol and/or other drugs.

High scorers on scale 4 tend to be rebellious toward authority figures and often are in conflict with authorities of one kind or another. They often have stormy relationships with families, and family members tend to be blamed for their difficulties. Underachievement in school, poor work history, and marital problems are characteristic of high scorers.

High scorers are very impulsive persons who strive for immediate gratification of impulses. They often do not plan their behavior very well, and they may act without considering the consequences of their actions. They are very impatient and have a limited frustration tolerance. Their behavior may involve poor judgment and considerable risk taking. They tend not to profit from experiences and may find themselves in the same difficulties time and time again.

High scorers are described by others as immature and childish. They are narcissistic, self-centered, selfish, and egocentric. Their behavior often is ostentatious and exhibitionistic. They are insensitive to the needs and feelings of other people and are interested in others in terms of how they can be used. Although they tend to be seen as likable and generally create good first impressions, their relationships tend to be shallow and superficial. This may be in part due to rejection on the part of the people they mistreat, but it also seems to reflect their own inability to form warm attachments with others.

In addition, high scorers typically are extroverted and outgoing. They are talkative, active, adventurous, energetic, and spontaneous. They are judged by others to be intelligent and self-confident. Although they have a wide range of interests and may become involved in many activities, they lack definite goals and their behavior lacks clear direction.

High scorers tend to be hostile and aggressive. They are resentful, rebellious, antagonistic, and refractory. Their attitude is characterized by sarcasm and cynicism. Both men and women with high scale 4 scores may act in aggressive ways, but women are likely to express aggression in more passive, indirect ways. Often there does not appear to be any guilt associated with the aggressive behavior. Whereas high scorers may feign guilt and remorse when their behaviors get them into trouble, such responses typically are short-lived, disappearing when the immediate crisis passes.

Although high scorers typically are not seen as being overwhelmed by emotional turmoil, at times they may admit feeling sad, fearful, and worried about the future. They may experience absence of deep emotional response, which may produce feelings of emptiness and boredom. Among psychiatric patients high scorers tend to receive personality disorder diagnoses, with antisocial personality disorder or passive-aggressive personality disorder occurring most frequently.

Because of their verbal facility, outgoing manner, and apparent intellectual resources, high scorers often are perceived as good candidates for psychotherapy or counseling. Unfortunately, the prognosis for change is poor. Although they may agree to treatment to avoid something more unpleasant (e.g., jail or divorce), they generally are unable to accept blame for their own problems, and they terminate treatment as soon as possible. In therapy they tend to intellectualize excessively and to blame others for their difficulties.

Summary of Descriptors for Scale 4

High scores on scale 4 indicate persons who:

1. have difficulty incorporating the values and standards of society
2. may engage in asocial and antisocial acts, including lying, cheating, stealing, sexual acting out, and excessive use of alcohol and/or drugs (especially if T > 75)
3. are rebellious toward authority figures
4. have stormy relationships with families
5. blame family members for difficulties
6. have histories of underachievement
7. tend to experience marital problems
8. are impulsive and strive for immediate gratification of impulses
9. do not plan their behavior well
10. tend to act without considering the consequences of their actions
11. are impatient and have limited frustration tolerance
12. show poor judgment and take risks
13. tend not to profit from experiences
14. are seen by others as immature and childish
15. are narcissistic, self-centered, selfish, and egocentric
16. are ostentatious and exhibitionistic
17. are insensitive to the needs and feelings of others
18. are interested in others in terms of how they can be used
19. are likable and create good first impressions
20. have shallow, superficial relationships
21. seem unable to form warm attachments with others
22. are extroverted and outgoing
23. are talkative, active, adventurous, energetic, and spontaneous
24. are judged by others to be intelligent and self-confident
25. have a wide range of interests, but behavior lacks clear direction
26. tend to be hostile, aggressive, resentful, rebellious, antagonistic, and refractory
27. have sarcastic and cynical attitudes
28. may act in aggressive ways
29. if women, may express aggression in passive, indirect ways
30. may feign guilt and remorse when in trouble
31. are not seen as overwhelmed by emotional turmoil
32. may admit feeling sad, fearful, and worried about the future
33. experience absence of deep emotional response

34. feel empty and bored
35. if psychiatric patients, are likely to receive diagnoses of antisocial or passive-aggressive personality disorder
36. have poor prognosis for psychotherapy or counseling
37. may agree to treatment to avoid something more unpleasant
38. tend to terminate treatment prematurely
39. in treatment tend to intellectualize excessively and to blame others for difficulties

SCALE 5 (MASCULINITY-FEMININITY)

Scale 5 originally was developed by Hathaway and McKinley to identify homosexual invert males. The test authors identified only a very small number of items that differentiated homosexual from heterosexual men. Thus, items also were added to the scale if they differentiated between normal men and women in the standardization sample. Items from the Terman and Miles Attitude Interest Test also were added to the scale. Although Hathaway and McKinley considered this scale as preliminary, it has come to be used routinely in its original form.

The test authors attempted unsuccessfully to develop a corresponding scale for identifying sexual inversion in women. As a result, scale 5 has been used for both male and female subjects. Fifty-two of the items are keyed in the same direction for both genders, whereas four items, all dealing with frankly sexual material, are keyed in opposite directions for men and women. After obtaining raw scores, T-score conversions are reversed for the sexes so that a high raw score for men automatically is transformed by means of the profile sheet itself to a high T score, whereas a high raw score for women is transformed to a low T score. The result is that high T scores for both genders are indicative of deviation from one's own gender.

In the MMPI-2, fifty-six of the sixty items in the original scale 5 were maintained. Although a few of the items in scale 5 have frankly sexual content, most items are not sexual in nature and cover a diversity of topics, including work and recreational interests, worries and fears, excessive sensitivity, and family relationships.

Although the MMPI scale 5 scores were strongly related to the amount of formal education, the relationship is much more modest in the MMPI-2. Butcher (1990b) reported a correlation of .348 between scale 5 scores and years of education for men in the MMPI-2 normative sample. For women, the correlation was −.152. Thus, more educated men tend to obtain slightly higher T scores on scale 5 than less educated men. More educated women tend to obtain slightly lower T scores on scale 5 than less educated women. These differences probably reflect the broader interest patterns of more educated men and women and are not large enough to necessitate different scale 5 interpretations for persons with differing levels of education.

Interpretation of Scores on Scale 5

Very high scores on scale 5 (T > 65) for men and women suggest the possibility of sexual concerns and problems. These concerns and problems may be associated with homoerotic trends or homosexual behavior, but they also can center around sexual problems and behaviors of other kinds.

High scores (T > 60) for men on scale 5 are indicative of a lack of stereotypic masculine interests. High scorers tend to have aesthetic and artistic interests, and they are likely to participate in housekeeping and child-rearing activities to a greater extent than do most men.

High scores on scale 5 are very uncommon among female subjects. When they are encountered, they generally indicate rejection of the traditional female role. Women having high scale 5 scores are interested in sports, hobbies, and other activities that tend to be stereotypically more masculine than feminine.

Men who score low on scale 5 are presenting themselves as extremely masculine. They clearly have stereotypically masculine preferences in work, hobbies, and other activities. Women who score low on scale 5 are indicating that they have many stereotypically feminine interests. They are likely to derive satisfaction from their roles as spouses and mothers. Women with low scale 5 scores may be traditionally feminine or may have adopted an androgenous lifestyle.

Summary of Descriptors for Scale 5

High scale 5 scores for men indicate persons who:

1. may have sexual problems and concerns (if T > 65)
2. lack stereotypically masculine interests

High scale 5 scores for women indicate persons who:

1. are rejecting a traditional female role
2. have interests that tend to be stereotypically more masculine than feminine

Low scale 5 scores for men indicate persons who:

1. are presenting themselves as extremely masculine
2. have stereotypically masculine interests

Low scale 5 scores for women indicate persons who:

1. have many stereotypically feminine interests
2. are likely to derive satisfaction from their roles as spouses and mothers
3. may be traditionally feminine or androgenous

SCALE 6 (PARANOIA)

Scale 6 originally was developed to identify patients who were judged to have paranoid symptoms such as ideas of reference, feelings of persecution, grandiose self-concepts, suspiciousness, excessive sensitivity, and rigid opinions and attitudes. Although the scale was considered as preliminary because of problems in cross-validation, a major reason for its retention was that it produced relatively few false positives. Persons who score high on the scale usually have paranoid symptoms. However, some patients with clearly paranoid symptoms are able to achieve average scores on scale 6.

All forty of the items in the original scale were maintained in the MMPI-2. Although some of the items in the scale deal with frankly psychotic behaviors (e.g., suspiciousness, ideas of reference, delusions of persecution, and grandiosity), many items cover such diverse topics as sensitivity, cynicism, asocial behavior, excessive moral virtue, and complaints about other people. It is possible to obtain a T score greater than 65 on this scale without endorsing any of the frankly psychotic items.

Interpretation of High Scores on Scale 6

When T scores on scale 6 are above 70, and especially when scale 6 also is the highest scale in the profile, subjects may exhibit frankly psychotic behavior. Their thinking may be disturbed, and they may have delusions of persecution or grandeur. Ideas of reference also are common. They may feel mistreated and picked on; they may be angry and resentful; and they may harbor grudges. Projection is a common defense mechanism. Among psychiatric patients, diagnoses of schizophrenia or paranoid disorder are most frequent.

When scale 6 scores are within a T-score range of 60–70, frankly psychotic symptoms are not as common. However, persons with scores within this range are characterized by a variety of traits and behaviors that suggest a paranoid predisposition. They tend to be excessively sensitive and overly responsive to the opinions of others. They feel that they are getting a raw deal out of life and tend to rationalize and to blame others for their own difficulties. Also, they are seen as suspicious and guarded and commonly exhibit hostility, resentment, and an argumentative manner. They tend to be very moralistic and rigid in their opinions and attitudes. Rationality is likely to be greatly overemphasized. Women who score in this range may describe sadness, withdrawal, and anxiety, and they are seen by others as emotionally labile and moody. Prognosis for psychotherapy is poor because these subjects do not like to talk about emotional problems and are likely to rationalize most of the time. They have great difficulty in establishing rapport with therapists. In therapy, they are likely to reveal hostility and resentment toward family members.

Summary of Descriptors for Scale 6

Extreme elevations (T > 70) on scale 6 indicate persons who:

1. may exhibit frankly psychotic behavior
2. may have disturbed thinking, delusions of persecution or grandeur, and ideas of reference
3. feel mistreated and picked on
4. feel angry and resentful
5. harbor grudges
6. utilize projection as a defense mechanism
7. if psychiatric patients, often receive diagnoses of schizophrenia or paranoid disorder

Moderate elevations (T = 60–70) on scale 6 indicate persons who:

1. have a paranoid predisposition
2. tend to be excessively sensitive and overly responsive to the opinions of others
3. feel they are getting a raw deal out of life
4. tend to rationalize and blame others for difficulties
5. are suspicious and guarded
6. have hostility, resentment, and an argumentative manner
7. are moralistic and rigid in their opinions and attitudes
8. overemphasize rationality
9. if women, may describe sadness, withdrawal, and anxiety
10. if women, are seen by others as emotionally labile and moody
11. have poor prognosis for therapy
12. do not like to talk about emotional problems
13. rationalize excessively in therapy
14. have difficulty establishing rapport with therapists
15. in therapy reveal hostility and resentment toward family members

SCALE 7 (PSYCHASTHENIA)

Scale 7 originally was developed to measure the general symptomatic pattern labeled psychasthenia. Although this diagnostic label is not used commonly today, it was popular when the scale was developed. Among currently popular diagnostic categories, the obsessive-compulsive disorder probably is closest to the original psychasthenia label. Persons diagnosed as psychasthenic had thinking characterized by excessive doubts, compulsions, obsessions, and unreasonable fears. This symptom pattern was much more common among outpatients than among hospitalized patients, so the number of cases available for scale construction was small. All forty-eight items in the original scale were maintained in the MMPI-2. They cover a variety of symptoms and behaviors. Many of the items deal with uncontrollable or obsessive thoughts, feelings of fear and/or anxiety, and doubts about one's own ability. Unhappiness, physical complaints, and difficulties in concentration also are represented in the scale.

Interpretation of High Scores on Scale 7

Scale 7 is a reliable index of psychological turmoil and discomfort, with higher scorers experiencing greater turmoil. High scorers tend to be extremely anxious, tense, and agitated. They worry a great deal, even over small problems, and they are fearful and apprehensive. They are high-strung and jumpy and report difficulties in concentrating. High scorers on scale 7 often receive diagnoses of anxiety disorder.

High scorers tend to be very introspective, and they sometimes report fears that they are losing their minds. Obsessive thinking, compulsive and ritualistic behavior, and ruminations, often centering on feelings of insecurity and inferiority, are common among the highest scorers. They lack self-confidence, are self-critical, self-conscious, and self-degrading, and are plagued by self-doubts. High scorers tend to be very rigid and moralistic and to have high standards of behavior and performance for themselves and others. They are likely to be perfectionistic and conscientious; they may feel guilty about not living up to their own standards and depressed about falling short of goals.

In general, high scorers are neat, orderly, organized, and meticulous. They are persistent and reliable, but they lack ingenuity and originality in their approach to problems. They are seen by others as dull and formal. They have great difficulties in decision making. In addition, they are likely to distort the importance of problems and to be overreactive to stressful situations.

High scorers tend to be shy and do not interact well socially. They are described as hard to get to know, and they worry a great deal about popularity and social acceptance. Other people see them as sentimental, peaceable, soft-hearted, trustful, sensitive, and kind. Other adjectives used to describe them include dependent, unassertive, and immature.

Some high scorers express physical complaints that may center on the heart, the gastrointestinal system, or the genitourinary system. Complaints of fatigue, exhaustion, insomnia, and bad dreams are common.

Although high scorers may be motivated for therapy because they feel so uncomfortable and miserable, they are not very responsive to brief psychotherapy or counseling. In spite of some insight into their problems, they tend to rationalize and to intellectualize a great deal. They often are resistant to interpretations and may express much hostility toward the therapist. However, they tend to remain in therapy longer than most patients, and they may show very slow but steady progress. Problems presented in therapy may include difficulties with authority figures, poor work or study habits, or concern about homosexual impulses.

Summary of Descriptors for Scale 7

High scores on scale 7 indicate persons who:

1. experience psychological turmoil and discomfort
2. feel anxious, tense, and agitated
3. are worried, fearful, apprehensive, high-strung, and jumpy
4. report difficulties in concentrating
5. often receive diagnoses of anxiety disorder
6. are introspective
7. may report fears that they are losing their minds
8. have obsessive thinking, compulsive and ritualistic behavior, and ruminations
9. feel insecure and inferior
10. lack self-confidence
11. are self-critical, self-conscious, and self-degrading
12. are plagued by self-doubts
13. tend to be very rigid and moralistic
14. have high standards of performance for self and others
15. are perfectionistic and conscientious
16. feel depressed and guilty about falling short of goals
17. are neat, organized, and meticulous
18. are persistent and reliable
19. lack ingenuity in their approach to problems
20. are seen by others as dull and formal
21. have difficulties in making decisions
22. distort the importance of problems and overreact to stressful situations
23. tend to be shy and do not interact well socially
24. are described as hard to get to know
25. worry about popularity and social acceptance
26. are seen by others as sentimental, peaceable, soft-hearted, sensitive, and kind
27. are described as dependent, unassertive, and immature
28. may have physical complaints centering on:
    a. the heart
    b. the genitourinary system
    c. the gastrointestinal system
    d. fatigue, exhaustion, insomnia, and bad dreams
29. may be motivated for therapy because of inner turmoil
30. are not responsive to brief therapy or counseling
31. show some insight into their problems
32. rationalize and intellectualize excessively

33. are resistant to interpretations
34. may express hostility toward the therapist
35. remain in therapy longer than most patients
36. make slow but steady progress in therapy
37. discuss in therapy problems that include difficulty with authority figures, poor work or study habits, and concerns about homosexual impulses

SCALE 8 (SCHIZOPHRENIA)

Scale 8 was developed to identify patients diagnosed as schizophrenic. This category included a heterogeneous group of disorders characterized by disturbances of thinking, mood, and behavior. Misinterpretations of reality, delusions, and hallucinations may be present. Ambivalent or constricted emotional responsiveness is common. Behavior may be withdrawn, aggressive, or bizarre.

All seventy-eight of the items in the original scale were maintained in the MMPI-2. Some of the items deal with frankly psychotic symptoms, such as bizarre mentation, peculiarities of perception, delusions of persecution, and hallucinations. Other topics covered include social alienation, poor family relationships, sexual concerns, difficulties in impulse control and concentration, and fears, worries, and dissatisfactions.

Scores on scale 8 are related to age and to race. College students often obtain T scores in a range of 50 to 60, perhaps reflecting the turmoil associated with that period in life. African-American, Native-American, and Hispanic subjects in the MMPI-2 normative sample scored higher (approximately five T-score points higher) than Caucasian subjects. The elevated scores for minority subjects do not necessarily suggest greater overt psychopathology. They may simply be indicative of the alienation and social estrangement experienced by minority group members. Some elevations of scale 8 can be accounted for by subjects who are reporting a large number of unusual experiences, feelings, and perceptions related to the use of prescription and nonprescription drugs, especially amphetamines. Also, some persons with the disorders of epilepsy or stroke endorse sensory and cognitive items that lead to high scores on scale 8 (e.g., Dikmen, Hermann, Wilensky, & Rainwater, 1983; Dodrill, 1986; Gass & Lawhorn, 1991).

Interpretation of High Scores on Scale 8

Although one should be cautious about assigning a diagnosis of schizophrenia on the basis of only the score on scale 8, T scores in a range of 75 to 90 suggest the possibility of a psychotic disorder. Confusion, disorganization, and disorientation may be present. Unusual thoughts or attitudes, perhaps even delusional in nature, hallucinations, and extremely poor judgment may be evident. Extreme scores (T > 90) usually are not produced by psychotic subjects. They are more likely to be indicative of an individual who is in acute psychological turmoil or of a less disturbed person who is endorsing many deviant items as a cry for help. However, some recently hospitalized psychiatric patients obtain high-ranging scores on scale 8 that accurately reflect their severe psychopathology.

High scores on scale 8 may suggest a schizoid lifestyle. High scorers tend to feel as if they are not a part of their social environments. They feel isolated, alienated, misunderstood, and unaccepted by their peers. They are withdrawn, seclusive, secretive, and inaccessible and may avoid dealing with people and with new situations. They are described by others as shy, aloof, and uninvolved.

High scorers experience a great deal of apprehension and generalized anxiety, and they often report having bad dreams. They may feel sad or blue. They may feel very resentful, hostile, and aggressive, but they are unable to express such feelings. A typical response to stress is withdrawal into daydreams and fantasies, and some subjects may have a difficult time in separating reality and fantasy.

High scorers may be plagued by self-doubts. They feel inferior, incompetent, and dissatisfied. They give up easily when confronted with problem situations. Sexual preoccupation and sex-role confusion are common. Their behavior often is characterized by others as nonconforming, unusual, unconventional, and eccentric. Physical complaints may be present, and they usually are vague and long-standing in nature.

High scorers may at times be very stubborn, moody, and opinionated. At other times they are seen as generous, peaceable, and sentimental. Other adjectives used to describe high scorers include immature, impulsive, adventurous, sharp-witted, conscientious, and high-strung. Although they may have a wide range of interests and may be creative and imaginative in approaching problems, their goals generally are abstract and vague, and they seem to lack the basic information required for problem solving.

The prognosis for psychotherapy is not good because of the long-standing nature of high scorers' problems and their reluctance to relate in a meaningful way to the therapist. However, high scorers tend to stay in therapy longer than most patients, and eventually they may come to trust the therapist. Medical consultation to evaluate the appropriateness of chemotherapy may be indicated.

Summary of Descriptors for Scale 8

High scores on scale 8 indicate persons who:

1. may have a psychotic disorder (especially if T = 75–90)
2. may be confused, disorganized, and disoriented

3. may report unusual thoughts or attitudes and hallucinations
4. may show extremely poor judgment
5. may be in acute psychological turmoil
6. may be exaggerating deviance as a cry for help
7. tend to have a schizoid lifestyle
8. do not feel a part of their environments
9. feel isolated, alienated, misunderstood, and unaccepted
10. are withdrawn, seclusive, secretive, and inaccessible
11. avoid dealing with people and new situations
12. are described as shy, aloof, and uninvolved
13. experience apprehension and generalized anxiety
14. have bad dreams
15. may feel sad or blue
16. may feel resentful, hostile, and aggressive
17. are unable to express negative feelings
18. typically respond to stress by withdrawing into daydreams and fantasies
19. may have difficulty separating reality and fantasy
20. are plagued by self-doubts
21. feel inferior, incompetent, and dissatisfied
22. give up easily when confronted with problem situations
23. may have sexual preoccupation and/or sex-role confusion
24. are nonconforming, unusual, unconventional, and eccentric
25. have vague and long-standing physical complaints
26. may at times be stubborn, moody, and opinionated
27. may at times be seen as generous, peaceable, and sentimental
28. are described as immature, impulsive, adventurous, sharp-witted, conscientious, and high-strung
29. may have a wide range of interests
30. may be creative and imaginative in approaching problems
31. have abstract and vague goals
32. seem to lack the basic information required for problem solving
33. have a poor prognosis for psychotherapy because of the long-standing nature of their problems and reluctance to relate in a meaningful way to the therapist
34. tend to stay in therapy longer than most patients
35. may eventually come to trust the therapist
36. may require medical referral to evaluate the appropriateness of chemotherapy

SCALE 9 (HYPOMANIA)

Scale 9 originally was developed to identify psychiatric patients manifesting hypomanic symptoms. Hypomania is characterized by elevated mood, accelerated speech and motor activity, irritability, flight of ideas, and brief periods of depression.

All forty-six items in the original scale were maintained in the MMPI-2. Some of the items deal specifically with features of hypomanic disturbance (e.g., activity level, excitability, irritability, and grandiosity). Other items cover topics such as family relationships, moral values and attitudes, and physical or bodily concerns. No single dimension accounts for much of the variance in scores, and most of the sources of variance represented in the scale are not duplicated in other clinical scales.

Scores on scale 9 are related to age and race. Younger subjects (e.g., college students) typically obtain scores in a T-score range of 50 to 60, and for elderly subjects T scores below 50 are common. African-American, Native-American, and Hispanic subjects in the MMPI-2 normative samples scored somewhat higher (5–10 T-score points higher) than Caucasian subjects.

Scale 9 can be viewed as a measure of psychological and physical energy, with high scorers having excessive energy. When scale 9 scores are high, one expects that characteristics suggested by other aspects of the profile will be acted out. For example, high scores on scale 4 suggest asocial or antisocial tendencies. If scale 9 is elevated along with scale 4, these tendencies are likely to be expressed overtly in behavior.

Interpretation of High Scores on Scale 9

Extreme elevations (T > 80) on scale 9 may be suggestive of a manic episode. Patients with such scores are likely to show excessive, purposeless activity and accelerated speech; they may have hallucinations and/or delusions of grandeur; and they are very emotionally labile. Some confusion may be present, and flight of ideas is common.

Subjects with more moderate elevations are not likely to exhibit frankly psychotic symptoms, but there is a definite tendency toward overactivity and unrealistic self-appraisal. High scorers are energetic and talkative, and they prefer action to thought. They have a wide range of interests and are likely to have many projects going at once. However, they do not utilize energy wisely and often do not see projects through to completion. They may be creative, enterprising, and ingenious, but they have little interest in routine or details. High scorers tend to become bored and restless very easily, and their frustration tolerance is quite low. They have great difficulty in inhibiting expression of impulses, and periodic episodes of irritability, hostility, and aggressive outbursts are common. An unrealistic and unqualified optimism is also characteristic of high scorers. They seem to think that nothing is impossible, and they have grandiose aspirations. Also, they have an exaggerated appraisal of their own self-worth and self-importance and are not able to see their own limitations. High scorers have a greater-than-average likelihood of using nonprescription drugs and getting into trouble with the law.

High scorers are very outgoing, sociable, and gregarious. They like to be around other people and generally create good first impressions. They impress others as being friendly, pleasant, enthusiastic, poised, and self-confident. They tend to try to dominate other people. Their relationships are usually quite superficial, and as others get to know them better they become aware of their manipulations, deceptions, and unreliability.

In spite of the outward picture of confidence and poise, high scorers are likely to harbor feelings of dissatisfaction concerning what they are getting out of life. They may feel upset, tense, nervous, anxious, and agitated, and they describe themselves as prone to worry. Periodic episodes of depression may occur.

In psychotherapy, high scorers may reveal negative feelings toward domineering parents, may report difficulties in school or at work, and may admit to a variety of delinquent behaviors. High scorers are resistant to interpretations, are irregular in their attendance, and are likely to terminate therapy prematurely. They engage in a great deal of intellectualization and may repeat problems in a stereotyped manner. They do not become dependent on the therapist, who may be a target for hostility and aggression.

Summary of Descriptors for Scale 9

High scores on scale 9 indicate persons who:

1. if T > 80, may exhibit behavioral manifestations of a manic episode, including:
   a. excessive, purposeless activity
   b. accelerated speech
   c. hallucinations
   d. delusions of grandeur
   e. emotional lability
   f. confusion
   g. flight of ideas
2. are overactive
3. have unrealistic self-appraisal
4. are energetic and talkative
5. prefer action to thought
6. have a wide range of interests
7. may have many projects going at once
8. do not utilize energy wisely
9. often do not see projects through to completion
10. may be creative, enterprising, and ingenious
11. have little interest in routine or detail
12. tend to become bored and restless very easily
13. have a low frustration tolerance

14. have difficulty in inhibiting expression of impulses
15. have periodic episodes of irritability, hostility, and aggressive outbursts
16. are characterized by unrealistic and unqualified optimism
17. have grandiose aspirations
18. have an exaggerated appraisal of self-worth
19. are unable to see their own limitations
20. may use nonprescription drugs
21. may get into trouble with the law
22. are outgoing, sociable, and gregarious
23. like to be around other people
24. create good first impressions
25. impress others as friendly, pleasant, enthusiastic, poised, and self-confident
26. try to dominate other people
27. have quite superficial relationships with other people
28. eventually are seen by others as manipulative, deceptive, and unreliable
29. beneath an outward picture of confidence and poise, harbor feelings of dissatisfaction
30. may feel upset, tense, nervous, anxious, and agitated
31. may describe selves as prone to worry
32. may experience periodic episodes of depression
33. in psychotherapy may reveal negative feelings toward domineering parents; difficulties in school or at work; and a variety of delinquent behaviors
34. have poor prognosis for psychotherapy
35. are resistant to interpretations
36. are irregular in therapy attendance
37. are likely to terminate therapy prematurely
38. engage in a great deal of intellectualization
39. repeat problems in stereotyped manner
40. do not become dependent on therapists
41. may make therapists targets of hostility and aggression

SCALE 0 (SOCIAL INTROVERSION)

Although scale 0 was developed later than the other clinical scales, it has come to be used routinely. The scale was designed to assess a subject's tendency to withdraw from social contacts and responsibilities. Items were selected by contrasting high and low scorers on the Social Introversion-Extroversion scale of the Minnesota T-S-E Inventory. Only female subjects were used to develop the scale, but its use has been extended to male subjects as well.

All but one of the seventy items in the original scale were maintained in the MMPI-2. The items are of two general types. One group of items deals with social participation, whereas the other group deals with general neurotic maladjustment and self-depreciation. High scores can be obtained by endorsing either kind of item, or both. Scores on scale 0 are quite stable over extended periods of time.

Interpretation of Scores on Scale 0

The most salient characteristic of high scorers on scale 0 is social introversion. High scorers are very insecure and uncomfortable in social situations. They tend to be shy, reserved, timid, and retiring. They feel more comfortable when alone or with a few close friends, and they do not participate in many social activities. They may be especially uncomfortable around members of the opposite sex.

High scorers lack self-confidence, and they tend to be self-effacing. They are hard to get to know and are described by others as cold and distant. They are sensitive to what others think of them and they are likely to be troubled by their lack of involvement with other people. They are overcontrolled and are not likely to display their feelings directly. They are submissive and compliant in interpersonal relationships, and they are overly accepting of authority.

High scorers also are described as serious and as having a slow personal tempo. Although they are reliable and dependable, their approach to problems tends to be cautious, conventional, and unoriginal, and they give up easily. They are somewhat rigid and inflexible in their attitudes and opinions. They also have great difficulty in making even minor decisions. They seem to enjoy their work and get pleasure from productive personal achievement.

High scorers tend to worry, to be irritable, and to feel anxious. They are described by others as moody. Guilt feelings and episodes of depression may occur. High scorers seem to lack energy and do not have many interests.

Low scorers on scale 0 tend to be sociable and extroverted. They are outgoing, gregarious, friendly, and talkative. They have a strong need to be around other people, and they mix well with other people. They are seen by others as verbally fluent and expressive. They are active, energetic, and vigorous. They are interested in power, status, and recognition, and they tend to seek out competitive situations.

Summary of Descriptors for Scale 0

High scores on scale 0 indicate persons who:

1. are socially introverted
2. are very insecure and uncomfortable in social situations
3. tend to be shy, reserved, timid, and retiring
4. feel more comfortable alone or with a few close friends
5. do not participate in many social activities

6. may be especially uncomfortable around members of the opposite sex
7. lack self-confidence and tend to be self-effacing
8. are hard to get to know
9. are described by others as cold and distant
10. are sensitive to what others think of them
11. are likely to be troubled by their lack of involvement with other people
12. are quite overcontrolled and are not likely to display feelings openly
13. are submissive and compliant in interpersonal relationships
14. are overly accepting of authority
15. are described as serious and as having a slow personal tempo
16. are reliable and dependable
17. tend to have a cautious, conventional, and unoriginal approach to problems
18. tend to give up easily
19. are somewhat rigid and inflexible in attitudes and opinions
20. have great difficulty in making even minor decisions
21. seem to enjoy their work and get pleasure from productive personal achievement
22. tend to worry, to be irritable, and to feel anxious
23. are described by others as moody
24. may experience episodes of depression
25. seem to lack energy
26. do not have many interests

Low scores on scale 0 indicate persons who:

1. are sociable and extroverted
2. are outgoing, gregarious, friendly, and talkative
3. have a strong need to be around other people
4. mix well
5. are seen as expressive and verbally fluent

5

Profile Configurations

From the MMPI's inception, Hathaway and McKinley emphasized that configural interpretation of scores was diagnostically richer and thus more useful than interpretation that examined single scales without regard for relationships among the scales. Meehl (1951), Meehl and Dahlstrom (1960), Taulbee and Sisson (1957), and others also stressed configural approaches to MMPI interpretation. Thus, some of the earlier MMPI validity studies (e.g., Black, 1953; Guthrie, 1952; Meehl, 1951) grouped profiles according to the two highest clinical scales in the profile and tried to identify reliable extratest behaviors that were uniquely related to each such profile type. Other investigators (e.g., Gilberstadt & Duker, 1965; Marks & Seeman, 1963) developed complex rules for classifying profiles into homogeneous groups and tried to identify extratest correlates for each group. Although clinicians initially were quite enthusiastic about this complex approach to profile classification, they became disenchanted as accumulating research and clinical evidence indicated that only a small proportion of the MMPI protocols encountered in a typical clinical setting could be classified using the code types defined by these complex classification systems (Fowler & Coyle, 1968; Huff, 1965; Meikle & Gerritse, 1970).

Interest in complex rules for classifying profiles diminished, and there was a resurgence of interest in the simpler (two-scale, three-scale) approach to classification of profiles. Gynther and his colleagues (Gynther, Altman, & Sletten, 1973) and Lewandowski and Graham (1972) demonstrated that reliable extratest correlates can be identified for profiles that are classified according to their two highest clinical scales. An obvious advantage of the simpler approach is that a large proportion of the profiles encountered in most settings can be classified into code types that have been studied empirically. Marks et al. (1974), in their revision and extension of the earlier work by Marks and Seeman (1963), acknowledged that no appreciable loss in accuracy of extratest descriptions resulted when they used two-point code types instead of their more complex rules for classifying MMPI profiles.

This chapter presents interpretive information for certain configurations of MMPI-2 scales. Because only limited data are available concerning correlates of code types for the MMPI-2, the descriptions reported here are based mostly on studies of code types with the original MMPI. This approach seems justified because of the continuity between the original MMPI and the MMPI-2. Additionally, external correlates for the individual clinical scales of the MMPI-2 are markedly similar to those previously identified for the clinical scales of the original MMPI (Graham, 1988; Lachar, Hays, & Buckle, 1991). Several unpublished research studies have determined correlates of some of the MMPI-2 code types and concluded that they are consistent with those previously reported for the MMPI code types (Moreland & Walsh, 1991; Timbrook & Graham, 1992). Additional research is needed to verify that the same descriptors are appropriate for frequently occurring MMPI and MMPI-2 code types in various settings.

DEFINING CODE TYPES

Described very simply, code types are ways of classifying MMPI-2 profiles that take into account more than a single clinical scale at a time. Although code types based on both lowest and highest clinical scales were discussed for the original MMPI, the lack of information concerning the meaning of low MMPI-2 scores suggests that at the present time MMPI-2 code types based on low scores should not be interpreted. (See Chapter 4.)

The simplest code types are high-points. A high-point code type (e.g., high-point 2) tells us that the scale, in this example scale 2, is higher than any other clinical scale in the profile. The high-point code type does not tell us anything about the absolute level of the highest scale. There can only be one high-point code type for any particular profile. In an earlier chapter, we defined high scores in terms of absolute T-score levels and not in relation to other scales in the profile. In this earlier conceptualization, a subject could have several or more high scores. A careful examination of the MMPI literature suggests that basically the same descriptors apply to high-scoring individuals, whether we are dealing with high scores or with high-point code types. Thus, no additional interpretive information will be presented in this chapter for high-point code types.

Two-point code types tell us which two clinical scales are the highest ones in the profile. Thus, a 2–7 two-point code type tells us that scale 2 is the highest clinical scale in the profile and scale 7 is the second-highest clinical scale in the profile. For most two-point code types, the scales are interchangeable. For example, we often talk about code types such as 27/72, and we make basically the same interpretations for the 2–7 code type as we do for the 7–2 code type. Whenever order of the scales in the two-point code type makes a difference in interpretation, specific mention is made in the descriptive data for those particular code types in this chapter. As with high-point code types, two-point code types tell us nothing about the absolute level of scores for the two scales in the code type.

Three-point code types tell us which three clinical scales are the highest in the profile. For example, a 2–7–8 code type is one in which scale 2 is the highest in the profile, scale 7 is the second-highest, and scale 8 is the third-highest. For most three-point code types the order of scales is interchangeable. Whenever order is important in terms of the interpretations of code types, specific mention is made when the descriptive data for those code types are presented in this chapter.

Early approaches to code type interpretation did not take into account the extent to which the code types were defined. Definition refers to the difference in scores between the scales in the code type and those not in the code type. The T-score difference between the lowest scale in the code type and the next-highest scale in the profile should be considered. For example, for a two-point code type we would examine the T-score difference between the second- and third-highest scales. For a three-point code type we would examine the T-score difference between the third- and fourth-highest scales. Because the MMPI-2 clinical scales are not perfectly reliable, small differences between scores should not be considered meaningful. Given the standard errors of measurement of the clinical scales, T-score differences of less than five points should not be considered meaningful. Those who argue against restricting interpretation to defined code types (e.g., Dahlstrom, 1992) seem to be ignoring measurement error.

Thus, a defined code type is one in which the lowest scale in the code type has a score at least five T-score points higher than the next-highest scale in the profile. When such definition is present, we can be more confident that the descriptors associated with the code type are likely to characterize the person whose profile we are interpreting. Graham, Smith, and Schwartz (1986) found that code types of psychiatric patients were more likely to be stable over time when they were defined. Graham, Timbrook, Ben-Porath, and Butcher (1991) demonstrated that the congruence between the MMPI and MMPI-2 code types was greater when the code types were defined. (See Chapter 8.) When an MMPI-2 protocol does not have a defined code type, the code type approach to interpretation should not be used. Rather, inferences about the examinee should be based on the levels of scores on individual scales.

Levels of the scale scores in code types have been considered in different ways. Sometimes, the level of scores are not taken into account at all. For example, the interpretation of a 27/72 code type would be the same whether the T scores on scales 2 and 7 were above 90 or below 70. At other times, code types are interpreted only when T scores on the scales in the code type are significantly elevated (usually > 70 with the original MMPI).

This author believes that defined code types can be interpreted regardless of the level of scores on the scales in the code types. Typically, the list of descriptors for a particular code type includes both symptoms and personality characteristics. When the scores on the scales in the code types are quite high, inferences concerning both symptoms and personality characteristics should be considered. When the scores on the scales in the code types are not very high, inferences about symptoms should not be made (or should be made with considerable caution), but inferences about personality characteristics would apply. It is difficult to specify a T-score level above which scores would be considered to be very high and below which they would not be considered to be very high. As a rough rule of thumb, the inferences concerning symptoms probably should not be included (or included cautiously) unless T scores on the scales in the code type are greater than 65. Obviously, when T scores are considerably higher than 65 there is an even greater likelihood that inferences concerning symptoms will be appropriate.

To illustrate, consider a protocol in which there is a defined 27/72 code type. The list of descriptors associated with this code type (which is presented later in this chapter) includes statements about anxiety and depression as well as statements about personality characteristics, including insecurity, perfectionism, and passive-dependent relationships with other people. If the scores on scales 2 and 7 are very high (T > 65), we would include inferences about symptoms of anxiety and depression and about the personality characteristics listed above. However, if the scores on scales 2 and 7 are not very high (T < 65), we would not make the inferences about anxiety and depression but would make inferences about insecurity, perfectionism, and passive-dependent relationships.

As more aspects of the MMPI-2 profile are taken into consideration in determining code types, the profiles that fit into each type are more homogeneous. Likewise, the persons producing those profiles are more likely to be similar to each other. Thus, we expect that the descriptors for more complex code types are more likely to apply to a particular examinee with one of those code types than will be the case for descriptors derived from less complex code types.

If the ten clinical scales are used interchangeably, 90 two-point code types and 720 three-point code types are possible. The code types for which interpretive information is provided in this chapter are those that occur reasonably frequently in a variety of settings and for which an adequate amount of interpretive information is available in the literature.[1] Few code types including scales 5 and 0 are included. The reason is that in many code type studies scales 5 and 0 were excluded because they were added after the original publication of the MMPI and thus were not available for subjects in some of the early studies. Some investigators have reported correlates of infrequently occurring code types (e.g., Kelley & King, 1979c; Tanner, 1990). The results

---

[1]Sources consulted in preparing Chapter 5: Anderson and Bauer (1985); Carson (1969); Dahlstrom et al. (1972); Davis and Sines (1971); Drake and Oetting (1959); Duckworth and Anderson (1986); Gilberstadt and Duker (1965); Good and Brantner (1961); Gynther et al. (1973); Hovey and Lewis (1967); Kelley and King (1979a, 1979b); Lachar (1974b); Lewandowski and Graham (1972); Marks et al. (1974); Moreland and Walsh (1991); Nelson and Marks (1985); Persons and Marks (1971); Schubert (1973); Timbrook and Graham (1992).

of these studies are not included in this chapter because the samples were relatively small and cross-validational studies have not been reported.

Descriptors for the various MMPI and MMPI-2 code types were determined in research studies by comparing persons with a particular code type with other persons in the same setting without that code type. It should be emphasized that the descriptors for two- and three-point code types were not rationally generated by combining descriptors for the individual scales included in the code types. It should be understood that the descriptions that follow are modal patterns and obviously do not describe unfailingly each and every person with a specific code type. Rather, the descriptors presented for a particular code type are more likely to apply to persons with that code type than to persons who do not have the code type. The descriptive information is not presented separately by gender because most earlier research did not analyze data by gender. However, some analyses have indicated gender differences in descriptors associated with certain code types. Future research should analyze and report data separately by gender. For profiles that do not have defined two- or three-point code types, interpretations should be based on scores on the individual clinical scales. (See Chapter 4.)

TWO-POINT CODE TYPES

12/21

The most prominent features of the 12/21 code type are somatic discomfort and pain. Individuals with this code type present themselves as physically ill, although there may be no clinical evidence of an organic basis for their symptoms. They are very concerned about health and bodily functions and they are likely to overreact to minor physical dysfunction. They may present multiple somatic complaints, or the symptoms may be restricted to a particular system. Although headaches and cardiac complaints may occur, the digestive system is more likely to be involved. Ulcers, particularly of the upper gastrointestinal tract, are common, and anorexia, nausea, and vomiting may be present. Individuals with the 12/21 code type also may complain of dizziness, insomnia, weakness, fatigue, and tiredness. They tend to react to stress with physical symptoms, and they resist attempts to explain their symptoms in terms of emotional or psychological factors.

12/21 individuals are generally anxious, tense, and nervous. Also, they are high-strung and prone to worry about many things, and they tend to be restless and irritable. Although pronounced clinical depression is not common for persons with the 12/21 code type, they may report feelings of unhappiness or dysphoria, brooding, and loss of initiative.

Persons with the 12/21 code type report feeling very self-conscious. They are introverted and shy in social situations, particularly with members of the opposite sex, and they tend to be somewhat withdrawn and seclusive. They harbor many doubts about their own abilities, and they are indecisive about even minor, everyday matters. They are hypersensitive concerning what other people think about them, and they may be somewhat suspicious and untrusting in interpersonal relations. They also tend to be passive-dependent in their relationships, and they may harbor hostility toward people who are perceived as not offering enough attention and support.

Excessive use of alcohol may be a problem for 12/21 individuals, especially among psychiatric patients. Their histories may include blackouts, job loss, arrests, and family problems associated with alcohol abuse. Persons with the 12/21 code type most often are given diagnoses of anxiety disorders, depressive disorders, or somatoform disorders, although a small proportion of individuals with this code type may be diagnosed as schizophrenic. In this latter group (schizophrenic), usually scale 8 also is elevated along with scales 1 and 2.

Individuals with the 12/21 code type are not seen as good risks for traditional psychotherapy. They can tolerate high levels of discomfort before becoming motivated to change. They utilize repression and somatization excessively, and they lack insight and self-understanding. In addition, their passive-dependent lifestyles make it difficult for them to accept responsibility for their own behavior. Although long-term change after psychotherapy is not likely, short-lived symptomatic changes often occur.

13/31

The 13/31 code type is more common among women and older persons than among men and younger persons. Psychiatric patients with the 13/31 code type usually receive somatoform disorder diagnoses. Classical conversion symptoms may be present, particularly if scale 2 is considerably lower than scales 1 and 3 (i.e., the so-called conversion V pattern). Whereas some tension may be reported by 13/31 persons, severe anxiety and depression usually are absent, as are clearly psychotic symptoms. Rather than being grossly incapacitated in functioning, the 13/31 individual is likely to continue functioning but at a reduced level of efficiency.

The somatic complaints presented by 13/31 persons include headaches, chest pain, back pain, and numbness or tremors of the extremities. Eating problems, including anorexia, nausea, vomiting, and obesity, are common. Other physical complaints include weakness, fatigue, dizziness, and sleep disturbance. The physical symptoms increase in times of stress, and often there is clear secondary gain associated with the symptoms.

Individuals with the 13/31 code type present themselves as normal, responsible, and without fault. They make excessive use of denial, projection, and rationalization, and they blame others for their difficulties. They prefer medical explanations for their symptoms, and they lack insight into the psychological factors underlying their symptoms. They manifest an overly optimistic and Pollyannaish view of their situations and of the world in general, and they do not show appropriate concern about their symptoms and problems.

13/31 persons tend to be rather immature, egocentric, and selfish. They are insecure and have a strong need for attention, affection, and sympathy. They are very dependent, but they are uncomfortable with the dependency and experience conflict because of it. Although they are outgoing and socially extroverted, their social relationships tend to be shallow and superficial, and they lack genuine emotional involvement with other people. They tend to exploit social relationships in an attempt to fulfill their own needs. They lack skills in dealing with the opposite sex, and they may be deficient in heterosexual drive.

13/31 individuals harbor resentment and hostility toward other people, particularly those who are perceived as not fulfilling their needs for attention. Most of the time they are overcontrolled and likely to express their negative feelings in indirect, passive ways, but they occasionally lose their tempers and express themselves in angry, but not violent, ways. Behaving in a socially acceptable manner is important to 13/31 persons. They need to convince other people that they are logical and reasonable, and they are conventional and conforming in their attitudes and values.

Because of their unwillingness to acknowledge psychological factors underlying their symptoms, 13/31 persons are difficult to motivate in traditional psychotherapy. They are reluctant to discuss psychological factors that might be related to somatic symptoms, and if therapists insist on doing so, people with this code type are likely to terminate therapy prematurely. Sometimes it is possible to get these persons to discuss problems as long as no direct link to somatic symptoms is suggested. In therapy they expect therapists to provide definite answers and solutions to their problems, and they may terminate therapy when therapists fail to respond to their demands. Because 13/31 persons tend to be suggestible, they often will try activities suggested by their therapists.

14/41

The 14/41 code type is not encountered frequently in clinical practice and is much more likely to be found for men than for women. Persons with the 14/41 code type frequently report severe somatic symptoms, particularly nonspecific headaches. They also may appear to be indecisive and anxious. Although they are socially extroverted, they lack skills with members of the opposite sex. They may feel rebellious toward home and parents, but direct expression of these feelings is not likely. Excessive use of alcohol may be a problem, and 14/41 persons may have a history of alcoholic benders, job loss, and family problems associated with their drinking behavior. In school or on the job, 14/41 persons lack drive and do not have well-defined goals. They are dissatisfied and pessimistic in their outlook toward life, and they are demanding, grouchy, and referred to as "bitchy" in interpersonal relationships. Because they are likely to deny psychological problems, they tend to be resistant to traditional psychotherapy.

18/81

Persons with the 18/81 code type harbor feelings of hostility and aggression, and they are not able to express these feelings in a modulated, adaptive manner. Either they inhibit expression almost completely, which results in feelings of being "bottled up," or they are overly belligerent and abrasive.

18/81 persons feel socially inadequate, especially around members of the opposite sex. They lack trust in other people, keep them at a distance, and feel generally isolated and alienated. A nomadic lifestyle and a poor work history are common.

Psychiatric patients with the 18/81 code type most often are diagnosed as schizophrenic, although diagnoses of anxiety disorders and schizoid personality disorder are sometimes given. 18/81 persons tend to be unhappy and depressed, and they may display flat affect. They present somatic concerns (including headaches and insomnia), which at times are so intense that they border on being delusional. 18/81 persons also may be confused in their thinking, and they are very distractible.

19/91

Persons with the 19/91 code type are likely to be experiencing a great deal of distress and turmoil. They tend to be very anxious, tense, and restless. Somatic complaints, including gastrointestinal problems, headaches, and exhaustion, are common, and these people are reluctant to accept psychological explanations of their symptoms. Although on the surface 19/91 individuals appear to be verbal, socially extroverted, aggressive, and belligerent, they are basically passive-dependent persons who are trying to deny this aspect of their personalities.

19/91 persons have a great deal of ambition. They expect a high level of achievement from themselves, but they lack clear and definite goals. They are frustrated by their inability to achieve at a high level. The 19/91 code type is sometimes found for brain-damaged individuals who are experiencing difficulties in coping with their limitations and deficits.

23/32

Although persons with the 23/32 code type typically do not experience disabling anxiety, they report feeling nervous, agitated, tense, and worried. They also report feeling sad, unhappy, and depressed. Fatigue, exhaustion, and weakness are common. They lack interest and involvement in their life situations, and they have difficulty in getting started on a project. Decreased physical activity is likely, and somatic complaints, usually gastrointestinal in nature, may occur.

23/32 individuals are rather passive, docile, and dependent. They are plagued by self-doubts, and they harbor feelings of inadequacy, insecurity, and helplessness. They tend to elicit helping behaviors from other people. However, persons with the 23/32 code type are very interested in achievement, status, and power. They may appear to be competitive, industrious, and driven, but they do not really place themselves in directly competitive situations where they might experience failure. They seek increased responsibility, but they dread the stress and pressure associated with it. They often feel that they do not get adequate recognition for their accomplishments, and they are easily hurt by even mild criticism.

23/32 persons are extremely overcontrolled. They have difficulty expressing their feelings, and they may feel bottled up much of the time. They tend to deny unacceptable impulses, and when denial fails they feel anxious and guilty. Persons with the 23/32 code type feel socially inadequate, and they tend to avoid social involvement. They are especially uncomfortable with members of the opposite sex, and sexual maladjustment, including frigidity and impotence, is common.

The 23/32 code type is much more common for women than for men. Rather than indicating incapacitating symptoms, it suggests a lowered level of efficiency for prolonged periods. Problems are long-standing, and the 23/32 person has learned to tolerate a great deal of unhappiness. Among psychiatric patients, the diagnosis most frequently assigned to persons with the 23/32 code type is depressive disorder. Antisocial personality disorder diagnoses are very rare among persons with this code type.

Response to traditional psychotherapy is likely to be poor for the 23/32 persons. They are not introspective; they lack insight into their own behavior; they resist psychological formulations of their problems; and they tolerate a great deal of unhappiness before becoming motivated to change.

24/42

When persons with the 24/42 code type come to the attention of professionals, it usually is after they have been in trouble with their families or with the law. They are impulsive and unable to delay gratification of their impulses. They have little respect for social standards and often find themselves in direct conflict with societal values. Their acting-out behavior is likely to involve excessive use of alcohol, and their histories often include alcoholic benders, arrests, job loss, and family discord associated with drinking.

24/42 persons feel frustrated by their own lack of accomplishment and are resentful of demands placed on them by other people. They may react to stress by drinking excessively or by using addictive drugs. After periods of acting out, they express a great deal of remorse and guilt about their misdeeds. They may report feeling depressed, anxious, and worthless, but their expressions do not seem to be sincere. In spite of their resolutions to change, they are likely to act out again in the future. It has been noted in the literature that when scales 2 and 4 are grossly elevated, suicidal ideation and attempts are quite possible. Often the suicide attempts are directed at making other people feel guilty.

When they are not in trouble, 24/42 persons may seem to be energetic, sociable, and outgoing. They create favorable first impressions, but their tendencies to manipulate others produce feelings of resentment in long-term relationships. Beneath the outer facade of competent, comfortable persons, 24/42 individuals tend to be introverted, self-conscious, and passive-dependent. They harbor feelings of inadequacy and self-dissatisfaction, and they are uncomfortable in social interactions, particularly those involving the opposite sex. At times they appear to be rigid and overly intellectualized.

Personality disorder diagnoses often are given to persons with the 24/42 code type. Although persons with this code type may express the need for help and the desire to change, the prognosis for traditional psychotherapy is not good. They are likely to terminate therapy prematurely when the situational stress subsides or when they have extracted themselves from their current difficulties. Even when they stay in therapy, not much improvement is likely.

27/72

27/72 individuals tend to be anxious, nervous, tense, high-strung, and jumpy. They worry excessively, and they are vulnerable to real and imagined threat. They tend to anticipate problems before they occur and to overreact to minor stress. Somatic symptoms are common among 27/72 persons. They usually involve rather vague complaints of fatigue, tiredness, and exhaustion, but insomnia, anorexia, and cardiac pain may be reported. Depression also is an important feature of the 27/72 code type. Although 27/72 persons may not report feeling especially sad or unhappy, they show symptoms of clinical depression, including weight loss, slow personal tempo, and retarded thought processes. They are extremely pessimistic about the world in general and more specifically about the likelihood of overcoming their problems, and they brood and ruminate about their problems much of the time.

Individuals with the 27/72 code type have a strong need for achievement and for recognition for their accomplishments. They have high expectations for themselves, and they feel guilty when they fall short of their goals. They tend to be indecisive, and they harbor feelings of inadequacy, insecurity, and inferiority. They are intropunitive, blaming themselves for all the problems in their life situations. 27/72 individuals are rigid in their thinking and problem solving, and they are meticulous and perfectionistic in their daily activities. They also may be very religious and extremely moralistic.

Persons with the 27/72 code type tend to be docile and passive-dependent in their relationships with other people. In fact, they often find it difficult to be even appropriately assertive. They have the capacity for forming deep, emotional ties, and in times of stress they become overly clinging and dependent. Not aggressive or belligerent, they tend to elicit nurturance and helping behavior from other people. Because of the intense discomfort they experience, they are motivated for psychotherapy. They tend to remain in psychotherapy longer than many patients, and slow but steady progress can be expected.

Psychiatric patients with the 27/72 code type are likely to receive a diagnosis of anxiety disorder, depressive disorder, or obsessive-compulsive disorder. Diagnoses of antisocial personality disorder are very rare among persons with this code type.

28/82

Persons with the 28/82 code type report feeling anxious, agitated, tense, and jumpy. Sleep disturbance, inability to concentrate, confused thinking, and forgetfulness also are characteristic of 28/82 people. Such persons are inefficient in carrying out their responsibilities, and they tend to be unoriginal in their thinking and problem solving. They are likely to present themselves as physically ill, and somatic complaints include dizziness, blackout spells, nausea, and vomiting. They resist psychological interpretations of their problems, and they are resistant to change. They underestimate the seriousness of their problems, and they tend to be unrealistic about their own capabilities.

28/82 individuals are basically dependent and ineffective, and they have problems in being assertive. They are irritable and resentful much of the time; they fear loss of control and do not express themselves directly. They attempt to deny undesirable impulses, and cognitive dissociative periods during which negative emotions are expressed may occur. Such periods are followed by guilt and depression. 28/82 persons are rather sensitive to the reactions of others, and they are quite suspicious of the motivations of others. They may have a history of being hurt emotionally, and they fear being hurt again. They avoid close interpersonal relationships, and they keep people at a distance emotionally. This lack of meaningful involvement with other people increases their feelings of despair and worthlessness.

If both scales 2 and 8 are very elevated, the 28/82 code type is suggestive of serious psychopathology. The most common diagnoses given to psychiatric patients with this code type are bipolar disorder and schizoaffective disorder. 28/82 individuals have chronic, incapacitating symptomatology. They are guilt-ridden and appear to be clinically depressed. Withdrawal, flat affect, soft and reduced speech, retarded stream of thought, and tearfulness also are common. Psychiatric patients with the 28/82 code type may be preoccupied with suicidal thoughts, and they may have a specific plan for doing away with themselves.

29/92

29/92 persons tend to be self-centered and narcissistic, and they ruminate excessively about self-worth. Although they may express concern about achieving at a high level, it often appears that they set themselves up for failure. In younger persons, the 29/92 code type may be suggestive of an identity crisis characterized by lack of personal and vocational direction.

29/92 persons report feeling tense and anxious, and somatic complaints, often centering in the upper gastrointestinal tract, are common. Although they may not appear to be clinically depressed at the time they are examined, their histories typically suggest periods of serious depression. Excessive use of alcohol may be employed as an escape from stress and pressure.

The 29/92 code type is found primarily among individuals who are denying underlying feelings of inadequacy and worthlessness and defending against depression through excessive activity. Alternating periods of increased activity and fatigue may occur. Although the most common diagnosis for psychiatric patients with the 29/92 code type is bipolar disorder, it sometimes is found for patients with brain damage who have lost emotional control or who are trying to cope with deficits through excessive activity.

34/43

The most salient characteristic of 34/43 persons is chronic, intense anger. They harbor hostile and aggressive impulses, but they are unable to express their negative feelings appropriately. If scale 3 is higher than scale 4, passive, indirect expression of anger is likely. Persons with scale 4 higher than scale 3 appear to be overcontrolled most of the time, but brief episodes of aggressive acting out may occur. Prisoners with the 4–3 code type have histories of assaultive, violent crimes. In some rare instances, individuals with the 34/43 code type successfully dissociate themselves from their aggressive acting-out behavior. 34/43 individuals lack insight into the origins and consequences of their behavior. They tend to be extrapunitive and to blame other people for their difficulties. Other people may define the 34/43 person's behavior as problematic, but he or she is not likely to view it in the same way.

Persons with the 34/43 code type are reasonably free of disabling anxiety and depression, but complaints of headaches, upper gastrointestinal discomfort, and other somatic distress may occur. Although these persons may feel upset at times, the upset does not seem to be related directly to external stress.

Most of the 34/43 person's difficulties stem from deep, chronic feelings of hostility toward family members. They demand attention and approval from others. They are very sensitive to rejection, and they become hostile when criticized. Although they appear outwardly to be socially conforming, inwardly they are quite rebellious. They may be sexually maladjusted, and marital instability and sexual promiscuity are common. Suicidal thoughts and attempts are characteristic of 34/43 individuals; these are most likely to follow episodes of excessive drinking and acting-out behavior. Personality disorder diagnoses are most commonly associated with the 34/43 code type, with passive-aggressive personality being most common.

36/63

Individuals with the 36/63 code type may report moderate tension and anxiety and may have physical complaints, including headaches and gastrointestinal discomfort, but their problems do not seem to be acute or incapacitating. Most of their difficulties stem from deep, chronic feelings of hostility toward family members. They do not express these feelings directly, and much of the time they may not even recognize the hostile feelings. When they become aware of their anger, they try to justify it in terms of the behavior of others. In general, 36/63 individuals are defiant, uncooperative, and hard to get along with. They may express mild suspiciousness and resentment about others, and they are very self-centered and narcissistic. They deny serious psychological problems and express a very naive, Pollyannaish attitude toward the world.

38/83

Persons with the 38/83 code type appear to be in a great deal of psychological turmoil. They report feeling anxious, tense, and nervous. Also, they are fearful and worried, and phobias may be present. Depression and feelings of hopelessness are common among 38/83 individuals, and they have difficulties in making even minor decisions. A wide variety of physical complaints (gastrointestinal and musculoskeletal discomfort, dizziness, blurred vision, chest pain, genital pain, headaches, insomnia) may be presented. 38/83 persons tend to be vague and evasive when talking about their complaints and difficulties.

38/83 persons are rather immature and dependent and have strong needs for attention and affection. They display intropunitive reactions to frustration. They are not involved actively in their life situations, and they are apathetic and pessimistic. They approach problems in an unoriginal, stereotyped manner.

The 38/83 code type suggests the presence of disturbed thinking. Individuals with this code type complain of not being able to think clearly, of problems in concentration, and of lapses of memory. They express unusual, unconventional ideas, and their ideational associations may be rather loose. Obsessive ruminations, blatant delusions and/or hallucinations, and irrelevant, incoherent speech may be present. The most common diagnosis for psychiatric patients with the 38/83 code type is schizophrenia, but they are sometimes diagnosed as having somatoform disorders. Although response to insight-oriented psychotherapy is not likely to be good for 38/83 persons, they often benefit from a supportive psychotherapeutic relationship.

46/64

Persons with the 46/64 code type are immature, narcissistic, and self-indulgent. They are passive-dependent individuals who make excessive demands on others for attention and sympathy, but they are resentful of even the most mild demands made on them by others. Women with the 46/64 code type seem overly identified with the traditional female role and are very dependent on men. Both 46/64 men and women do not get along well with others in social situations, and they are especially uncomfortable around members of the opposite sex. They are suspicious of the motivations of others and avoid deep emotional involvement. They generally have poor work histories, and marital problems are quite common. Repressed hostility and anger are characteristic of 46/64 persons. They appear to be irritable, sullen, argumentative, and generally obnoxious. They seem to be especially resentful of authority and may derogate authority figures.

Individuals with the 46/64 code type tend to deny serious psychological problems. They rationalize and transfer blame to others, accepting little or no responsibility for their own behavior. They are somewhat unrealistic and grandiose in their self-appraisals. Because they deny serious emotional problems, they generally are not receptive to traditional counseling or psychotherapy.

Among psychiatric patients, diagnoses associated with the 46/64 code type are about equally divided between passive-aggressive personality disorder and schizophrenia, paranoid type. In general, as the elevation of scales 4 and 6 increases and as scale 6 becomes higher than scale 4, a prepsychotic or psychotic disorder becomes more likely. 46/64 individuals present vague emotional and physical complaints. They report feeling nervous and depressed, and they are indecisive and insecure. Physical symptoms may include asthma, hay fever, hypertension, headaches, blackout spells, and cardiac complaints.

47/74

Persons with the 47/74 code type may alternate between periods of gross insensitivity to the consequences of their actions and excessive concern about the effects of their behavior. Episodes of acting out, which may include excessive drinking and sexual promiscuity, may be followed by temporary expressions of guilt and self-condemnation. However, the remorse does not inhibit further episodes of acting out. 47/74 individuals may present vague somatic complaints, including headaches and stomach pain. They also may report feeling tense, fatigued, and exhausted. They are rather dependent, insecure individuals who require almost constant reassurance of their self-worth. A diagnosis of passive-aggressive personality disorder often is given to persons with the 47/74 code type. In psychotherapy they tend to respond symptomatically to support and reassurance, but long-term changes in personality are unlikely.

48/84

48/84 individuals do not seem to fit into their environments. They are seen by others as odd, peculiar, and queer. They are nonconforming and resentful of authority, and they often espouse radical religious or political views. Their behavior is erratic and unpredictable, and they have marked problems with impulse control. They tend to be angry, irritable, and resentful, and they act out in asocial or antisocial ways. When crimes are committed by 48/84 persons, they tend to be vicious and assaultive and often appear to be senseless, poorly planned, and poorly executed. Prostitution, promiscuity, and sexual deviation are fairly common among 48/84 individuals. This is the most common code type among male rapists. Excessive drinking and drug abuse (particularly involving hallucinogens) may also occur. Histories of 48/84 individuals usually indicate underachievement, uneven performance, and marginal adjustment.

Persons with the 48/84 code type harbor deep feelings of insecurity, and they have exaggerated needs for attention and affection. They have poor self-concepts, and it seems as if they set themselves up for rejection and failure. They may have periods during which they become obsessed with suicidal ideation. 48/84 persons are quite distrustful of other people, and they avoid close relationships. When they are involved interpersonally, they have impaired empathy and try to manipulate others into satisfying their needs. They lack basic social skills and tend to be socially withdrawn and isolated. The world is seen as a threatening and rejecting place, and their response is to withdraw or to strike out in anger as a defense against being hurt. They accept little responsibility for their own behavior, and they rationalize excessively, blaming their difficulties on other people. 48/84 persons tend to harbor serious concerns about their masculinity or femininity. They may be obsessed with sexual thoughts, but they are afraid that they cannot perform adequately in sexual situations. They may indulge in antisocial sexual acts in an attempt to demonstrate sexual adequacy.

Psychiatric patients with the 48/84 code type tend to be diagnosed as schizophrenic (paranoid type) or as having an antisocial, schizoid, or paranoid personality disorder. If both scales 4 and 8 are very elevated, and particularly if scale 8 is much higher than scale 4, the likelihood of psychosis and bizarre symptomatology, including unusual thinking and paranoid suspiciousness, increases.

49/94

The most salient characteristic of 49/94 individuals is a marked disregard for social standards and values. They frequently get in trouble with the authorities because of antisocial behavior. They have a poorly developed conscience, easy morals, and fluctuating ethical values. Alcoholism, fighting, marital problems, sexual acting out, and a wide array of delinquent acts are among the difficulties in which they may be involved. This is a common code type among persons who abuse alcohol and other substances.

49/94 individuals are narcissistic, selfish, and self-indulgent. They are quite impulsive and are unable to delay gratification of their impulses. They show poor judgment, often acting without considering the consequences of their acts, and they fail to learn from experience. They are not willing to accept responsibility for their own behavior, rationalizing shortcomings and failures and blaming difficulties on other people. They have a low tolerance for frustration, and they often appear to be moody, irritable, and caustic. They harbor intense feelings of anger and hostility, and these feelings are expressed in occasional emotional outbursts.

49/94 persons tend to be ambitious and energetic, and they are restless and overactive. They are likely to seek out emotional stimulation and excitement. In social situations they tend to be uninhibited, extroverted, and talkative, and they create a good first impression. However, because of their self-centeredness and distrust of people, their relationships are likely to be superficial and not particularly rewarding. They seem to be incapable of forming deep emotional ties and keep others at an emotional distance. Beneath the facade of self-confidence and security, the 49/94 individuals are immature, insecure, and dependent persons who are trying to deny these aspects of themselves. A diagnosis of antisocial personality disorder is usually associated with the 49/94 code type, although patients with this disorder occasionally are diagnosed as having a bipolar disorder.

68/86

Persons with the 68/86 code type harbor intense feelings of inferiority and insecurity. They lack self-confidence and self-esteem, and they feel guilty about perceived failures. Withdrawal from everyday activities and emotional apathy are common, and suicidal ideation may be present. 68/86 persons are not emotionally involved with other people. They are suspicious and distrustful of others, and they avoid deep emotional ties. They are seriously deficient in social skills, and they are most comfortable when alone. They are quite resentful of demands placed on them, and other people see them as moody, irritable, unfriendly, and negativistic. In general, their lifestyles can be characterized as schizoid.

Although some persons with the 68/86 code type are diagnosed as having paranoid or schizoid personality disorders, among psychiatric patients this configuration usually is associated with a diagnosis of schizophrenia, paranoid type, particularly if scales 6 and 8 are considerably higher than scale 7. 68/86 individuals are likely to manifest clearly psychotic behavior. Thinking is described as autistic, fragmented, tangential, and circumstantial, and thought content is likely to be bizarre. Difficulties in concentrating and attending, deficits in memory, and poor judgment are common. Delusions of persecution and/or grandeur and hallucinations may be present, and feelings of unreality may be reported. Persons with the 68/86 code type often are preoccupied with abstract or theoretical matters to the exclusion of specific, concrete aspects of their life situations. Affect may be blunted, and speech may be rapid and at times incoherent. Effective defenses seem to be lacking, and these persons respond to stress and pressure by withdrawing into fantasy and daydreaming. Often it is difficult for 68/86 persons to differentiate between fantasy and reality. Medical consultation to determine appropriateness of psychotropic medication should be considered.

69/96

69/96 individuals are rather dependent and have strong needs for affection. They are vulnerable to real or imagined threat, and they feel anxious and tense much of the time. In addition, they may appear to be tearful and trembling. A marked overreaction to minor stress also is characteristic of persons with the 69/96 code type. A typical response to severe stress is withdrawal into fantasy. 69/96 individuals are unable to express emotions in an adaptive, modulated way, and they may alternate between overcontrol and direct, undercontrolled emotional outbursts.

Psychiatric patients with the 69/96 code type frequently receive a diagnosis of schizophrenia, paranoid type, and they are likely to show signs of a thought disorder. They complain of difficulties in thinking and concentrating, and their stream of thought is retarded. They are ruminative, overideational, and obsessional. They may have delusions and hallucinations, and their speech seems to be irrelevant and incoherent. They appear to be disoriented and perplexed, and they may show poor judgment.

78/87

78/87 individuals typically are in a great deal of turmoil. They are not hesitant to admit to psychological problems, and they seem to lack adequate defenses to keep them reasonably comfortable. They report feeling depressed, worried, tense, and nervous. When first seen professionally, they may appear to be confused and in a state of panic. They show poor judgment and do not seem to profit from experience. They are introspective and are characterized as ruminative and overideational.

Persons with the 78/87 code type harbor chronic feelings of insecurity, inadequacy, and inferiority, and they tend to be indecisive. They lack even an average number of socialization experiences and are not socially poised or confident. As a result, they withdraw from social interactions. They are passive-dependent individuals who are unable to take a dominant role in interpersonal relationships. Mature heterosexual relationships are especially difficult for 78/87 persons. They feel quite inadequate in the traditional sex role, and sexual performance may be poor. In an apparent attempt to compensate for these deficits, they engage in rich sexual fantasies.

Diagnoses of schizophrenia, depressive disorders, obsessive-compulsive disorders, and personality disorders are all represented among individuals with the 78/87 code type. Schizoid is the most common personality disorder diagnosis assigned to persons with this code type. The relative elevations of scales 7 and 8 are important in differentiating psychotic from nonpsychotic disorders. As scale 8 becomes greater than scale 7, the likelihood of a psychotic disorder increases. Even when a psychotic label is applied, blatant psychotic symptoms may not be present.

89/98

Persons with the 89/98 code type tend to be rather self-centered and infantile in their expectations of other people. They demand a great deal of attention and may become resentful and hostile when their demands are not met. Because they fear emotional involvement, they avoid close relationships and tend to be socially withdrawn and isolated. They seem especially uncomfortable in heterosexual relationships, and poor sexual adjustment is common.

89/98 persons also are characterized as hyperactive and emotionally labile. They appear to be agitated and excited, and they may talk excessively in a loud voice. They are unrealistic in self-appraisal, and they impress others as grandiose, boastful, and fickle. They are vague, evasive, and denying in talking about their difficulties, and they may state that they do not need professional help.

Although 89/98 persons have a high need to achieve and may feel pressured to do so, their actual performance tends to be mediocre. Their feelings of inferiority and inadequacy and their low self-esteem limit the extent to which they involve themselves in competitive or achievement-oriented situations.

The 89/98 code type is suggestive of serious psychological disturbance, particularly if scales 8 and 9 are grossly elevated. The modal diagnosis for 89/98 persons is schizophrenia. Severe disturbance in thinking is likely. 89/98 individuals are confused, perplexed, and disoriented, and they report feelings of unreality. They have difficulty concentrating and thinking, and they are unable to focus on issues. Thinking also may appear to be odd, unusual, autistic, and circumstantial. Speech may be bizarre and may involve clang associations, neologisms, and echolalia. Delusions and hallucinations may be present.

THREE-POINT CODE TYPES

As stated earlier in this chapter, three-point code types tell us which three clinical scales are the highest in the profile. Far less research has been conducted concerning three-point code types than concerning two-point code types or single-scale scores. The three-point code types included in this section are those that occur reasonably frequently in a variety of clinical settings and for which some research data are available. Because profiles classified according to three-point code types result in rather homogeneous groupings, the descriptors presented for any particular code type are likely to fit many individuals with that code type rather well. However, it must again be emphasized that the descriptions provided represent modal patterns. Not every descriptor will apply to every person with a particular three-point code type.

123/213/231

Persons with this code type usually are diagnosed as having a somatoform disorder, anxiety disorder, or depressive disorder. Somatic complaints, particularly those associated with the gastrointestinal system, are common, and often there appears to be clear secondary gain associated with the symptoms. Sleep disturbance, perplexity, despondency, and feelings of hopelessness occur. Persons with this code type are in conflict about dependency and self-assertion, and they often keep other people at an emotional distance. They tend to have a low energy level and are lacking in sex drive. Such persons often show good work and marital adjustment, but they rarely take risks in their lives.

132/312

This configuration, in which scales 1 and 3 often are significantly higher than scale 2, has been referred to as the "conversion valley." Persons with this code type may show classic conversion symptoms, and diagnoses of conversion disorder or somatoform pain disorder are common. Stress is often converted into physical symptoms. Persons with this code type use denial and repression excessively, lack insight into the causes of their symptoms, and resist psychological explanations of their problems. Although these individuals tend to be rather sociable, they tend to be passive-dependent in relationships. It is important for them to be liked and approved of by others, and their behavior typically is conforming and conventional. They typically are seeking medical treatment for their symptoms and are likely to terminate treatment prematurely if they are pressed to deal with psychological matters.

138

Persons with this code type usually are diagnosed with schizophrenic disorder (paranoid type) or paranoid personality disorder. They are likely to have rather bizarre somatic symptoms that may be delusional in nature. Depressive episodes, suicidal ideation, and sexual and religious preoccupation may occur. Clear evidence of thought disorder may be observed. These individuals are agitated, excitable, loud, and short-tempered. They often have histories of excessive use of alcohol and feel restless and bored much of the time. They are ambivalent about forming close relationships, and they often feel suspicious and jealous.

139

Persons with this code type often are diagnosed as having a somatoform disorder or organic brain syndrome. If they are given the latter diagnosis, they may show spells of irritation, assaultiveness, and outbursts of temper.

247/274/472

The modal diagnosis for persons with this code type is passive-aggressive personality disorder, and symptoms of depression and anxiety may be present. This is a very common code type among patients who abuse alcohol and/or other substances. Family and marital problems are common among these individuals. They may feel fearful, worried, and high-strung. They overreact to stress and undercontrol impulses. They tend to be angry, hostile, and immature, with strong unfilled needs for attention and support. They are in conflict about dependency and sexuality. They tend to be phobic, ruminative, and overideational, and they experience guilt associated with anger. Although they often have strong achievement needs, they are afraid to compete for fear of failing. They have difficulty enduring anxiety during treatment, and they may respond best to directive, goal-oriented treatment.

278/728

Persons with this code type are experiencing a great deal of emotional turmoil, and they tend to have a rather schizoid lifestyle. They tend to feel tense, nervous, and fearful, and they have problems in concentrating and attending. They feel depressed, despondent, and hopeless, and they often ruminate about suicide. Affect appears to be blunted or otherwise inappropriate. Eating problems often are reported by women with this code type. These persons lack basic social skills and are shy, withdrawn, introverted, and socially isolated. They tend to be passive in relationships. They feel inadequate and inferior. They tend to set high standards for themselves and to feel guilty when these standards are not met. They tend to show interest in obscure, esoteric subjects. They may use alcohol or other drugs as a way of coping with stress.

The diagnostic picture for 278/728 persons is mixed. They may receive either neurotic or psychotic diagnoses. In making a differential diagnosis it often is helpful to try to understand why they have high scores on scale 8. If examination of the Harris-Lingoes subscales indicates that the scale 8 elevation is accounted for primarily by items in the Sc3 (Lack of Ego Mastery, Cognitive) or Sc6 (Bizarre Sensory Experiences) subscales, a psychotic disorder is more likely than if items in the Sc 4 (Lack of Ego Mastery, Conative) seem to account for much of the scale 8 elevation.

687/867

This code type, in which scales 6 and 8 typically are much more elevated than scale 7, has been referred to as the "psychotic valley." It suggests very serious psychopathology, and the most common diagnosis for persons with the code type is schizophrenia, paranoid type. Hallucinations, delusions, and extreme suspiciousness are common. Affect tends to be blunted. Persons with this code type tend to be shy, introverted, and socially withdrawn, but they may become quite aggressive when drinking. They tend to have problems with memory and concentration. Although persons with this code type may not be experiencing disabling emotional turmoil, they often are unable to handle the responsibilities of everyday life and require inpatient treatment. Psychotropic medications often are prescribed.

OTHER CONFIGURAL ASPECTS

Regardless of their absolute elevations and whether or not they are the highest scales in the profile, the relative elevations of scales 1, 2, and 3 provide important interpretive information. When scales 1 and 3 are ten or more T-score points higher than scale 2, individuals probably are using denial and repression excessively. They tend to have little or no insight into their own needs, conflicts, or symptoms. They are reasonably free of depression, anxiety, and other emotional turmoil, but somatic symptoms are likely. These persons want medical explanations for their problems, resisting psychological explanations. When scale 2 is equal to or higher than scales 1 and 3, the individuals are not likely to be so well defended, and they may report emotional turmoil and a wide variety of symptoms.

The relationship between scales 3 and 4 gives important information about impulse control. Even when these two scales are not the most elevated ones in the profile, their relative positions are meaningful. When scale 4 is ten or more T-score points higher than scale 3, we expect problems with impulse control. Such persons tend to act without adequately considering the consequences of their actions. When scale 3 is ten or more T-score points higher than scale 4, we expect persons to have adequate control and not to act impulsively. When scores on scales 3 and 4 are about equally elevated, and especially when they are both above T scores of 65, persons may be overly controlled and not even appropriately assertive most of the time, but periods of impulsive acting out may occur.

Scale 5's position in the profile of male subjects also tells us something about control. Regardless of the scores on other scales in the profile, elevation of scale 5 suggests an element of control. High scale 5 men are not likely to act out impulsively. Hathaway and Monachesi (1953) found that even in environments where the base rate for delinquency was very high, adolescent boys who scored high on scale 5 did not become involved in delinquent acts.

A profile configuration in which scales 4 and 6 are at above-average levels suggests rather intense anger that is expressed in rather passive-aggressive ways. This configuration occurs for both men and women, but it is more common for women. Women with this configuration often have mislabeled their feelings and present themselves as depressed rather than angry. Women with this configuration may feel trapped in a role (e.g., housewife, mother) that is not satisfying to them.

The relationship between scales 7 and 8 gives important information about chronicity of problems and about the likelihood of thought disorder. When scale 7 is ten or more T-score points greater than scale 8, problems tend to be acute rather than chronic and thought disorder is not likely. As scale 8 becomes greater than scale 7, problems tend to be more chronic, and the likelihood of thought disorder increases. When both scales 7 and 8 are elevated, persons may be rather confused, but a well-developed delusional system is not to be expected.

6

Content Interpretation

Hathaway and McKinley utilized empirical keying procedures to construct the original MMPI scales. Items were included in a scale if they empirically differentiated between external criterion groups. No emphasis was placed on the content of the items identified in this manner, and only for scale 7 were attempts made to ensure internal consistency. In fact, early in the history of the MMPI some clinicians seemed to believe that examination of the content of the items endorsed by test subjects would spoil the empirical approach to assessment.

More recently, clinicians and researchers have become increasingly aware that consideration of item content adds significantly to MMPI interpretation. The purpose of this chapter is to discuss some approaches to the interpretation of content dimensions of the MMPI-2. It should be emphasized that these approaches are viewed as supplementary to interpretation of the standard MMPI-2 scales and should not be used instead of the standard scales.

THE HARRIS-LINGOES SUBSCALES

Subscale Development

As mentioned above, the standard MMPI clinical scales were constructed by empirical keying procedures. Because little attention was given by Hathaway and McKinley to scale homogeneity, most of the standard clinical scales are quite heterogeneous in terms of item content. The same total raw score on a clinical scale can be obtained by individuals endorsing combinations of very different kinds of items. A number of investigators have suggested that systematic analysis of these subgroups of items within the standard clinical scales can add significantly to the interpretation of protocols (e.g., Comrey, 1957a, 1957b, 1957c, 1958b, 1958c, 1958d, 1958e; Comrey & Marggraff, 1958; Graham, Schroeder, & Lilly, 1971; Harris & Lingoes, 1955, 1968; Pepper & Strong, 1958). The subscales developed by Harris and Lingoes represent the most comprehensive effort of this kind. Their scales have come to be widely used clinically and are routinely scored and reported by some of the automated scoring and interpretation services.

Harris and Lingoes (1955, 1968) reported the construction of subscales for six of the ten standard clinical scales (scales 2, 3, 4, 6, 8, and 9). They did not develop subscales for scales 1 or 7 because they considered them homogeneous in content. Whereas a factor-analytic study by Comrey (1957b) suggests that Harris and Lingoes are correct about the unidimensionality of scale 1, factor analyses of scale 7 items have not been as conclusive. Comrey (1958d) factor-analyzed scale 7 item responses and identified several factors, but Strenger (1989) was not able to develop reliable and valid subscales for scale 7, largely because of the homogeneity of the scale. Harris and Lingoes did not develop subscales for scales 5 and 0 because these scales were not considered to be standard clinical scales. Subsequent efforts to develop subscales for scales 5 and 0 will be discussed later in this chapter.

Each of the Harris-Lingoes subscales was constructed logically by examining the content of items within a standard clinical scale and grouping together items that seemed similar in content or that were judged to reflect a single attitude or trait. A label was assigned to each subscale on the basis of the investigators' clinical judgments of the content of items in the subscale. Although it was assumed that the resulting subscales would be more homogeneous than their parent scales, no statistical estimates of homogeneity were provided by Harris and Lingoes. Although thirty-one subscales were developed, three subscales that are obtained by summing scores on other subscales generally are not used in clinical interpretation.

Because the MMPI-2 includes most of the items in the standard clinical scales, the Harris-Lingoes subscales can be scored. Several changes were made in the subscales when the MMPI-2 was developed. First, several items that were scored on some of the subscales were deleted, so the MMPI-2 has fewer items for those scales than does the original MMPI. Although only a few items were deleted, some of the subscales were already so short that the deletions are of serious concern. Second, Harris and Lingoes included items in some of the subscales that were not on the parent scales, apparently because they were using preliminary versions of some of the clinical scales. In the MMPI-2 the items that were in subscales but not in the parent scales were deleted from the subscales. Finally, the Harris-Lingoes subscales were renumbered to eliminate the lettered subscripts for some of the subscales.

Using samples of psychiatric inpatients and nonclinical subjects, Levitt, Browning, and Freeland (1992) compared MMPI and MMPI-2 raw scores for the scale 4 subscales. As expected, scores were lower for the MMPI-2 subscales, which have fewer items than the corresponding MMPI subscales. However, had the investigators converted raw scores to T scores, using the appropriate norms, they would have found the differences between the MMPI and MMPI-2 scores to be quite small and probably not clinically meaningful.

The names of the twenty-eight Harris-Lingoes subscales and the number of items in each subscale are presented in Table 6.1. The MMPI-2 booklet

Table 6.1. Internal Consistency (Alpha) and Test-Retest Coefficients for the Harris-Lingoes Subscales

	Subscale	No. Items	Internal Consistency[a]		Test-retest Reliability[b]	
			Men (*n*=1138)	Women (*n*=1462)	Men (*n*=82)	Women (*n*=111)
D1	Subjective Depression	32	.71	.74	.85	.84
D2	Psychomotor Retardation	14	.24	.28	.74	.76
D3	Physical Malfunctioning	11	.23	.29	.64	.74
D4	Mental Dullness	15	.60	.63	.76	.81
D5	Brooding	10	.63	.65	.81	.77
Hy1	Denial of Social Anxiety	6	.73	.74	.86	.85
Hy2	Need for Affection	12	.64	.62	.76	.80
Hy3	Lassitude-Malaise	15	.67	.73	.86	.88
Hy4	Somatic Complaints	17	.59	.68	.81	.77
Hy5	Inhibition of Aggression	7	.17	.11	.61	.68
Pd1	Familial Discord	9	.51	.57	.81	.73
Pd2	Authority Problems	8	.31	.23	.68	.62
Pd3	Social Imperturbability	6	.57	.55	.85	.77
Pd4	Social Alienation	13	.48	.52	.81	.75
Pd5	Self-alienation	12	.62	.67	.78	.78
Pa1	Persecutory Ideas	17	.59	.64	.78	.69
Pa2	Poignancy	9	.39	.43	.66	.69
Pa3	Naiveté	9	.56	.57	.58	.73
Sc1	Social Alienation	21	.66	.69	.78	.75
Sc2	Emotional Alienation	11	.29	.34	.69	.74
Sc3	Lack of Ego Mastery, Cognitive	10	.66	.68	.68	.64
Sc4	Lack of Ego Mastery, Conative	14	.58	.62	.76	.81
Sc5	Lack of Ego Mastery, Defective Inhibition	11	.50	.56	.72	.70
Sc6	Bizarre Sensory Experiences	20	.64	.66	.78	.67
Ma1	Amorality	6	.29	.32	.72	.78
Ma2	Psychomotor Acceleration	11	.53	.50	.81	.68
Ma3	Imperturbability	8	.43	.46	.65	.70
Ma4	Ego Inflation	9	.32	.48	.74	.58

[a] Cronbach's coefficient alpha.

[b] Average retest interval was 9 days.

Source: Unpublished data from the MMPI restandardization project. John R. Graham, Department of Psychology, Kent State University, Kent, OH 44242.

numbers of the items in each subscale, and the scored response for each item are presented in Appendix C. NCS offers scoring templates and profile sheets for the Harris-Lingoes subscales. Linear T-score transformations for raw scores on each of the subscales are presented in Appendix D.

Harris and Lingoes did not avoid placing an item in more than one subscale. Thus, item overlap among the subscales is considerable and may account for the high correlations between certain subscales. Table 6.2 summarizes these intercorrelations for men and women in the MMPI-2 normative sample.

Table 6.2. Intercorrelations for the Harris-Lingoes Subscales for 1,138 Men and 1,462 Women in the MMPI-2 Normative Sample

	D	D1	D2	D3	D4	D5	
D	—	88	57	58	77	67	
D1	85	—	44	45	84	83	
D2	61	45	—	09	40	20	
D3	52	41	09	—	41	34	
D4	70	82	36	34	—	70	
D5	57	79	19	25	65	—	

	Hy	Hy1	Hy2	Hy3	Hy4	Hy5	
Hy	—	35	39	52	53	31	
Hy1	39	—	31	−25	−22	16	
Hy2	54	31	—	−31	−32	28	
Hy3	43	−27	−28	—	55	−16	
Hy4	45	−20	−24	53	—	−14	
Hy5	47	16	36	−09	−09	—	

	Pd	Pd1	Pd2	Pd3	Pd4	Pd5	
Pd	—	68	45	13	68	73	
Pd1	63	—	16	−15	46	49	
Pd2	52	19	—	32	15	15	
Pd3	13	−12	24	—	−10	−22	
Pd4	66	39	21	−07	—	71	
Pd5	67	41	21	−26	69	—	

	Pa	Pa1	Pa2	Pa3			
Pa	—	51	62	38			
Pa1	48	—	36	−39			
Pa2	57	32	—	−18			
Pa3	43	−37	−18	—			

	Sc	Sc1	Sc2	Sc3	Sc4	Sc5	Sc6
Sc	—	83	56	76	74	73	74
Sc1	81	—	43	51	53	54	46
Sc2	52	39	—	37	70	27	31
Sc3	74	47	33	—	69	50	58
Sc4	72	50	68	69	—	40	43
Sc5	71	50	24	46	38	—	60
Sc6	68	38	26	49	32	55	—

	Ma	Ma1	Ma2	Ma3	Ma4		
Ma	—	51	74	26	66		
Ma1	57	—	25	03	21		
Ma2	72	33	—	−05	41		
Ma3	28	01	−13	—	−14		
Ma4	64	21	41	−06	—		

Note: Correlations for women are above diagonal; correlations for men are below diagonal.

Source: Unpublished data from the MMPI restandardization project. John R. Graham, Department of Psychology, Kent State University, Kent, OH 44242.

Subscale Norms

Harris and Lingoes (1955) did not present normative data when their subscales were first described, but a later paper (Harris & Lingoes, 1968)
reported means and standard deviations for psychiatric patients at the Langley Porter Clinic. Gocka and Holloway (1963) presented means and standard
deviations for sixty-eight male veteran psychiatric patients. Appendix D presents T-score transformations for the Harris-Lingoes subscale raw scores
based on data from the MMPI-2 normative sample. These are the T scores
that were used to construct the profile sheets available from NCS for the Harris-Lingoes subscales.

Subscale Reliability

Table 6.1 reports internal consistency (Alpha) coefficients for men and
women in the MMPI-2 normative sample. Although several of the subscales
have rather low Alpha coefficients, most have a high degree of internal consistency. The table also reports test-retest coefficients for subsamples of the
normative subjects who took the MMPI-2 twice. The test-retest data suggest
that the temporal stability of the subscales is less than the parent scales, but
stability is adequate for most scales. As one would expect, the subscales with
fewer items have the lower test-retest coefficients.

Subscale Validity

Although the Harris-Lingoes subscales have been in existence for more than
thirty years and have gained fairly wide usage among clinicians (largely
because they are scored routinely by some automated scoring and interpretation services), only limited empirical research concerning the subscales has
been published. The factor-analytic work of Comrey (1957a, 1957b, 1957c,
1958b, 1958c, 1958d, 1958e; Comrey & Marggraff, 1958) indirectly offers support for the construct validity of the Harris-Lingoes subscales. Comrey
reported factor analyses of the intercorrelations of items within each scale,
separately for each of the clinical scales of the MMPI (excluding scales 5 and
0). Although there are some significant differences between the logically
derived Harris-Lingoes subscales and the corresponding factor-analytically
derived Comrey factors, in general the Comrey studies revealed factors within
each clinical scale that are similar to the Harris-Lingoes subscales and that
supported Harris's notion that the clinical scales are not homogeneous and
unidimensional.

Lingoes (1960) factor-analyzed scores on the Harris-Lingoes subscales and on the Wiener Subtle-Obvious subscales of the MMPI (Wiener, 1948) in an attempt to determine the statistical factor structure of the MMPI. He concluded that the dimensionality of the MMPI was more complex than the six standard scales from which the various subscales were derived but simpler than the thirty-six subscales (Harris-Lingoes and Wiener) included in his own factor analysis. Harris and Lingoes (1955) reasoned intuitively that the subscales should be more homogeneous than the parent scales from which they were drawn, but they did not offer any evidence in this regard. Calvin (1974) statistically examined the homogeneity of the five Harris-Lingoes subscales for scale 2 (Depression). He separately factor analyzed inter-item correlations for each of the five subscales and concluded that four of the subscales appeared to be unidimensional, whereas one subscale (Psychomotor Retardation) was two-dimensional (loss of interest in life activities and inhibition of hostility). Examination of the internal consistency coefficients in Table 6.1 indicates that there is considerable variability among the Harris-Lingoes subscales. Some of them, such as D1 (Subjective Depression) and Hy1 (Denial of Social Anxiety), have good internal consistency, whereas others, such as D2 (Psychomotor Retardation) and Hy5 (Inhibition of Aggression), have very poor internal consistency.

Harris and Christiansen (1946) studied pretherapy MMPI differences between neurotic patients who were judged to have been successful in psychotherapy and similar patients who were judged to have been unsuccessful. They found that the successful patients scored lower on scales 4, 6, 8, and 9 of the MMPI, suggesting that they had more ego strength. Significant differences between successful and unsuccessful patients also were identified for eight Harris-Lingoes subscales. Successful patients scored lower on the Familial Discord, Authority Problems, and Social Alienation subscales of scale 4, on the Persecutory Ideas subscale of scale 6, and on the Defective Inhibition and Bizarre Sensory Experiences subscales of scale 8. Harris and Christiansen did not address the question of whether greater accuracy of prediction of psychotherapy outcome was possible with the subscales than with only the standard clinical scales. They felt, however, that the subscale information could lead to a better understanding of how successful therapy patients view themselves and the environments in which they live.

Gocka and Holloway (1963) correlated scores of psychiatric patients on the Harris-Lingoes subscales with other MMPI scales assessing social desirability, introversion-extroversion, and dissimulation, with a number of demographic variables (intelligence, occupational level, marital status), with legal competency status at the time of hospital admission, and with number of days of hospitalization. Most of the Harris-Lingoes subscales were related to the social desirability scale, and some were related to the introversion-extroversion and dissimulation scales. Few significant correlations were found between the Harris-Lingoes subscale scores and demographic variables. Two subscale scores correlated significantly with competency status, and no subscale correlated significantly with length of hospitalization.

Panton (1959) compared the Harris-Lingoes subscale scores of African-American and Caucasian prison inmates. He found that Caucasians scored higher on the Authority Problems subscale, suggesting that they had more authority problems and aggressive tendencies than African Americans. African Americans scored higher on the Persecutory Ideas, Social Alienation, and Ego Inflation subscales, suggesting more psychotic trends for African Americans than for Caucasians. Panton also compared the Harris-Lingoes subscale scores of prison inmates with the psychiatric norms presented by Harris and Lingoes (1968). He found that prison inmates were higher than the psychiatric patients on the Social Alienation, Self-Alienation, and Amorality subscales. Prisoners scored lower than the psychiatric patients on the Subjective Depression, Psychomotor Retardation, Mental Dullness, Need for Affection, Lassitude-Malaise, Inhibition of Aggression, Lack of Cognitive Ego Mastery, Lack of Conative Ego Mastery, and Psychomotor Acceleration subscales.

Calvin (1975) attempted to identify empirical behavioral correlates for the Harris-Lingoes subscales for a sample of hospitalized psychiatric patients. He compared high scorers on each Harris-Lingoes subscale with other scorers on that subscale on a number of extratest variables, including psychiatric diagnosis, reasons for hospitalization, nurses' ratings, and psychiatrists' ratings. Although ten of the twenty-eight subscales were determined to have reliable behavioral correlates, Calvin concluded that in most cases the subscales are not likely to add significantly to interpretation based on the standard clinical scales for psychiatric patients. The results of Calvin's study are included in the interpretive descriptions that are presented in the following section.

Several studies have demonstrated that the Harris-Lingoes subscales add important information to that provided by the clinical scales. Prokop (1986) found that a group of chronic-pain patients obtained high scores on scale 3 by endorsing items assessing somatic concerns rather than items indicating histrionic tendencies. Moore, McFall, Kivlahan, and Capestany (1988) compared the MMPI results of chronic-pain patients and psychotic patients and found that the chronic-pain patients obtained elevated scale 8 scores because they endorsed items dealing with somatic symptoms and depression. By contrast, the psychotic patients obtained elevated scale 8 scores because they endorsed items indicative of bizarre thinking, social alienation, and defective inhibition. Moore et al. (1988) compared groups of psychotic and nonpsychotic patients. Although both groups had similarly elevated scores on scale 8, they had different patterns of scores on the Harris-Lingoes subscales for scale 8. The psychotic patients scored higher on subscales indicating bizarre thinking and loss of control of impulses and emotions, whereas the nonpsychotic patients scored higher on subscales assessing depression, anxiety, and thinking difficulties.

Interpretation of the Harris-Lingoes Subscales

Scores on the Harris-Lingoes subscales provide information concerning the kinds of items that subjects endorsed in the scored direction in obtaining a particular score on a clinical scale. Because some of the subscales have very few items and are relatively unreliable and because there is only limited research concerning extratest correlates of the subscales, they should not be interpreted independently of their parent scales. The subscales generally should not be interpreted unless their parent scales are significantly elevated (T > 65) and interpretation should be limited to trying to understand why test subjects have obtained high scores on the parent scales.

There are two circumstances in which the subscales may be especially helpful. First, they can help to explain why a subject receives an elevated score on a clinical scale when that elevation was not expected from history and other information available to the clinician. For example, a patient whose primary symptom is depression could produce a profile with elevations on scales 2, 7, and 8. The scale 2 and scale 7 elevations are consistent with the patient's history and with clinical observation. However, the scale 8 elevation is somewhat troublesome. Why does this patient, for whom there is no history or clinical indication of schizophrenia or thought disorder, score relatively high on scale 8? Reference to the Harris-Lingoes subscale scores might reveal that most of the scale 8 elevation is coming from items in the Lack of Ego Mastery, Conative (Sc4) subscale. This subscale assesses depression and despair and whether the subject feels life is a strain much of the time. These characteristics would be highly consistent with those based on the rest of the profile and with the patient's history.

Second, the Harris-Lingoes subscales can be very useful in interpreting clinical scale scores that are marginally elevated (T = 60–65). Often it does not seem that many of the interpretations suggested for a high score on a scale are appropriate for the less elevated scores. For example, a subject might receive a T score of 67 on scale 4, and there would be some reluctance to attribute to that subject the antisocial characteristics often suggested for high scores on scale 4. A high score on Pd1 (Familial Discord) of the Harris-Lingoes subscales, for example, could explain the moderately elevated score on scale 4 without requiring inferences about more deviant asocial or antisocial behaviors.

One can also consider the Harris-Lingoes subscales in relation to the empirical correlates for the clinical scales. For any of the clinical scales, many different kinds of behaviors and characteristics have been associated with higher scores. For example, elevated scores on scale 4 have been associated with family problems, antisocial behavior, and absence of social anxiety. Usually, not all of the descriptors associated with elevated scores on a scale will be characteristic of any specific individual with an elevated score on that scale. Examination of the Harris-Lingoes subscales can be helpful in determining which of the many different descriptors should be emphasized. In the above example, if the person had an elevated score on the Familial Discord subscale and not on other scale 4 subscales, in our interpretation we would emphasize the correlates associated with family problems. On the other hand, if the Authority Problems subscale was the only elevated scale 4 subscale, we would emphasize the correlates having to do with acting-out behaviors.

As with many of the other supplementary scores discussed in this book, it is not possible to establish absolutely firm cutoff scores to define high and low scorers on the subscales. As clinicians gain experience with the subscales, they will come to establish cutoff scores for the settings in which the MMPI-2 is used. The individual who is just beginning to use the subscales for MMPI-2 interpretation should find it useful to consider T scores greater than 65 as high scores.

The descriptions that follow for the subscales are based on information provided by Harris and Lingoes (1955, 1968), on the validity studies reviewed in the previous section, on the author's own clinical experience, and on examination of the content of the items in each subscale. The resulting descriptions can be used to generate hypotheses concerning why persons have obtained elevated scores on the parent clinical scales. Because not every descriptor will be characteristic of every person who has an elevated score on a subscale, the hypotheses need to be evaluated in relation to other information available about the person being assessed. It should be emphasized again that the Harris-Lingoes subscales should be used to supplement the standard validity and clinical scales. Some of the subscales are too short, and therefore probably too unreliable, to be used as independent scales on which to base clinical interpretation. Descriptions are not provided for low scorers on the subscales, because it is difficult to know what low scores mean about subjects. It could mean that they do not have the characteristics reported by high scorers or that they do not want to admit to having them. Therefore, it is recommended that low scores on the Harris-Lingoes subscales not be interpreted.

Subjective Depression (D1)

High scorers on the D1 subscale are reporting that they:

1. feel unhappy, blue, or depressed much of the time
2. lack energy for coping with the problems of their everyday lives
3. are not interested in what goes on around them
4. feel nervous or tense much of the time
5. have difficulties in concentrating and attending
6. have poor appetite and trouble sleeping
7. brood and cry frequently
8. lack self-confidence
9. feel inferior and useless
10. are easily hurt by criticism

11. feel uneasy, shy, and embarrassed in social situations
12. tend to avoid interactions with other people, except for relatives and close friends

Psychomotor Retardation (D2)

High scorers on the D2 subscale are reporting that they:

1. feel immobilized and withdrawn
2. lack energy to cope with everyday activities
3. avoid other people
4. do not have hostile or aggressive impulses

Physical Malfunctioning (D3)

High scorers on the D3 subscale are reporting that they:

1. are preoccupied with their own physical functioning
2. do not have good health
3. experience a wide variety of specific somatic symptoms that may include weakness, hay fever or asthma, poor appetite, nausea or vomiting, and convulsions

Mental Dullness (D4)

High scorers on the D4 subscale are reporting that they:

1. lack energy to cope with the problems of everyday life
2. feel tense
3. experience difficulties in concentrating
4. have poor memory and/or show poor judgment
5. lack self-confidence
6. feel inferior to others
7. get little enjoyment out of life
8. have concluded that life is no longer worthwhile

Brooding (D5)

High scorers on the D5 subscale are reporting that they:

1. brood, ruminate, and cry much of the time
2. lack energy to cope with problems
3. have concluded that life is no longer worthwhile
4. feel inferior, unhappy, and useless
5. are easily hurt by criticism
6. feel that they are losing control of their thought processes

Denial of Social Anxiety (Hy1)

High scorers on the Hy1 subscale are reporting that they:

1. are socially extroverted
2. feel quite comfortable in interacting with other people
3. find it easy to talk with other people
4. are not easily influenced by social standards and customs

Need for Affection (Hy2)

High scorers on the Hy2 subscale are reporting that they:

1. have strong needs for attention and affection from others and fear that those needs will not be met if they are more honest about their feelings and attitudes
2. have optimistic and trusting attitudes toward other people
3. see others as honest, sensitive, and reasonable
4. do not have negative feelings about other people
5. try to avoid unpleasant confrontations whenever possible

Lassitude-Malaise (Hy3)

High scorers on the Hy3 subscale are reporting that they:

1. feel uncomfortable and are not in good health
2. feel weak, fatigued, or tired
3. do not have specific somatic complaints
4. have difficulties in concentrating, poor appetite, and sleep disturbance
5. feel unhappy and blue
6. see their home environments as unpleasant and uninteresting

Somatic Complaints (Hy4)

High scorers on the Hy4 subscale are reporting that they:

1. have many somatic complaints
2. experience pain in the heart and/or chest
3. have fainting spells, dizziness, or balance problems
4. experience nausea and vomiting, poor vision, shakiness, or feeling too hot or too cold
5. express little or no hostility toward other people

Inhibition of Aggression (Hy5)

High scorers on the Hy5 subscale are reporting that they:

1. do not experience hostile and aggressive impulses

2. are not interested in reading about crime and violence
3. are sensitive about how others respond to them
4. are decisive

Familial Discord (Pd1)

High scorers on the Pd1 subscale are reporting that they:

1. see their home and family situations as quite unpleasant
2. have felt like leaving their home situations
3. see their homes as lacking in love, understanding, and support
4. feel that their families are critical, quarrelsome, and refuse to permit adequate freedom and independence

Authority Problems (Pd2)

High scorers on the Pd2 subscale are reporting that they:

1. resent societal and parental standards and customs
2. have been in trouble in school or with the law
3. have definite opinions about what is right and wrong
4. stand up for what they believe
5. are not greatly influenced by the values and standards of others

Social Imperturbability (Pd3)

High scorers on the Pd3 subscale are reporting that they:

1. are comfortable and confident in social situations
2. like to interact with other people
3. experience no difficulty in talking with other people
4. tend to be somewhat exhibitionistic and show-offish
5. have strong opinions about many things and are not reluctant to defend their opinions vigorously

Social Alienation (Pd4)

High scorers on the Pd4 subscale are reporting that they:

1. feel alienated, isolated, and estranged
2. feel that other people do not understand them
3. feel lonely, unhappy, and unloved
4. feel that they get a raw deal from life
5. see other people as responsible for their problems and shortcomings
6. are concerned about how other people react to them
7. experience regret, guilt, and remorse for their actions

Self-alienation (Pd5)

High scorers on the Pd5 subscale are reporting that they:

1. are uncomfortable and unhappy
2. have problems in concentrating
3. do not find daily life interesting or rewarding
4. experience regret, guilt, and remorse for past deeds but are vague about the nature of this misbehavior
5. find it hard to settle down
6. may use alcohol excessively

Persecutory Ideas (Pa1)

High scorers on the Pa1 subscale are reporting that they:

1. view the world as a threatening place
2. feel that they are getting a raw deal from life
3. feel misunderstood
4. feel that others have unfairly blamed or punished them
5. are suspicious and untrusting of other people
6. blame others for their own problems and shortcomings
7. feel that others are trying to influence or control them
8. believe that others are trying to poison or otherwise harm them

Poignancy (Pa2)

High scorers on the Pa2 subscale are reporting that they:

1. are more high-strung and more sensitive than other people
2. feel more intensely than others
3. feel lonely and misunderstood
4. look for risky or exciting activities to make them feel better

Naiveté (Pa3)

High scorers on the Pa3 subscale are reporting that they:

1. have very optimistic attitudes about other people
2. see others as honest, unselfish, generous, and altruistic
3. are trusting
4. have high moral standards
5. do not experience hostility and negative impulses

Social Alienation (Sc1)

High scorers on the Sc1 subscale are reporting that they:

1. are getting a raw deal from life
2. believe that other people do not understand them
3. believe that other people have it in for them
4. believe that other people are trying to harm them
5. feel that their family situations are lacking in love and support
6. feel that their families treat them more as children than as adults
7. feel lonely and empty
8. have never had love relationships with anyone
9. harbor hostility and hatred toward family members
10. avoid social situations and interpersonal relationships whenever possible

Emotional Alienation (Sc2)

High scorers on the Sc2 subscale are reporting that they:

1. experience feelings of depression and despair and may wish that they were dead
2. are apathetic and frightened
3. have sadistic and/or masochistic needs

Lack of Ego Mastery, Cognitive (Sc3)

High scorers on the Sc3 subscale are reporting that they:

1. feel that they might be losing their minds
2. have strange thought processes and feelings of unreality
3. have problems with concentration and memory

Lack of Ego Mastery, Conative (Sc4)

High scorers on the Sc4 subscale are reporting that they:

1. feel that life is a strain and that they experience depression and despair
2. have difficulty in coping with everyday problems and worry excessively
3. respond to stress by withdrawing into fantasy and daydreaming
4. do not find their daily activities interesting and rewarding
5. have given up hope of things getting better
6. may wish that they were dead

Lack of Ego Mastery, Defective Inhibition (Sc5)

High scorers on the Sc5 subscale are reporting that they:

1. feel that they are not in control of their emotions and impulses and are frightened by this perceived loss of control
2. tend to be restless, hyperactive, and irritable
3. have periods of laughing and crying that they cannot control
4. have experienced episodes during which they did not know what they were doing and later could not remember what they had done

Bizarre Sensory Experiences (Sc6)

High scorers on the Sc6 subscale are reporting that they:

1. experience feelings that their bodies are changing in strange and unusual ways
2. experience skin sensitivity, feeling hot or cold, voice changes, muscle twitching, clumsiness, problems in balance, ringing or buzzing in the ears, paralysis, and weakness
3. experience hallucinations, unusual thought content, and ideas of external influence

Amorality (Ma1)

High scorers on the Ma1 subscale are reporting that they:

1. perceive other people as selfish, dishonest, and opportunistic and because of these perceptions feel justified in behaving in similar ways
2. derive vicarious satisfaction from the manipulative exploits of others

Psychomotor Acceleration (Ma2)

High scorers on the Ma2 subscale are reporting that they:

1. experience acceleration of speech, thought processes, and motor activity
2. feel tense and restless
3. feel excited or elated without cause
4. become bored easily and seek out risk, excitement, or danger as a way of overcoming the boredom
5. have impulses to do something harmful or shocking

Imperturbability (Ma3)

High scorers on the Ma3 subscale are reporting that they:

1. do not experience social anxiety

2. feel comfortable around other people
3. have no problem in talking with others
4. are not concerned about the opinions, values, and attitudes of other people
5. feel impatient and irritable toward others

Ego Inflation (Ma4)

High scorers on the Ma4 subscale are reporting that they:

1. are important persons
2. are resentful when others make demands on them, particularly if the persons making the demands are perceived as less capable
3. have been treated unfairly

SUBSCALES FOR SCALES 5 AND 0

As stated previously, Harris and Lingoes (1955, 1968) did not develop subscales for scales 5 and 0. Their omission of these scales was consistent with other early research efforts that did not consider scales 5 and 0 as standard clinical scales. An early effort by Pepper and Strong (1958), who used clinical judgment in forming subgroups of items for scale 5, received little attention among MMPI users. Graham et al. (1971) factor analyzed scale 5 and 0 item responses of psychiatric inpatients, psychiatric outpatients, and normal subjects. For each of the two scales, seven factors emerged, one of which represented demographic variables included in the analyses.

Serkownek (1975) utilized the data from the Graham et al. (1971) factor analyses to develop subscales for scales 5 and 0. Items that loaded higher than .30 on a factor were selected for the scale to assess that factor dimension. Labels were assigned to the subscales on the basis of an examination of the content of the items included in the scale. Prior to the publication of the MMPI-2 the Serkownek subscales gained popularity among MMPI users. However, several concerns about the subscales led to a decision not to include them in the MMPI-2.

One major concern about Serkownek's scale 5 subscales was that the factor analysis on which they were based may have had methodological problems. Graham et al. (1971) combined male and female data in their analysis, and this could have produced artificially the masculine interest and feminine interest factors. Another concern was that some of the items in the Serkownek subscales were scored in the opposite direction from the parent scales. Finally, for women the raw scores on the scale 5 subscales were transformed to T scores in the opposite direction from the parent scale. High raw scores on scale 5 subscales yielded high T scores, whereas for scale 5 itself, higher raw scores yield lower T scores.

Development of the Scale 0 Subscales for the MMPI-2

Ben-Porath, Hostetler, Butcher, and Graham (1989) developed scale 0 subscales for the MMPI-2 to replace the Serkownek subscales for that scale. These investigators also tried unsuccessfully to develop subscales for scale 5. Factor analyses of scale 0 item responses of normal college men and women were used to construct provisional subscales. Internal consistency procedures were then used to refine the subscales. The three subscales resulting from these procedures are mutually exclusive, internally consistent, moderately independent, and representative of the major content dimensions of scale 0. Item numbers and scored directions for each of the subscales are reported in Appendix E. Linear T-score values for raw scores on the subscales can be found in Appendix F.

Reliability and Validity of the Scale 0 Subscales

Internal consistency (Alpha) coefficients were computed for the subscales for college and normative samples (Ben-Porath, Hostetler, Butcher, & Graham, 1989). These coefficients are reported in Table 6.3. Sieber and Meyers (1992) reported comparable coefficients for other samples of college students. The internal consistency of the subscales compares quite favorably with that of other MMPI-2 scales and subscales.

Test-retest reliability coefficients were computed for the subscales using a subsample of 82 men and 111 women from the MMPI normative samples who took the test twice with approximately a one-week interval between testings (Ben-Porath, Hostetler, Butcher, & Graham, 1989). The test-retest coefficients are reported in Table 6.3. The temporal stability of the subscales seems to be greater than for most of the other MMPI-2 scales and subscales.

Table 6.3. Internal Consistency and Test-Retest Reliability Coefficients for the Scale 0 Subscales

	Alpha Coefficients				Test-Retest Coefficients	
	College		Normative		Normative	
	Men	Women	Men	Women	Men	Women
Subscale	($n = 525$)	($n = 797$)	($n = 1138$)	($n = 1462$)	($n = 82$)	($n = 111$)
Si1 Shyness/Self-Consciousness	.82	.82	.81	.84	.91	.90
Si2 Social Avoidance	.77	.75	.77	.75	.88	.87
Si3 Self/Other Alienation	.77	.77	.75	.78	.77	.88

Source: Ben-Porath, Y.S., Hostetler, K., Butcher, J.N., & Graham, J.R. (1989). New subscales for the MMPI-2 Social Introversion (Si) Scale. *Psychological Assessment: A Journal of Consulting and Clinical Psychology, 1,* 169–174. Copyright © 1989 by the American Psychological Association. Adapted and reproduced by permission of the publisher.

Ben-Porath, Hostetler, Butcher, and Graham (1989) reported some preliminary validity data for the subscales. Scores on the subscales were correlated with behavioral ratings for a sample of 822 couples from the normative sample who participated in the study together and independently rated each other. The patterns of correlations were judged to offer support for the convergent and discriminant validity of the subscales.

Sieber and Meyers (1992) examined the validity of the scale 0 subscales by correlating them with other self-report measures of constructs that were believed to be differentially related to the three subscales. The results were very similar for men and women. The authors concluded that persons with elevated Si1 scale scores may be more socially anxious, less social, and have lower self-esteem; those with elevated Si2 scale scores may be more shy and less social; and persons with elevated Si3 scale scores may possess lower self-esteem and have a more external locus of control.

Interpretation of the Scale 0 Subscales

Examination of the content of the items in the subscales and of the correlations between scores on the subscales and other measures can be used to generate tentative hypotheses concerning persons who score high or low on the subscales. Ben-Porath, Hostetler, Butcher, and Graham (1989) recommended that T scores of 65 or greater be considered high scores on the subscales. Although the subscale developers did not recommend any cutoff for low scores, it seems reasonable to use the same cutoff (T < 40) that has been recommended for the other MMPI-2 scales.

Shyness/Self-consciousness (Si1)

High scores on the Si1 subscale indicate persons who:

1. feel shy, anxious, and uncomfortable in social situations
2. feel easily embarrassed
3. feel ill at ease in new situations
4. are not talkative or friendly
5. lack self-confidence and give up easily

Low scores on the Si1 subscale indicate persons who:

1. are extroverted
2. initiate social contact with other people
3. are talkative and friendly
4. are self-confident and do not give up easily

Social Avoidance (Si2)

High scores on the Si2 subscale indicate persons who:

1. do not enjoy being involved with groups or crowds of people
2. actively avoid getting involved with other people

Low scores on the Si2 subscale indicate persons who:

1. enjoy being involved with groups or crowds of people
2. initiate social contact with other people

Self/Other Alienation (Si3)

High scores on the Si3 subscale indicate persons who:

1. have low self-esteem
2. lack interest in activities
3. feel unable to effect changes in their life situations
4. have a more external locus of control

Low scores on the Si3 subscale indicate persons who:

1. have high self-esteem
2. appear to be interested in activities
3. feel able to effect changes in their life situations

THE MMPI-2 CONTENT SCALES

Whereas Harris and Lingoes formed content subscales within individual clinical scales, Wiggins (1969) used the entire MMPI item pool to form content scales. Starting with twenty-six content categories suggested by Hathaway and McKinley (1940), Wiggins used a combination of rational and statistical procedures to develop his scales. The resulting thirteen scales were psychometrically sound and seemed to represent well the content dimensions of the original MMPI. Unfortunately, when the MMPI was revised in 1989, Wiggins's scales were no longer adequate. One of the scales, Religious Fundamentalism, could no longer be scored because of item deletions. In addition, the Wiggins scales did not represent adequately the new content dimensions introduced into the MMPI-2 by the addition of new items.

Content scales based on the MMPI-2 item pool were developed by Butcher, Graham, Williams, and Ben-Porath (1990) to assess the content dimensions of the revised instrument. The content scales were developed using a combination of rational and statistical procedures.

Development of the Content Scales

The first step in the development of the MMPI-2 content scales was to define clinically relevant content areas represented by the items in Form AX of the MMPI. Twenty-two categories were rationally identified, and a definition was written for each. Three clinical psychologists served as judges and assigned items to the content categories. Judges were free to add categories, and items could be assigned to more than one category. Items assigned to a category by two or three of the judges were placed into provisional scales. Raters then met and reviewed all the item placements. Any disagreements were discussed until there was full agreement by all three raters concerning item placement. For one of the original categories, sufficient items could not be identified, so it was dropped from further consideration.

In the next step of scale development, item responses for two samples of psychiatric patients and two samples of college students were used to identify items in the provisional scales that did not correlate highly with total scores for the scales and that detracted from internal consistency. Such items were dropped. At this stage, four additional scales were eliminated from further consideration because of unacceptably low internal consistencies. Also, the data indicated that another content category, cynicism, which had been previously identified by item factor analysis, was not represented in the content scales. Thus, a twenty-item cynicism scale was added.

Another way of ensuring appropriate item placement was to examine correlations between each item in the inventory and total scores on the provisional content scales. Items that correlated higher with a score from a scale other than the one on which it was placed were deleted or moved to the other scale.

A final step involved examination of the content of the items in each content scale to determine rationally whether the items fit conceptually with the definition of the content domain. A number of items that were statistically related to the total score for a scale but whose content did not seem appropriate for that scale were eliminated.

The multistage procedures used to develop the scales yielded a set of fifteen that were judged to be internally consistent, relatively independent, and representative of clinically relevant content dimensions in the MMPI-2 item pool. Although item overlap between scales was kept to a minimum, some overlap was permitted when the constructs assessed by the scales were conceptually related. Table 6.4 presents a listing of the fifteen content scales. Item numbers and scored directions for each of the content scales are reported in Appendix G.

Table 6.4. Reliability of the MMPI-2 Content Scales

Scale		No. Items	Internal[a] Consistency		Test-Retest[b] Reliability	
			Men (n = 1138)	Women (n = 1462)	Men (n = 82)	Women (n = 111)
ANX	Anxiety	23	.82	.83	.90	.87
FRS	Fears	23	.72	.75	.81	.86
OBS	Obsessiveness	16	.74	.77	.83	.85
DEP	Depression	33	.85	.86	.87	.88
HEA	Health Concerns	36	.76	.80	.81	.85
BIZ	Bizarre Mentation	24	.73	.74	.78	.81
ANG	Anger	16	.76	.73	.85	.82
CYN	Cynicism	23	.86	.85	.80	.89
ASP	Antisocial Practices	22	.78	.75	.81	.87
TPA	Type A Behavior	19	.72	.68	.82	.79
LSE	Low Self-esteem	24	.79	.83	.84	.86
SOD	Social Discomfort	24	.83	.84	.91	.90
FAM	Family Problems	25	.73	.77	.84	.83
WRK	Work Interference	33	.82	.84	.90	.91
TRT	Negative Treatment Indicators	26	.78	.80	.79	.88

[a] Cronbach's coefficient alpha.

[b] Average retest interval was 9 days.

Source: Butcher, J.N., Graham, J.R., Williams, C.L., & Ben-Porath, Y.S. (1990). *Development and use of the MMPI-2 content scales.* Minneapolis: University of Minnesota Press. Copyright © 1990 by University of Minnesota. Reproduced by permission.

Norms for the Content Scales

Data from men and women in the MMPI-2 normative sample (Butcher et al., 1989) were used to generate T-score conversions for the raw scores of the content scales. The same uniform T scores used for the validity and clinical scales of MMPI-2 are used with the content scales. Raw scores for the content scales were regressed on percentile-corresponding T scores from the uniform distribution derived for the clinical scales. This procedure permits scores for the content scales to be expressed on the same metric as the clinical scales, thus ensuring comparability within the set of content scales and between the content scales and the clinical scales. Uniform T-score transformations for the content scales are reported separately for men and women in Appendix H.

Reliability of the Content Scales

Table 6.4 reports internal consistency (Alpha) coefficients for the content scales, based on responses of the male and female normative subjects. As would be expected, the internal consistency of the content scales is quite high. In general, the content scales are more internally consistent than the clinical scales and similar in internal consistency to the Wiggins scales, which they were developed to replace.

Table 6.4 also reports test-retest reliability coefficients for the content scales for 82 men and 111 women in the normative sample. The average retest interval was approximately 9 days. These coefficients indicate that the content scales are quite stable over this short time interval. In fact, the content scales appear to be more reliable than the basic clinical scales.

Validity of the Content Scales

Butcher, Graham, Williams, and Ben-Porath (1990) reported several kinds of preliminary validity data for the content scales. Correlations between the content scales and other MMPI-2 scales are reported in Table 6.5. These correlational data contribute significantly to our understanding of the construct validity of the content scales. Some of the content scales correlate highly with the standard scales, suggesting that they can be interpreted in similar ways. For example, the HEA (Health Concerns) scale and the Hs (Hypochondriasis) scale correlate .89 for men and .91 for women, suggesting that both are measures of health concern. Likewise, the SOD (Social Discomfort) scale and the Si (Social Introversion) scale correlate .85 for men and .84 for women. However, other content scales are not so highly correlated with standard scales with similar labels, suggesting that these scales are assessing unique characteristics as well as common ones. For example, the correlation between DEP (Depression) and D (Depression) was .52 for men and .63 for women, suggesting that these two measures of depression are not interchangeable.

Butcher, Graham, Williams, and Ben-Porath (1990) also presented data concerning behavioral correlates for the content scales. More than eight hundred couples participated in the MMPI-2 standardization project (Butcher et al., 1989). In addition to responding to the MMPI items, these couples, most of whom were married to each other, independently rated each other on 110 items concerning personality and behavior. Butcher et al. correlated ratings on these items and on factor scales derived from the items with scores on the content scales. The resulting correlations were used to generate behavioral descriptors for high scorers on each of the content scales.

Butcher, Graham, Williams, and Ben-Porath (1990) reported data concerning scores of chronic-pain patients, psychiatric patients, and normal subjects on the HEA (Health Concerns) scale. As expected, chronic-pain patients scored significantly higher than the other groups on the HEA scale. A T-score cutoff of 65 on the HEA scale correctly classified most of the chronic-pain patients and incorrectly classified very few of the other subjects.

Butcher, Graham, Williams, and Ben-Porath (1990) also presented data concerning WRK (Negative Work Attitudes) scale scores for several groups of men who would be expected to differ on this scale. The scores of pilot applicants, military personnel, alcoholics in treatment, and psychiatric inpatients

Table 6.5. Correlations of the MMPI-2 Content Scales with the Validity and Clinical Scales for Men and Women in the MMPI-2 Normative Samples

	?	L	F	K	Hs	D	Hy	Pd	Mf	Pa	Pt	Sc	Ma	Si
						Men ($n = 1138$)								
ANX	−04	−27	47	−61	50	45	04	50	20	33	80	69	31	43
FRS	00	−07	24	−29	34	22	02	16	01	09	37	35	06	28
OBS	−02	−30	40	−63	40	26	−16	29	12	18	77	64	31	44
DEP	01	−17	57	−56	48	52	02	58	16	38	80	75	27	48
HEA	02	−06	47	−29	89	45	39	35	10	25	50	55	18	29
BIZ	03	−14	51	−44	38	03	−09	36	08	33	51	62	48	11
ANG	−03	−38	34	−66	33	01	−21	36	−02	15	55	53	42	19
CYN	−01	−17	39	−71	33	07	−43	26	−17	−16	51	53	42	32
ASP	−03	−34	42	−60	26	01	−36	37	−15	−12	45	50	51	18
TPA	−06	−37	29	−68	29	05	−30	22	−05	04	53	48	38	25
LSE	−01	−19	48	−52	42	42	−11	27	07	18	72	61	11	59
SOD	−01	−07	31	−31	24	39	−19	04	11	09	40	36	−20	85
FAM	02	−27	53	−55	32	21	−11	57	18	21	59	66	43	31
WRK	00	−26	53	−63	49	44	−09	41	14	21	81	73	23	59
TRT	00	−19	54	−57	46	40	−12	40	02	19	72	68	20	56
						Women ($n = 1462$)								
ANX	−05	−24	47	−67	58	60	17	51	10	39	83	71	34	48
FRS	−05	02	16	−41	34	20	−01	13	00	06	39	33	11	32
OBS	−04	−26	39	−69	45	40	−06	36	08	25	79	65	36	47
DEP	00	−19	58	−63	54	63	12	61	01	44	83	77	31	55
HEA	01	−07	42	−43	91	45	48	33	00	25	55	59	29	31
BIZ	00	−07	49	−46	36	11	−03	39	−14	31	51	65	50	15
ANG	−04	−34	38	−69	39	20	−06	44	00	25	62	60	44	27
CYN	−05	−08	41	−70	41	17	−24	32	−24	−06	51	54	46	35
ASP	−05	−30	41	−57	30	09	−25	37	−28	−09	44	51	51	23
TPA	−03	−29	31	−65	32	15	−16	23	−05	13	53	49	36	28
LSE	−04	−16	44	−60	44	53	−04	31	01	23	74	61	14	65
SOD	−03	−04	28	−38	24	43	−17	06	07	19	43	35	−17	84
FAM	−01	−21	56	−57	40	32	04	61	04	33	60	72	45	33
WRK	−04	−23	50	−69	52	58	02	44	04	29	82	72	27	63
TRT	−01	−15	50	−63	45	50	−05	42	−04	26	72	69	24	61

Source: Butcher, J.N., Graham, J.R., Williams, C.L., & Ben-Porath, Y.S. (1990). *Development and use of the MMPI-2 content scales.* Minneapolis: University of Minnesota Press. Copyright © 1990 by the University of Minnesota. Reproduced by permission.

were compared. The pilot applicants, who would be expected to have the most positive work attitudes, scored lowest on the WRK scale, whereas the alcoholics and psychiatric patients obtained the highest scores.

Ben-Porath, Butcher, and Graham (1991) investigated the contribution of the MMPI-2 content scales to the differential diagnosis of schizophrenia and major depression in an inpatient psychiatric setting. They found that both the clinical scales and the content scales were related to the differential diagnosis of these two conditions and that the content scales contained information relevant to this diagnostic question beyond that available from the clinical scales. For male patients, the Depression and Bizarre Mentation content scales added to the diagnostic discrimination. For female patients, the Bizarre Mentation content scale added to the diagnostic discrimination.

Several correlational studies have suggested that the MMPI-2 content scales are related to other relevant measures. Schill and Wang (1990) reported that the Anger content scale correlated positively with Spielberger's anger expression measure and negatively with anger control. For men the Anger content scale was correlated significantly with verbal expression of aggression, and for women with physical expression of aggression.

Dwyer, Graham, and Ott (1992) examined correlations between the MMPI-2 content scales and ratings of symptoms in an inpatient psychiatric setting. The results suggested both convergent and discriminant validity for the content scales for which construct-relevant ratings were available. Patients with higher Anxiety scale scores were rated as more anxious ($r = .36$) and depressed ($r = .39$), and patients with higher Depression scale scores were rated as more depressed ($r = .49$). Patients scoring higher on the Bizarre Mentation scale were rated as more likely to have hallucinations ($r = .38$) and unusual thought content ($r = .33$). Patients with higher scores on the Health Concerns scale were rated as having more somatic symptoms ($r = .45$).

Clearly, additional empirical data are needed before the validity of the content scales can be judged adequately. However, the data presented to date are impressive and encouraging.

Interpretation of the Content Scales

Although the validity data for the content scales are limited at this time, the data summarized above and examination of the content of items in each content scale can be used to offer some tentative interpretive inferences about persons who score relatively high on each scale. As with the Harris-Lingoes subscales discussed earlier, until additional data are available users of the content scales should consider T scores greater than 65 as high scores. Because of limited data, it is not recommended that low scores on the content scales be used to generate interpretive statements. As additional validity data accumulate, it may be possible to state more precise cutoff scores when using some of the content scales for specific purposes.

Test-taking attitude must be taken into account when interpreting the content scales. Because the scales contain items with obvious content, scores on the scales are susceptible to distortion related to test-taking attitude. Persons who approach the MMPI-2 in a defensive manner obtain low scores on most of the scales, and persons who exaggerate problems in taking the MMPI-2 obtain elevated scores on most of the scales. Clearly, the content scales are most useful when test subjects have approached the MMPI-2 in a cooperative, open manner.

Clinicians should view scores on the content scales as direct communication between test subjects and examiners. Characteristics reflected by elevated scores on the content scales are those that the test subjects want examiners to know about. Rapport with clients often is increased when the content scale results are used to give feedback to clients that we are aware of the things they were trying to communicate to us when they completed the MMPI-2.

Anxiety (ANX)

High scores on the ANX scale are indicative of persons who:

1. feel anxious, nervous, worried, and apprehensive
2. have problems with concentration
3. complain of sleep disturbance
4. are uncomfortable making decisions
5. may report feeling sad, blue, or depressed
6. feel that life is a strain and are pessimistic about things getting better
7. lack self-confidence
8. feel overwhelmed by the responsibilities of daily life
9. if female, may appear to be irritable and hostile
10. if psychiatric patients, are likely to have been given anxiety disorder diagnoses

Fears (FRS)

High scores on the FRS scale are indicative of persons who:

1. feel fearful and uneasy much of the time
2. report multiple specific fears or phobias

Obsessiveness (OBS)

High scores on the OBS scale are indicative of persons who:

1. have great difficulty making decisions
2. are rigid and dislike change
3. engage in compulsive behaviors such as counting or hoarding
4. fret, worry, and ruminate about trivial things
5. may feel dysphoric and despondent and lack interest in things
6. lack self-confidence

Depression (DEP)

High scores on the DEP scale are indicative of persons who:

1. feel depressed, sad, blue, or despondent
2. feel fatigued and lack interest in things
3. are pessimistic and feel hopeless
4. may recently have been preoccupied with thoughts of death and suicide
5. cry easily

6. are indecisive and lack self-confidence
7. feel guilty
8. have health concerns
9. feel lonely and empty much of the time
10. if female, may be resentful and demanding
11. if psychiatric patients, are likely to have been given depressive disorder diagnoses

Health Concerns (HEA)

High scores on the HEA scale are indicative of persons who:

1. deny good physical health
2. are preoccupied with bodily functioning
3. feel worn out and lack energy
4. report a variety of specific somatic symptoms, including some that could be suggestive of a neurological disorder

Bizarre Mentation (BIZ)

High scores on the BIZ scale are indicative of persons who:

1. may have psychotic thought processes
2. may report auditory, visual, or olfactory hallucinations
3. report feelings of unreality
4. feel that other people say bad things about them
5. may believe that other people are trying to harm them
6. may believe that other people can read their minds or control their thinking or behavior
7. if psychiatric patients, are likely to have been given psychotic diagnoses

Anger (ANG)

High scores on the ANG scale are indicative of persons who:

1. feel angry and hostile much of the time
2. are seen by others as irritable, grouchy, impatient, and stubborn
3. may feel like swearing or smashing things
4. have temper tantrums
5. may lose control and be physically abusive

Cynicism (CYN)

High scores on the CYN scale are indicative of persons who:

1. see other people as dishonest, selfish, and uncaring

2. question the motives of others
3. are guarded and untrusting in relationships
4. may be hostile and overbearing
5. may be demanding of themselves but resent even mild demands placed on them by others
6. are not friendly or helpful

Antisocial Practices (ASP)

High scores on the ASP scale are indicative of persons who:

1. are likely to have been in trouble in school or with the law
2. believe that there is nothing wrong with getting around laws as long as they are not broken
3. may enjoy hearing about the antics of criminals
4. have generally cynical attitudes about other people, seeing them as selfish and dishonest
5. resent authority
6. if men, may express anger and hostility by cursing, swearing, or having temper tantrums
7. if men, may use nonprescription drugs
8. if women, may express anger and hostility less directly than males
9. if women, may be seen by others as dishonest and not helpful or considerate of others

Type A Behavior (TPA)

High scores on the TPA scale are indicative of persons who:

1. are hard-driving, fast-moving, and work-oriented
2. feel there is never enough time to get things done
3. do not like to wait or be interrupted
4. frequently are hostile, irritable, and easily annoyed
5. tend to be overbearing and critical in relationships
6. tend to hold grudges and want to get even
7. if women, may be seen as tense, restless, nervous, and suspicious

Low Self-Esteem (LSE)

High scores on the LSE scale are indicative of persons who:

1. have very poor self-concepts
2. anticipate failure and give up easily
3. are overly sensitive to criticism and rejection
4. find it difficult to accept compliments
5. are passive in relationships

6. have difficulty making decisions
7. may have many worries and fears

Social Discomfort (SOD)

High scores on the SOD scale are indicative of persons who:

1. are shy and socially introverted
2. would rather be alone than around other people
3. dislike parties and other group activities
4. do not initiate conversations

Family Problems (FAM)

High scores on the FAM scale are indicative of persons who:

1. describe considerable discord in their current families and/or families of origin
2. describe their families as lacking in love, understanding, and support
3. resent the demands and advice of their families
4. feel angry and hostile toward their families
5. see marital relationships as involving unhappiness and lack of affection

Work Interference (WRK)

High scores on the WRK scale are indicative of persons who:

1. are reporting a wide variety of attitudes and behaviors that are likely to contribute to poor work performance
2. may be questioning their own career choices
3. say that their families have not approved of their career choices
4. are not ambitious and are lacking in energy
5. express negative attitudes toward co-workers
6. have poor self-concepts
7. are obsessive and have problems concentrating
8. have difficulty making decisions and may show poor judgment
9. feel tense, worried, and fearful

Negative Treatment Indicators (TRT)

High scores on the TRT scale are indicative of persons who:

1. have negative attitudes toward doctors and mental health treatment
2. feel that no one can understand them
3. believe that they have problems that they cannot share with anyone
4. give up easily when problems are encountered
5. feel unable to make significant changes in their lives

6. are poor problem solvers
7. often show poor judgment

CRITICAL ITEMS

Critical items are those whose content is judged to be indicative of serious psychopathology. The first set of MMPI critical items was identified by Grayson (1951) based on subjective clinical judgment. The thirty-eight items dealt primarily with severe psychotic symptoms and overlapped considerably with scales F and 8. Grayson believed that responses in the scored direction to any of these items suggested potentially serious emotional problems that should be studied further. Caldwell (1969) also generated intuitively a more comprehensive set of critical items that he intended for use with computerized scoring and interpretive services. Koss, Butcher, and Hoffman (1976) investigated the validity of the Grayson and Caldwell critical items as indicators of crises, and they concluded that both sets of items performed poorly as indexes of serious malfunctioning.

Koss-Butcher Critical Items

Koss et al. (1976) asked clinicians to nominate MMPI items that seemed to be related to six crisis areas (acute anxiety state, depressed suicidal ideation, threatened assault, situational stress due to alcoholism, mental confusion, and persecutory ideas). The nominated items were then compared with criterion measures of the crises, resulting in a list of seventy-three valid critical items. Following the revision of the MMPI in 1989, the Koss-Butcher critical item set was revised to reflect changes in the item pool (Butcher et al., 1989). The revised Koss-Butcher critical items are reprinted in Appendix I.

Lachar-Wrobel Critical Items

Lachar and Wrobel (1979) used a similar approach in identifying 111 critical items related to fourteen problem areas frequently encountered in inpatient and outpatient samples. All but 4 of the original Lachar-Wrobel critical items are included in the MMPI-2. These items are listed in Appendix I. It should be noted that the Lachar-Wrobel critical items have not been revised to include new items that were added to the original MMPI item pool.

Recommendations Concerning Use of the Critical Items

Koss (1979, 1980) summarized the usefulness of critical items. She reviewed research suggesting that the Koss-Butcher and Lachar-Wrobel critical items are more valid than the Grayson or Caldwell critical items. However, Koss also pointed out some cautions in using critical items. All of the critical-item sets overlap considerably with scales F and 8, and most critical items are keyed in the true direction. Thus, critical-item endorsements can be misleading for persons who are displaying an acquiescence response set or exaggerating their symptoms and problems.

MMPI-2 users who interpret critical-item endorsements should seriously consider Koss's (1980) cautions and recommendations. She concluded that critical items should not be used as a quick assessment of level of maladjustment. Data indicate that the critical items perform poorly in separating normal and psychiatric subjects. Also, critical-item lists are not as reliable as scales because of the vulnerability of single-item responses to error. A test subject can misinterpret and/or mismark a single item, leading the test interpreter to an erroneous conclusion, whereas that same mistake in the context of a longer scale would not have much impact on the individual's total score on that scale.

The potential value of using critical items is the same as with the other content approaches discussed in this chapter. Examination of the content of subjects' responses can clarify the kinds of things they are telling us about themselves. However, critical-item responses should not be overinterpreted. In a valid MMPI-2 protocol, endorsement of critical items should lead the clinician to inquire further into the areas assessed by the items.

7

Supplementary Scales

In addition to its utilization in the construction of the standard validity and clinical scales, the original MMPI item pool was used to develop numerous other scales by variously recombining the 566 items using item-analytic, factor-analytic, and intuitive procedures. Dahlstrom, Welsh, and Dahlstrom (1972, 1975) presented more than 450 supplementary scales. The scales had quite diverse labels, ranging from more traditional ones, such as "Dominance" and "Suspiciousness," to more unusual ones, such as "Success in Baseball." The scales varied considerably in terms of what they were supposed to measure, the manner in which they were constructed, their reliabilities, the extent to which they were cross-validated, the availability of normative data, and the amount of additional validity data that were generated. The scales also varied in terms of how frequently they were used in clinical and research settings. Some scales were used only by their constructors, whereas others were employed extensively in research studies and used routinely in clinical interpretation of the MMPI.

Caldwell (1988) offered interpretive information for many of the supplementary scales of the original MMPI. Not all supplementary scales were maintained in the MMPI-2 (Butcher et al., 1989) because that would have increased the length of the booklet beyond a point judged acceptable for routine clinical use. For the most part, the extent to which existing research data supported a scale's reliability and validity determined which scales were retained. However, a number of scales were maintained on the basis of less scientific criteria. For example, the Harris-Lingoes subscales were judged to be a very helpful supplementary source of information in interpreting the clinical scales. The Wiener Subtle-Obvious subscales were maintained because some persons believe that they are useful in detecting certain response sets that invalidate profiles.

Levitt (1990) provided information concerning how many of the items needed to score some commonly used supplementary scales are included in the MMPI-2. He concluded that certain scales included in the MMPI-2 manual can still be scored. However, no MMPI-2 norms are readily available for use with these additional scales.

In addition to maintaining some of the existing supplementary scales, new scales were developed for the MMPI-2. Several new validity scales were developed (described in Chapter 3), and new content scales were developed (described in Chapter 6). Other scales also were developed as part of the restandardization project and will be described in this chapter. Subsequent to the publication of the MMPI-2, several additional scales for assessing substance abuse and marital distress were published. They too will be described in this chapter.

The same format will be used for discussing each supplementary scale. Scale development information will be presented, and, to the extent that they are available, reliability and validity data will be reported. Interpretive information for each scale also will be summarized. As with the clinical and validity scales, no absolute cutoff scores can be specified. In general, T scores greater than 65 should be considered as high scores. Whenever information about specific cutoff scores for a scale is available, the information will be presented. The higher the scores are, the more likely it is that the interpretive information presented will apply. As with other scales discussed earlier in this book, low scores should not be interpreted for most of the supplementary scales. There is not enough research information available to have confidence in interpretive statements based on low scores. For several scales where data concerning the meaning of low scores are available, the interpretation of low scores is discussed in this chapter.

Although an attempt was made to rely on research studies for interpretive information, in some cases examining item content was necessary in generating descriptors. For supplementary scales that were developed from the original MMPI and maintained in the MMPI-2, research data from the original MMPI were used to generate interpretive statements. Because these scales are essentially the same in the two versions of the test, this approach seems appropriate. It should be emphasized that the supplementary scales are not intended to replace the standard validity and clinical scales. Rather, they are to be used in addition to them.

The composition and scoring of each supplementary scale are presented in Appendix J. It should be noted that most of the supplementary scales can be scored only if the entire 567-item MMPI-2 is administered. The test publisher provides scoring keys for the supplementary scales discussed in this chapter. Appendix K presents T-score conversions for the supplementary scales, and the test publisher provides profile sheets for plotting scores on most of the supplementary scales. The norms used to transform raw scores to T scores are the same ones used for the standard validity and clinical scales. Linear T-score transformations are used for all of the supplementary scales discussed in this chapter, except for the Marital Distress Scale (MDS) which has uniform T-score transformations.

ANXIETY (A) AND REPRESSION (R) SCALES

Scale Development

Whenever the basic validity and clinical scales of the MMPI or MMPI-2 have been factor analyzed to reduce them to their most common denominators, two basic dimensions have emerged consistently (e.g., Butcher et al., 1989; Block, 1965; Eichman, 1961, 1962; Welsh, 1956). Welsh (1956) developed the Anxiety (A) and Repression (R) scales to assess these two basic dimensions.

By factor analyzing MMPI scores for male Veterans Administration patients, Welsh identified a factor that he originally labeled "general maladjustment." A scale was developed to assess this factor by identifying items that were most highly associated with the factor. After being administered to new groups of psychiatric patients, this scale was refined by using internal consistency procedures. The original A scale included thirty-nine items, all of which are included in the MMPI-2 version of the scale. Welsh suggested from an examination of the items that the content of the A-scale items falls into four categories: thinking and thought processes; negative emotional tone and dysphoria; pessimism and lack of energy; and malignant mentation. The items are keyed in such a way that high scores on the A scale are associated with greater psychopathology.

The R scale was constructed by Welsh (1956) to measure the second major dimension emerging from factor analyses of the basic validity and clinical scales of the MMPI. A procedure similar to that used in developing the A scale also was employed with the R scale. It resulted in a final scale containing forty items, thirty-seven of which are included in the MMPI-2 version of the scale. Welsh suggested the following clusters based on the content of the R-scale items: health and physical symptoms; emotionality, violence, and activity; reactions to other people in social situations; social dominance, feelings of personal adequacy, and personal appearance; and personal and vocational interests.

Reliability and Validity

Welsh (1965) reported reliability data for the A and R scales based on unpublished research by Kooser and Stevens. For 108 college undergraduates, the split-half reliability coefficients for the A and R scales were .88 and .48, respectively. Gocka (1965) reported Kuder-Richardson 21 (internal-consistency) values of .94 and .72 for the A and R scales, respectively, for 220 male Veterans Administration psychiatric patients. For the MMPI-2 normative sample, internal-consistency coefficients for the A scale were .89 for men and .90 for women. Corresponding internal-consistency coefficients for the R scale were .67 and .57 (Butcher et al., 1989).

When 60 college sophomores were given the scales on two occasions, separated by 4 months, test-retest reliability coefficients for the A and R scales were .70 and .74, respectively (Welsh, 1956). Test-retest coefficients for the A scale for college students, with a 6-week interval, were .90 for men and .87 for women. Corresponding values for the R scale were .85 and .84, respectively (Moreland, 1985b). For the MMPI-2 normative sample, test-retest reliabilities (with an average interval of 1 week) for the A scale were .91 for men and .91 for women. Test-retest reliabilities for the R scale were .79 for men and .77 for women (Butcher et al., 1989). The stability of scores on the A and R scales over these relatively short periods of time is quite high.

It has been suggested by researchers that the major sources of variance in MMPI responses are associated with response sets. A response set exists when persons taking a test answer the items from a particular perspective or attitude about how they would like the items to show themselves to be. Edwards (1964) argued that the first factor of the MMPI, the one assessed by the A scale, simply assesses examinees' willingness, while describing themselves on the test, to endorse socially undesirable items. Messick and Jackson (1961) suggested that R-scale scores simply indicate the extent to which examinees are willing to admit (acquiesce) on the test to many kinds of emotional difficulties. This interpretation appears to be supported by the fact that all of the items in the R scale are keyed in the false direction. Block (1965) refuted the response set or bias arguments by demonstrating that the same two major factor dimensions emerge even when the MMPI scales were altered to control for social desirability and acquiescence effects with the use of techniques developed by Edwards (1964) and others. Block also was able to identify through his research reliable extratest correlates for the two factor dimensions.

Welsh (1956) reported some unpublished data supplied by Gough for a group of normal subjects. Gough found that A-scale scores correlated negatively with the K and L scales and with scale 1 of the MMPI and correlated positively with the F scale and with scales 9 and 0. Gough also reported that high A-scale scorers showed slowness of personal tempo, pessimism, vacillation, hesitancy, and inhibitedness. Sherriffs and Boomer (1954) found that high A-scale scorers showed more self-doubts in examination situations. Welsh (1956) reported an unpublished study by Welsh and Roseman indicating that patients who showed the most positive change during insulin shock therapy also showed marked decreases in A-scale scores after such therapy. There also is evidence that A-scale scores tend to decrease during psychiatric hospitalization (Lewinsohn, 1965). Duckworth and Duckworth (1975) suggested that a high A-scale score indicates that a person is experiencing enough discomfort that he or she is likely to be motivated to change in psychotherapy. Block and Bailey (1955) reported reliable extratest correlates for scores on the A scale. These correlates are presented later, in the discussion of the interpretation of A-scale scores.

Welsh (1956) also reported that in the study by Welsh and Roseman the patients who were judged as most improved during their course of insulin shock therapy showed some decreases in R-scale scores in addition to the decreases in A-scale scores. Lewinsohn (1965) found that only small changes were found in R-scale scores during psychiatric hospitalization. Welsh (1956) reported data provided by Gough indicating that in a sample of normal subjects, R-scale scores were positively correlated with the L and K scales and with scales 1 and 2 of the MMPI and negatively correlated with scale 9. Duckworth and Duckworth (1975) described high R-scale scorers as denying, rationalizing, and lacking self-insight. Block and Bailey (1955) identified extratest correlates of R-scale scores. These correlates are included below in connection with the interpretation of scores on the R scale.

Interpretation of High A-Scale Scores

High scores on the A scale indicate persons who:

1. are anxious and uncomfortable
2. have a slow personal tempo
3. are pessimistic
4. are apathetic, unemotional, and unexcitable
5. are shy and retiring
6. lack confidence in their own abilities
7. are hesitant and vacillating
8. are inhibited and overcontrolled
9. are influenced by diffuse personal feelings
10. are defensive
11. rationalize and blame others for difficulties
12. lack poise in social situations
13. are conforming and overly accepting of authority
14. are submissive, compliant, and suggestible
15. are cautious
16. are fussy
17. if men, have behavior that tends to be seen as effeminate
18. are seen as cool, distant, and uninvolved
19. become confused, disorganized, and maladaptive under stress
20. are uncomfortable enough to be motivated to change in psychotherapy

In summary, persons scoring high on the A scale, if from a normal population, are rather miserable and unhappy people. High A-scale scorers in a psychiatric setting fit such summarizing rubrics as neurotic, maladjusted, submissive, and overcontrolled. Because of their discomfort, high A-scale scorers usually are highly motivated for counseling or psychotherapy.

Interpretation of High R-Scale Scores

High scores on the R scale indicate persons who:

1. are submissive
2. are unexcitable

3. are conventional and formal
4. are clear-thinking
5. are slow and painstaking

In summary, high R-scale scorers are internalizing individuals who have adopted careful and cautious lifestyles.

Conjoint Interpretation of the A and R Scales

Welsh (1956, 1965) suggested that a more complete understanding of an examinee is possible if scores on the A and R scales are considered conjointly. Welsh (1956) reported some preliminary work carried out by Welsh and Pearson in which protocols of Veterans Administration psychiatric inpatients were categorized as high A–low R, high A–high R, low A–low R, and low A–high R. Different diagnostic labels were associated with cases in the four quadrants (e.g., depressive diagnoses most often occurring in the high A–high R quadrant and personality disorder diagnoses most often occurring in the low A–low R quadrant). Gynther and Brillant (1968) reported that Welsh's results were not replicated when the quadrant approach was utilized by them with their own sample of psychiatric outpatients.

Subsequently, Welsh (1965) suggested dividing each scale (A and R) into high, medium, and low levels to form nine categories, or novants. Using male Veterans Administration patients, Welsh identified protocols associated with each of his novants. He then determined the typical profiles for the novants and inferred personality descriptions from the profile configurations. Welsh noted that the descriptions are biased toward patient groups rather than normal individuals and that the descriptions, which are intended to lead to hypotheses for further investigation, should not be taken literally and should not be used for "cookbook" interpretation of profiles. Duckworth and Duckworth (1975) reported that they did not find Welsh's descriptions of the novants to be very accurate for college counselees, except for the high A–high R interpretation.

EGO STRENGTH (Es) SCALE

Scale Development

The Ego Strength (Es) scale was developed by Barron (1953) specifically to predict the response of neurotic patients to individual psychotherapy. The original Es scale had sixty-eight items, of which the MMPI-2 version of the scale includes fifty-two. To identify items for the original Es scale, item responses of 17 patients who were judged independently as clearly improved after 6 months of psychotherapy were compared with item responses of 16 patients who were judged as unimproved after 6 months of psychotherapy. Items are scored in the direction most often chosen by the improved patients. The Es-scale items deal with physical functioning, seclusiveness, moral posture, personal adequacy, ability to cope, phobias, and anxieties.

Reliability and Validity

Barron (1953) reported that the odd-even reliability of the Es scale for a sample of 126 patients was .76. Gocka (1965) reported a Kuder-Richardson 21 (internal-consistency) value of .78 for the Es scale for 220 male Veterans Administration psychiatric patients. For men and women in the MMPI-2 normative sample, internal-consistency values for the Es scale were .60 and .65, respectively.

Barron (1953) reported a test-retest reliability coefficient of .72 for a group of 30 patients, using a test-retest interval of 3 months. Moreland (1985b) reported test-retest coefficients for male and female college students (with a 6-week interval) of .80 and .82, respectively. Test-retest coefficients for subsamples of male and female subjects in the MMPI-2 normative sample were .78 and .83, respectively (Butcher et al., 1989). Schuldberg (1992) reported that for a sample of college students the internal consistency of the MMPI-2 Es scale was slightly higher than that of the original MMPI Es scale.

The Es scale was cross-validated by Barron (1953) using three different samples of neurotic patients for whom ratings of improvement during brief, psychoanalytically-oriented psychotherapy were available. Because pretherapy Es-scale scores were positively related to assessed improvement for all three samples, Barron concluded that the Es scale is useful in predicting responsiveness to psychotherapy. Unfortunately, subsequent attempts by others to cross-validate the Es scale as a predictor of response to psychotherapy or other treatment approaches have yielded inconsistent findings. Some data indicate that psychiatric patients who change most during treatment have higher pretreatment Es-scale scores than patients who show less change (e.g., Wirt, 1955, 1956), whereas other data suggest that change in treatment is unrelated to pretreatment Es-scale scores (Ends & Page, 1957; Fowler, Teel, & Coyle, 1967; Getter & Sundland, 1962; Sullivan, Miller, & Smelser, 1958). Distler, May, and Tuma (1964) found that pretreatment Es-scale scores were positively related to hospitalization outcome for male psychiatric patients and negatively related to hospitalization outcome for female psychiatric patients. Sinnett (1962) reported that Veterans Administration psychiatric patients with higher pretreatment Es-scale scores showed more personality growth during treatment, which included psychotherapy, than did patients with lower scores, but pretreatment Es-scale scores were unrelated to assessed symptomatic change for these same patients. It should be noted that many of the studies failing to replicate Barron's finding that Es-scale scores were related to change after psychotherapy used change after hospitalization and therefore did not represent a true replication of his study.

Dahlstrom et al. (1975) tried to explain the inconsistent findings concerning the relationship between Es-scale scores and treatment outcome. They suggested that when high Es-scale scores occur for persons who obviously are having difficulties but who are denying them, the high Es-scale scores may not be predictive of a favorable treatment outcome. However, high Es-scale scores for persons who are admitting to emotional problems may suggest a favorable response to treatment. Clayton and Graham (1979) were not able to validate the Dahlstrom et al. hypothesis with a sample of hospitalized psychiatric patients. It is clear from the existing literature that the relationship between Es-scale scores and treatment outcome is not a simple one and that factors such as kind of patients, type of treatment, and nature of the outcome measure must be taken into account. In general, however, high Es-scale scores are predictive of positive personality change for neurotic patients who receive traditional, individual psychotherapy.

One must be very cautious in interpreting Es-scale scores in protocols that are suggestive of defensiveness. In such circumstances, Es-scale scores tend to be artificially high and are not predictive of positive response to therapy. Likewise, caution should be exercised in interpreting Es-scale scores in protocols suggestive of exaggeration of symptoms. In such circumstances, Es-scale scores tend to be artificially low and are not predictive of a negative response to therapy.

There also are research data indicating that the Es scale can be viewed as an indication of overall psychological adjustment. Higher scores on the Es scale are associated with more favorable adjustment levels as assessed by other MMPI-2 indexes and extratest criteria. Schuldberg (1992) found that MMPI-2 Es-scale scores of college students were positively related to other self-report measures of psychological health.

Es-scale scores tend to be lower for psychiatric patients than for nonpatients and for people receiving psychiatric or psychological treatment than for persons not involved in such treatment (Gottesman, 1959; Himelstein, 1964; Kleinmuntz, 1960; Quay, 1955; Spiegel, 1969; Taft, 1957). However, it has been reported that the Es scale fails to differentiate between delinquent and nondelinquent adolescents (Gottesman, 1959). Whereas Es-scale scores tend to be higher for neurotic patients than for psychotic patients, the scale fails to discriminate among more specific diagnostic categories (Hawkinson, 1961; Rosen, 1963; Tamkin, 1957; Tamkin & Klett, 1957).

There are some data indicating that Es-scale scores tend to increase during the course of psychotherapy or other treatment procedures. Lewinsohn (1965) reported that psychiatric patients showed an increase in level of Es-scale scores from hospital admission to discharge. However, Barron and Leary (1955) found that Es-scale scores did not change more for patients who received individual or group psychotherapy than for patients who remained on a waiting list for a similar period of time. It also was reported that psychotherapy patients who were self-referred scored higher on the Es scale than those who were referred by someone else (Himelstein, 1964), suggesting that high Es-scale scorers are more aware of internal conflicts than are low Es-scale scorers.

Scores on the Es scale are related positively to intelligence (Tamkin & Klett, 1957; Wirt, 1955) and to formal education (Tamkin & Klett, 1957). The relationship between Es-scale scores and age is less clear. Tamkin and Klett (1957) found no relationship between Es-scale scores and age, but Getter and Sundland (1962) reported that older persons tended to score lower on the Es scale. Consistent gender differences in Es-scale scores have been reported, with men obtaining higher raw scores than women (Distler et al., 1964; Getter & Sundland, 1962; Taft, 1957). This gender difference originally was interpreted as reflecting the greater willingness of women to admit to problems and complaints (Getter & Sundland, 1962). However, a more reasonable explanation of the gender difference on the Es scale is that men score higher than women because the scale contains a number of items dealing with masculine role identification (Holmes, 1967).

Interpretation of High Es-Scale Scores

From the above discussion, it may be concluded that high scorers on the Es scale generally tend to show more positive personality change during treatment than do low scorers. However, the relationship between Es-scale scores and treatment prognosis is not a simple one, and patient and treatment variables must be taken into account. Also, high Es-scale scorers tend to be better adjusted psychologically, and they are more able to cope with problems and stresses in their life situations. Among psychiatric patients, high Es-scale scores are likely to be associated with neurotic diagnoses and low Es-scale scores are more likely to be found for psychotic patients. In addition, high scores on the Es scale indicate persons who:

1. lack chronic psychopathology
2. are stable, reliable, and responsible
3. are tolerant and lack prejudice
4. are alert and adventuresome
5. are determined and persistent
6. are self-confident, outspoken, and sociable
7. are intelligent, resourceful, and independent
8. have a secure sense of reality
9. deal effectively with others
10. create favorable first impressions
11. gain acceptance of others
12. are opportunistic and manipulative
13. have strongly developed interests
14. if men, have an appropriately masculine style of behavior

15. are hostile and rebellious toward authority
16. are competitive
17. may be sarcastic and cynical
18. seek help because of situational problems
19. can tolerate confrontations in psychotherapy

In summary, people with high Es-scale scores appear to be fairly well put together emotionally. In nonpsychiatric settings, such people are not likely to have serious emotional problems. Among persons with emotional problems, high Es-scale scores suggest that problems are likely to be situational rather than chronic, that the individuals have psychological resources that can be drawn upon in helping them to solve the problems, and that the prognosis for positive change in counseling or psychotherapy is good.

MacANDREW ALCOHOLISM SCALE—REVISED (MAC-R)

Scale Development

The MacAndrew Alcoholism (MAC) scale (MacAndrew, 1965) was developed to differentiate alcoholic from nonalcoholic psychiatric patients. The scale was constructed by contrasting the MMPI responses of 200 male alcoholics seeking treatment at an outpatient clinic with responses of 200 male nonalcoholic psychiatric outpatients from the same facility. These analyses identified fifty-one items that differentiated the two groups. Because MacAndrew was interested in developing a subtle scale, two of the fifty-one items that deal directly with excessive drinking behavior were eliminated from the scale. The items are keyed in the direction selected most often by the alcoholic patients. Schwartz and Graham (1979) reported that the major content dimensions of the MAC scale are cognitive impairment, school maladjustment, interpersonal competence, risk taking, extroversion and exhibitionism, and moral indignation.

Four of the original MAC-scale items were among those eliminated from the MMPI-2 because of objectionable content. Because the MAC scale typically is interpreted in terms of raw scores, a decision was made to maintain a scale of forty-nine items in the MMPI-2. Thus, the four objectionable items were replaced with four new items that were selected because they differentiated alcoholic and nonalcoholic men (Butcher et al., 1989). Levitt et al. (1992) reported that psychiatric inpatients and nonclinical subjects scored somewhat lower on the MMPI-2 version of the scale than on the MMPI version. This finding is difficult to interpret because, based on an earlier study (Levitt, 1990), it would appear that the investigators used a fifty-one-item version of the original scale. In practice and in most research studies, a forty-nine-item version of the MAC has been used, and the MMPI-2 version of the scale also has forty-nine items.

Reliability and Validity

Internal-consistency data were not reported for the original MAC scale. The MAC-R scale does not seem to have particularly good internal consistency. Internal-consistency coefficients (Coefficient Alpha) for the MMPI-2 normative sample were .56 for men and .45 for women (Butcher et al., 1989). A factor analysis reported by Schwartz and Graham (1979) indicated that the original MAC scale was not unidimensional.

Moreland (1985b) reported test-retest reliability coefficients (6-week interval) for the MAC scale with samples of normal college men and women. The coefficients were .82 and .75, respectively. For subsamples of male and female subjects in the MMPI-2 normative sample, test-retest reliability coefficients (1-week interval) for the MAC-R scale were .62 and .78, respectively. These relatively modest test-retest reliability coefficients for normal samples can be explained, at least in part, by limited variability of scores in these groups.

Several studies reported that MAC-scale scores did not change significantly during treatment programs ranging in length from 28 to 90 days or during a 1-year follow-up period after treatment (Chang, Caldwell, & Moss, 1973; Gallucci, Kay, & Thornby, 1989; Huber & Danahy, 1975; Rohan, 1972; Rohan, Tatro, & Rotman, 1969). Hoffman, Loper, and Kammeier (1974) compared the MMPIs of male alcoholics at the time of treatment to MMPIs taken 13 years earlier when they had entered college, and found no significant changes in MAC-scale scores over this extended period of time. Apfeldorf and Hunley (1975) reported high MAC-scale scores for male veterans with histories of alcohol abuse who were no longer abusing alcohol when they completed the MMPI.

MacAndrew (1965) reported cross-validation data for his scale. A cutoff score of 24 correctly classified approximately 82 percent of the alcoholic and nonalcoholic subjects. Many subsequent studies have found that groups of alcoholic and nonalcoholic subjects in a variety of settings tend to obtain significantly different mean scores on the MAC scale (Apfeldorf & Hunley, 1975; Rhodes, 1969; Rich & Davis, 1969; Rohan, 1972; Rosenberg, 1972; Schwartz & Graham, 1979; Uecker, 1970; Williams, McCourt, & Schneider, 1971). There also are data suggesting that drug addicts score higher than other psychiatric patients but not differently from alcoholics on the MAC scale (Fowler, 1975; Kranitz, 1972). Graham (1978) reported that pathological gamblers scored similarly to alcoholics and heroin addicts on the MAC scale.

Hoffman et al. (1974) located MMPIs of male alcoholics in treatment; these MMPIs had been completed approximately 13 years previously when the men entered college. When the MAC scale scores of these men were com-

pared with those of their classmates who had not received treatment for alcoholism, significant differences were found. There also are data suggesting
that persons who drink excessively but who are not alcoholics score higher on
the MAC scale than persons who do not drink excessively (Apfeldorf & Hunley, 1975; Williams et al., 1971). Several studies have reported that the MAC
scale also seems to be effective in identifying adolescents who have significant
problems with alcohol and/or drug abuse (Gantner, Graham, & Archer,
1992; Wisniewski, Glenwick, & Graham, 1985; Wolfson & Erbaugh, 1984).
However, Colligan and Offord (1990) found that the MAC scale did not discriminate well between male adolescent substance abusers and nonabusers.

Several studies have suggested caution in using the MAC scale with African
Americans (Graham & Mayo, 1985; Walters, Greene, & Jeffrey, 1984; Walters,
Greene, Jeffrey, Kruzich, & Haskin, 1983). Although African-American alcoholics tend to obtain scores in the addictive range, classification rates in these
studies have not been very good because nonalcoholic African Americans
also tend to score rather high on the MAC scale. It should be noted that in all
of these MAC studies, the samples have involved military personnel or veterans. The extent to which these findings can be generalized to other kinds of
subjects remains to be determined. Likewise, additional research is needed to
determine to what extent the MAC-R scale is effective with other minority
subjects.

Although the MAC scale was developed using data from male substance
abusers and psychiatric outpatients, it has come to be used with both men
and women. However, data suggest that the scale does not work as well with
women as with men (Gottesman & Prescott, 1989; Schwartz & Graham,
1979).

Not all MAC-scale validity studies have reported positive results (Miller &
Streiner, 1990; Snyder, Kline, & Podany, 1985; Zager & Megargee, 1981).
After reviewing seventy-four empirical studies, Gottesman and Prescott
(1989) questioned the routine use of the MAC scale in clinical and employment settings. They indicated that the evidence for the use of the MAC scale
to identify substance abusers is not as compelling as many users assume. Further, accuracy of classification using the MAC scale is considerably lower
when the scale is used in settings where the base rate for substance abuse is
markedly different from that of the setting in which the scale was developed.
The observations made by Gottesman and Prescott are appropriate, but their
recommendation that the MAC scale not be used outside of research settings
seems too extreme. Most of the studies they reviewed used a cutoff score of
24 in discriminating abusers from nonabusers. Further, their recommendation seems to imply that one would decide that a person is or is not a substance abuser on the basis of MMPI (or MMPI-2) data only. This author recommends that no decisions be made on the basis of MAC (or MAC-R) scale
scores alone. High scores on the scale should alert clinicians to obtain corroborating information concerning the possibility of substance abuse.

At times the MAC scale has been described as assessing an addiction-prone personality. It has been suggested that higher MAC scorers are at greater risk for developing substance abuse problems even if they are not currently abusing substances. The existing literature simply does not support such interpretations. The relationships between MAC-scale scores and current substance abuse does not speak to the issue of predicting such abuse in persons who currently are not abusers.

It has been suggested that the MAC scale measures general antisocial tendencies and not specifically substance abuse. Data concerning this possibility are mixed. Schwartz and Graham (1979) concluded that the MAC scale was not measuring general antisocial behavior in a large sample of hospitalized psychiatric patients. However, Levenson et al. (1990) reported that older men with arrest histories but without drinking problems scored as high on the MAC scale as older men with drinking problems but without arrest histories. Zager and Megargee (1981) reported that young male prisoners scored relatively high on the MAC scale regardless of the extent to which they reported having drinking problems.

As Ward and Jackson (1990) suggested, there is evidence that MAC-scale scores may be a function of both psychiatric diagnosis and substance use/abuse. Substance-abusing patients who have other psychiatric diagnoses (i.e., MacAndrew's secondary alcoholics) tend to obtain relatively low scores on the MAC scale (e.g., Ward & Jackson, 1990) and are not easily discriminated from psychiatric patients who do not abuse substances. Patients with diagnoses of antisocial personality disorder often obtain relatively higher scores on the MAC scale whether or not they abuse substances (e.g., Wolf, Schubert, Patterson, Grande, & Pendleton, 1990). Thus, nonabusing patients with diagnoses of antisocial personality disorder are often misidentified as having substance abuse problems.

Interpretation of Scores on the MAC-R Scale

Except for four objectionable items that have been replaced, the MAC-R scale is essentially the same as the original scale. Thus, the interpretation of the scale can be similar to the interpretation of the original MAC scale. High scores on the MAC-R scale suggest the *possibility* of alcohol or other substance abuse problems. Obviously, it would not be responsible clinical practice to reach conclusions about substance abuse without obtaining corroborating information from other sources. In general, raw scores of 28 and above on the MAC-R scale are suggestive of substance abuse problems. In such cases, additional information about alcohol and drug use should be obtained. Scores between 24 and 27 are suggestive of such abuse, but at this level there will be many false positives (i.e., persons who are identified as abusers because of their scores who really are not abusers). Scores below 24 suggest that substance abuse problems are not very likely. Incorrect classification of nonabusers as abusers is especially likely to occur for individuals who have many of the characteristics often associated with the diagnosis of antisocial personality disorder. Substance-abusing patients who also have other psychiatric diagnoses, such as schizophrenia or major affective disorders, are likely to have relatively low scores on the MAC-R scale and may not be identified as having substance abuse problems on the basis of this scale. African Americans who abuse substances are likely to obtain elevated MAC-R scale scores, but the tendency for nonabusing African Americans to have elevated scores on the scale will lead to more false positives than with Caucasian subjects. It should be noted that persons who previously abused substances but no longer do so may still obtain high scores on the MAC-R scale.

Although most items on the MAC-R scale are not obviously related to substance use/abuse, data exist to suggest that alcoholics who took the MMPI under instructions to hide any problems or shortcomings produced lower scores on the MAC-R scale than when they took the MMPI under standard instructions (Otto et al., 1988). Using a MAC-scale cutoff score of 24, 83.8 percent of alcoholics who took the test with standard instructions were identified correctly, whereas only 65 percent of alcoholics who took the test with instructions to hide problems and shortcomings were correctly identified. However, almost all of the subjects who were trying to hide problems and shortcomings were identified by the MMPI validity scales and indexes. These data suggest that one should not interpret scores on the MAC-R scale when the validity scales of the MMPI-2 indicate that the subject has approached the MMPI-2 in an overly defensive manner.

In addition to the possibility of substance abuse, high scores on the MAC-R scale may indicate persons who:

1. are socially extroverted
2. are exhibitionistic
3. may experience blackouts
4. have difficulties in concentrating
5. may have histories of behavior problems in school or with the law
6. are self-confident and assertive
7. enjoy competition and risk taking

ADDICTION ACKNOWLEDGMENT SCALE (AAS)

Scale Development

Weed, Butcher, McKenna, and Ben-Porath (1992) developed the Addiction Acknowledgment Scale (AAS), using items in the MMPI-2 that have obvious content related to substance abuse (e.g., "I have a drug or alcohol problem."; "I can express my true feelings only when I drink."). A tentative scale was refined using internal-consistency procedures. The final AAS has thirteen items. Raw scores on the AAS are transformed to linear T scores using the MMPI-2 normative data.

Reliability and Validity

Weed et al. reported an internal-consistency coefficient (Alpha) of .74 for a combined sample of substance abusers, psychiatric patients, and normative subjects. They also reported test-retest reliability coefficients of .89 and .84, respectively, for men and women in the MMPI-2 normative sample.

The utility of the AAS was examined using samples of substance abusers (832 men, 380 women), psychiatric inpatients (232 men, 191 women), and MMPI-2 normative subjects (1,138 men and 1,462 women). For both men and women, the substance abusers had the highest mean AAS scores and the normative subjects the lowest mean scores. The mean scores of the psychiatric patients were between the means of the other two groups.

Although Weed et al. (1992) did not recommend cutoff scores, examination of their data suggests that a T score of 60 yielded optimal classification. When men with AAS T scores greater than 60 were considered to be substance abusers and those with T scores equal to or below 60 nonabusers, 72 percent of the substance abusers, 51 percent of the psychiatric patients, and 87 percent of the normative subjects were correctly classified. A T-score cutoff of 60 for women yielded correct classification of 58 percent of substance abusers, 66 percent of psychiatric patients, and 95 percent of normative subjects.

Weed et al. (1992) reported that the AAS and the Addiction Potential Scale (APS) (described below) covaried considerably. Their correlation in the total sample was .57. For the substance abuse, psychiatric, and normative samples, the correlations were .36, .33, and .33, respectively. In spite of these correlations, joint use of the two scales discriminated better between substance abusers and normative subjects than did either scale alone. The AAS alone did somewhat better than did the APS alone. Using both the APS and AAS did not increase discrimination between the psychiatric and substance abuse patients. In this analysis the APS alone did much better than the AAS alone.

In the Weed et al. study, the AAS discriminated much better between normative and substance abuse subjects than did the MAC-R (MacAndrew, 1965) or Substance Abuse Proclivity (SAP) (MacAndrew, 1986) scales. Although the AAS better discriminated between psychiatric and substance abuse patients than did the MAC-R or SAP scales, none of these three scales discriminated as well as the APS.

Greene, Weed, Butcher, Arrendondo, and Davis (1992) cross-validated the AAS, using inpatient substance abusers (82 men, 44 women) and psychiatric inpatients (71 men, 85 women). As in the Weed et al. (1992) study, both male and female substance abusers scored higher on the AAS than did male and female psychiatric patients. The AAS was slightly more effective than the MAC-R scale and much more effective than the SAP scale. A T-score cutoff of 60 for men correctly classified 69 percent of the substance abusers and 58 percent of the psychiatric patients. Using a T-score cutoff of 60 correctly classified 52 percent of female substance abusers and 73 percent of female psychiatric patients.

Based on the results of the Weed et al. (1992) and Greene et al. (1992) studies, the only relevant ones published to date, it would appear that the AAS has promise in discriminating between substance abusers and nonabusing normal subjects. However, the AAS seems to be much less useful in discriminating between substance abusers and general psychiatric patients. Using the AAS for this purpose leads to many psychiatric patients being misclassified as substance abusers. However, this may reflect the nature of the psychiatric samples used in the two studies. Neither study eliminated from the psychiatric samples patients who also had alcohol and/or drug problems. In fact, Greene et al. reported that between 10 to 20 percent of psychiatric patients in the study's setting typically received a diagnosis of alcohol or drug dependence.

Interpretation of AAS Scores

At this time it is premature to judge just how useful the AAS will be in identifying substance abusers in a variety of settings. What seems clear, however, is that persons who obtain high scores (T > 60) on the AAS are openly acknowledging substance abuse problems, and additional assessment in this area is indicated. The meaning of low scores on the AAS is less clear. Because the content of the items in the AAS is obviously related to substance abuse, persons not wanting to reveal substance abuse problems can easily obtain lower scores. Therefore, it is difficult to determine if low scores indicate the absence of substance abuse problems or simply reveal denial of such problems in persons who actually abuse substances. Examination of the validity scales of the MMPI-2 could be helpful in this regard. One should not interpret low AAS scores in a defensive profile. Validity patterns suggesting exaggeration are of less concern, because persons are not often motivated to appear to have alcohol or drug problems when they really do not have them. However, in certain forensic cases this motive should be considered (e.g., a case in which a defendant could receive a more favorable outcome if judged to be drug-dependent).

ADDICTION POTENTIAL SCALE (APS)

Scale Development

The Addiction Potential Scale (APS) was developed by Weed et al. (1992) using the MMPI-2 item pool. The thirty-nine items in the scale are those that 434 men and 164 women in an inpatient chemical-dependency program answered differently from 120 male and 90 female psychiatric inpatients and 584 men and 706 women in the MMPI-2 normative sample. The substance abusers included persons who abused alcohol only, other drugs only, or both alcohol and other drugs. Several tentative items were eliminated from the scale because their content obviously related to substance abuse. The items are keyed in the direction most often chosen by the substance abusers. Raw scores on the APS are transformed to linear T scores using the MMPI-2 normative data.

The content of the items in the APS is quite heterogenous, and many of the items do not seem to have obvious relevance to substance use or abuse. Some items have to do with extroversion, excitement seeking, and risk taking. Other items are related to self-doubts, self-alienation, and cynical attitudes about other people.

Reliability and Validity

Although items included in the APS were based, in part, on their contribution to internal consistency, no internal-consistency coefficients were reported. The test-retest reliability coefficients for men and women in the MMPI-2 normative sample (1-week interval) were .69 and .77, respectively.

In cross-validating the APS, Weed et al. (1992) drew different subjects from the same settings used in scale development. They reported data suggesting that the APS discriminated quite well between substance abusers and normative subjects and between substance abusers and psychiatric patients. Although the scale developers did not recommend cutoff scores for the APS, examination of their data reveals that the optimal T-score cutoff for both male and female subjects seemed to be 60. When men with T scores greater than 60 were considered to be substance abusers, and those with T scores equal to or less than 60 nonabusers, 71 percent of the substance abusers, 86 percent of the normative subjects, and 82 percent of the psychiatric patients were correctly classified. A T-score cutoff of 60 for women correctly classified 70 percent of the substance abusers, 87 percent of the normative subjects, and 81 percent of the psychiatric patients. By lowering the cutoff score, greater proportions of substance abusers could be correctly classified, but more normative subjects and psychiatric patients were incorrectly classified as substance abusers.

Weed et al. (1992) also examined the ability of the MacAndrew Alcoholism Scale–Revised (MAC-R) (MacAndrew, 1965) and the Substance Abuse Proclivity (SAP) scale (MacAndrew, 1986) to discriminate between subjects in the various groups in their study. They concluded that the SAP scale was quite ineffective. The mean MAC-R-scale scores of substance abusers were higher than those of the normative subjects, but the substance abusers and psychiatric patients had very similar mean scores in the MAC-R scale.

Greene et al. (1992) cross-validated the APS in settings different from those in which the scale had been developed. APS scores of 82 male and 44 female substance abusers were compared with those of 71 male and 85 female psychiatric patients. They also examined scores of their subjects on the MAC-R and SAP scales. The results indicated that a T-score cutoff of 60 on the APS seemed to yield optimal classification, but accuracy was not as great as had been reported by Weed et al. (1992). This cutoff correctly classified 54 percent of the male substance abusers and 76 percent of the male psychiatric patients. The corresponding rates for female substance abusers and female psychiatric patients were 54 percent and 79 percent, respectively. Although accuracy of classification using the MAC-R scale was not as great as when using the APS, the MAC-R worked better in the Greene et al. study than in the Weed et al. study. As in the Weed et al. study, the SAP scale was quite ineffective in discriminating between substance abusers and psychiatric patients.

Interpretation of APS Scores

The limited data available concerning the APS suggest that it has promise for discriminating between persons who abuse substances and those who do not. However, more research with the scale is needed to determine the generalizability of the original findings. Given the findings of Greene et al. (1992), it may well be that the scale will work better in some settings than others. As has been the case with other substance abuse scales, it is quite possible that the same APS cutoff score will not be equally effective in all settings. In addition, as Weed et al. (1992) have acknowledged, it will be necessary to determine positive predictive power and negative predictive power in settings with differing base rates for substance abuse.

Because of the manner in which the APS was constructed, its items should be related to substance abuse but not to many other dimensions of psychopathology. Therefore, it may be better able than scales such as the MAC-R to discriminate between psychiatric patients who are also substance abusers and those who are not. However, examination of the content of the APS items suggests the scale may reveal aspects of antisocial personality. If this is the case, this scale, like the MAC-R, may have difficulty in discriminating between persons who abuse substances and have other features of antisocial personality disorder and those who have features of antisocial personality disorder but who do not abuse substances.

The label of Addiction Potential suggests that the scale assesses a potential for or vulnerability to substance abuse whether or not that abuse is currently taking place. At this time there are no data concerning this very important issue. Available data address the ability of the scale to identify persons who currently are abusing substances and are in treatment settings where it is unlikely that they would be denying the abuse. The extent to which the scale can predict future abuse and can identify current abuse by persons who are denying abuse remains to be investigated.

The extent to which scores on the APS are affected by defensive test-taking attitudes has not been directly studied. Greene et al. (1992) have demonstrated that scores on the APS scale are less affected by random responding and/or exaggeration than are scores on the more face-valid Addiction Acknowledgement Scale (AAS). Given data discussed in relation to the MAC-R scale (Otto et al., 1988), it would not be surprising if persons who approach the MMPI-2 in a defensive manner would obtain lower APS scale scores than those who approach the test more honestly.

In spite of the limited data available concerning the APS scale, it should be considered as an indicator of possible substance abuse problems. As was discussed in relation to the MAC-R scale, it is not appropriate to reach conclusions about substance abuse on the basis of MMPI-2 scores alone. High APS scores (T > 60) should alert clinicians that additional information concerning possible substance abuse should be obtained.

MARITAL DISTRESS SCALE (MDS)

Scale Development

The Marital Distress Scale (MDS) of the MMPI-2 was developed by Hjemboe, Almagor, and Butcher (1992) to assess distress in marital relationships. Tentative items were selected by correlating MMPI-2 item responses with scores on the Spanier Dyadic Adjustment Scale (DAS) (Spanier, 1976) for 150 couples involved in marital counseling and 392 couples from the MMPI-2 normative sample. The DAS is a thirty-one-item inventory assessing relationship consensus, cohesion, affection, and satisfaction. Additional items were added to the tentative MDS on the basis of correlations between other MMPI-2 items and scores on the tentative scale. Items were later eliminated if their content was judged not to be specifically related to marital distress or if their removal led to improved discriminative ability. The final scale consists of fourteen items. The content of certain items seems to be related obviously to marital distress (e.g., "I have very few quarrels with members of my family (F)"; "I believe my home life is as pleasant as that of most people I know (F)"). Several items in the scale have less obvious relationships to marital distress (e.g., "I have lost out on things because I couldn't make up my mind soon enough (T)"; "My main goals in life are within my reach (F)"). The items are scored in the direction most often chosen by persons suffering greater marital distress. Raw scores are transformed to uniform T scores using the MMPI-2 normative data.

Reliability and Validity

Hjemboe et al. (1992) reported internal-consistency coefficients (Alphas) of .65 for the developmental sample and .60 for a cross-validational sample of couples from the MMPI-2 normative sample. No test-retest reliability coefficients were reported for the MDS.

The MDS was validated by correlating it with DAS scores for the combined marital counseling and normative groups. The correlation between the two measures was -.55. The ability of the MDS to discriminate between persons with high and low levels of marital distress was compared with that of several other relevant MMPI-2 scales: Pd (Psychopathic Deviate); Harris-Lingoes Pd1 subscale (Familial Discord); and the FAM (Family Problems) content scale. Regression analysis indicated that the MDS accounted for more of the variance in DAS scores than did any of the other measures. When marital adjustment groups were formed on the basis of extremely high or low scores on the DAS, a T-score cutoff of 60 on the MDS correctly classified slightly more subjects overall (88.0%) than did the Pd1 subscale (85.3%), the Pd scale (83.0%) or the FAM scale (84.0%). However, the MDS correctly classified a higher percentage of the high-marital-distress subjects (63.0%) than did the other scales. The results of the Hjemboe et al. (1992) study are somewhat difficult to evaluate, because most of the subjects used in the validity analyses were also used for scale development purposes.

Hjemboe et al. (1992) pointed out that a lower MDS cutoff score would correctly classify a greater proportion of high-distress subjects but at the cost of incorrectly classifying a greater proportion of low-distress subjects. They also pointed out that the utility of the MDS is likely to depend on the base rate of marital distress in a particular setting. In outpatient counseling settings, where the base rate is higher than in Hjemboe et al.'s study, the MDS may identify distressed subjects with greater accuracy.

Interpretation of MDS Scores

Because the MDS has only recently been developed and no data are available except that reported by its developers in their initial study, the scale should be interpreted cautiously. High scores (T > 60) may be indicative of significant marital distress and additional assessment in this area is recommended. Obviously, the scale may not be of much help when assessing persons who are readily admitting to marital problems and seeking help for them. However, when the MMPI-2 is used as part of a more general assessment, high MDS scores should alert clinicians that marital problems may be underlying other symptoms such as anxiety or depression.

Several important questions about the MDS have not yet been addressed by empirical research. Is the scale assessing marital distress specifically, or might it be assessing relationship problems more generally? How does test-taking attitude affect scores on the MDS? Since the content of items in the scale obviously relates to relationship problems, do persons who are having such problems but denying them obtain lower scores on the scale? How effective will the scale be in identifying marital distress in a variety of clinical settings? Are scores on the scale related to the severity of distress among persons seeking marital counseling and/or to the success of marital counseling? These questions remain to be addressed by future research with the scale.

OVERCONTROLLED-HOSTILITY (O-H) SCALE

Scale Development

Megargee, Cook, and Mendelsohn (1967) suggested that there are two major types of persons who commit acts of extreme physical aggression. Habitually aggressive (undercontrolled) persons have not developed appropriate controls against the expression of aggression so that when they are provoked they respond with aggression of an intensity proportional to the degree of provocation. Chronically overcontrolled persons have very rigid inhibitions against the expression of any form of aggression. Most of the time the overcontrolled individuals do not respond even with aggression appropriate to provocation, but occasionally, when the provocation is great enough, they may act out in an extremely aggressive manner. Megargee and his associates believed that the most aggressive acts typically are committed by overcontrolled rather than undercontrolled persons.

The original O-H scale was constructed by identifying items that were answered differently by extremely assaultive prisoners, moderately assaultive prisoners, prisoners convicted of nonviolent crimes, and men who had not been convicted of any crime. Items were scored so that higher scores on the O-H scale were indicative of more assaultive (overcontrolled) persons. The original O-H scale had thirty-one items, and the MMPI-2 version of the scale includes twenty-eight of the items.

Reliability and Validity

Megargee et al. (1967) reported a coefficient of internal consistency (Kuder-Richardson 21) of .56 for the O-H scale for a combined group of criminals and college students. Internal-consistency coefficients (Alpha) for men and women in the MMPI-2 normative sample were .34 and .24, respectively (Butcher et al., 1989). Clearly, the O-H scale is not very internally consistent. Moreland (1985b) reported test-retest coefficients for male and female college students of .72 and .56, respectively. Test-retest coefficients for men and women from the MMPI-2 normative sample were .68 and .69, respectively (Butcher et al., 1989).

Although some studies (e.g., Deiker, 1974; Fredericksen, 1976; Megargee et al., 1967) found that more violent criminals scored higher on the O-H scale, other studies (e.g., Fisher, 1970) failed to find O-H scale differences between assaultive and nonassaultive prisoners. Lane (1976) suggested that some of the negative findings could be due to a confounding of race and the assaultiveness criterion and to the manner in which the O-H scale was administered. Although no data were presented to support the contention, Lane stated that the O-H scale must be administered in the context of the entire MMPI. Data also suggest that when the O-H scale is used to identify assaultive prisoners, cutoff scores should be established individually for each setting in which the scale is used. There is little evidence to suggest that high scores on the O-H scale in groups other than prisoners are associated with violent acts.

Interpretation of High Scores on the O-H Scale

In correctional settings, high scores on the O-H scale tend to be associated with aggressive and violent acts. However, the validity of the O-H scale is such that individual predictions of violence from scores are not likely to be very accurate. In addition, cutoff scores for predicting violence should be established separately in each setting where the scale is used. The O-H scale has potential use in other settings because it tells clinicians something about how subjects typically respond to provocation. Higher scorers on the O-H scale tend not to respond to provocation appropriately most of the time, but occasional exaggerated aggressive responses may occur. High scores also may be indicative of persons who:

1. are impunitive
2. tend not to express angry feelings
3. tend not to express verbal hostility in reaction to frustration
4. are socialized and responsible
5. have strong needs to excel
6. are dependent on others
7. are trustful
8. describe nurturant and supportive family backgrounds

Interpretation of Low Scores on the O-H Scale

Relatively little data exist concerning the interpretation of low scores on the O-H scale. One does not expect low scorers to display the overcontrolled-hostility syndrome described for high scorers. Low scorers may be either chronically aggressive persons or persons who are quite appropriate in the expression of their aggression.

DOMINANCE (Do) SCALE

Scale Development

The Dominance (Do) scale was developed by Gough, McClosky, and Meehl (1951) as part of a larger project concerned with political participation. Because the sixty-item scale included twenty-eight MMPI items, it was possible to score an abbreviated version of the scale from standard administration of the MMPI. The MMPI-2 version of the Do scale has twenty-five of the twenty-eight items. To develop the Do scale, high school and college students were given a definition of dominance ("strength" in face-to-face personal situations; ability to influence others; not readily intimidated or defeated; feeling safe, secure, and confident in face-to-face situations) and were asked to nominate peers who were most and least dominant. High- and low-dominance criterion groups were defined on the basis of these peer nominations, and both groups were given a 150-item questionnaire, which included some MMPI items. Analyses of the responses identified items that differentiated between high- and low-dominance criterion groups. The items are keyed in such a way that a high score on the Do scale is suggestive of high dominance. The Do-scale items deal with a number of different content areas, including concentration, obsessive-compulsive behaviors, self-confidence, discomfort in social situations, concern about physical appearance, perseverance, and political opinions.

Reliability and Validity

Gough et al. (1951) reported an internal-consistency coefficient (Kuder-Richardson 21) of .79 for the sixty-item Do scale, and Gocka (1965) reported a Kuder-Richardson 21 value of .60 for the twenty-eight-item Do scale for 220 male Veterans Administration psychiatric patients. Internal-consistency coefficients (Alpha) for men and women in the MMPI-2 normative sample were .74 and .79, respectively (Butcher et al. 1989).

A test-retest coefficient of .86 for Marine Corps officers was reported by Knapp (1960). Moreland (1985b) reported test-retest coefficients (6-week interval) for male and female college students of .85 and .83, respectively. Test-retest coefficients (1-week interval) for subsamples of men and women in the MMPI-2 normative sample were .84 and .86, respectively (Butcher et al., 1989).

Gough et al. (1951) found that a raw-score cutoff of 36 on the sixty-item scale identified 94 percent of their high- and low-dominance high school subjects, whereas a raw-score cutoff of 39 on the sixty-item scale identified 92 percent of high- and low-dominance college students. Correlations between Do-scale scores based on the twenty-eight MMPI items and peer ratings and self-ratings of dominance were .52 and .65, respectively, for college students and .60 and .41, respectively, for high school students.

Knapp (1960) found that Marine Corps officer pilots scored significantly higher on the twenty-eight-item Do scale than did enlisted men. The mean scores for the officers and enlisted men were quite similar to mean scores reported for high- and low-dominance high school and college students. Knapp interpreted his data as supporting the use of the Do scale as a screening device in selecting officers. However, Olmstead and Monachesi (1956) reported that the MMPI Do scale was not able to differentiate between fire fighters and fire captains. Eschenback and Dupree (1959) found that Do-scale scores did not change as a result of situational stress (a realistic survival test). It would be interesting to know whether Do-scale scores change as individuals change their dominance roles (e.g., when an enlisted person becomes an officer). Unfortunately, no data of this kind are currently available.

Interpretation of High Do Scale Scores

High scorers on the Do scale see themselves and are seen by others as stronger in face-to-face personal situations, as not readily intimidated, and as feeling safe, secure, and self-confident. Although there is limited evidence to suggest that high scores on the Do-scale are more common among persons holding positions of greater responsibility and leadership, no data are available concerning the adequacy of performance in such positions as a function of Do-scale scores. Also, high Do scale scores may indicate persons who:

1. appear poised and self-assured
2. are self-confident
3. appear to feel free to behave in a straightforward manner
4. are optimistic
5. are resourceful and efficient
6. are realistic and task-oriented
7. feel adequate to handle problems
8. are persevering
9. have a dutiful sense of morality
10. have a strong need to face reality

In summary, high scorers on the Do scale are people who are confident of their abilities to cope with problems and stresses in their life situations.

SOCIAL RESPONSIBILITY (Re) SCALE

Scale Development

The Social Responsibility (Re) scale was developed by Gough, McClosky, and Meehl (1952) as part of a larger project concerning political participation.

The original scale contained fifty-six items, with thirty-two items coming from the MMPI item pool. A score based on the thirty-two MMPI items could be obtained, and normative data were available for the thirty-two-item Re scale. In the MMPI-2, thirty of the original thirty-two MMPI items were retained.

The samples used in constructing the Re scale consisted of 50 college fraternity men, 50 college sorority women, 123 social science students from a high school, and 221 ninth-grade students. In each sample, the most and least responsible individuals were identified. Responsibility was defined as willingness to accept the consequences of one's own behavior; dependability; trustworthiness; integrity; and sense of obligation to the group. For the high school and college samples, peer nominations were used to identify subjects high and low in responsibility. Teachers provided ratings of responsibility for the ninth-grade sample. The responses of the most responsible and least responsible subjects in each sample to the items in the MMPI item pool and to a questionnaire containing rationally generated items were examined. Items that revealed the best discrimination between most and least responsible subjects in all samples were included in the Re scale. The content of the MMPI-2 version of the Re scale deals with concern for social and moral issues; disapproval of privilege and favor; emphasis on duties and self-discipline; conventionality versus rebelliousness; trust and confidence in the world in general; and poise, assurance, and personal security (Gough et al., 1952).

Reliability and Validity

Gough et al. (1952) reported an uncorrected split-half coefficient of .73 for the fifty-six-item scale for a sample of ninth-grade students. Gocka (1965) reported a Kuder-Richardson 21 (internal-consistency) value of .63 for the thirty-two-item Re scale for 220 male Veterans Administration psychiatric patients. Internal-consistency values (Coefficient Alpha) for men and woman in the MMPI-2 normative sample were .67 and .61, respectively (Butcher et al., 1989).

Moreland (1985b) reported test-retest coefficients (6-week interval) for male and female college students of .85 and .76, respectively. Test-retest coefficients (with 1-week interval) for subsamples of men and women in the MMPI-2 normative sample were .85 and .74, respectively (Butcher et al., 1989). Gough and his colleagues reported correlations of .84 and .88 between Re-scale scores based on all fifty-six items and scores based on the thirty-two MMPI items in the Re scale for their college and high school samples.

Correlations between MMPI Re-scale scores and criterion ratings of responsibility in the derivation samples were .47 for college students and .53 for high school students. For college students, the correlation between MMPI Re-scale scores and self-ratings of responsibility was .20, and the correlation between these two variables for high school students was .23. Optimal cutoff scores for the MMPI Re scale yielded correct classification of 78 percent and 87 percent, respectively, of the most and least responsible individuals in the various derivation samples. Gough et al. (1952) reported some limited cross-validational data for the total (56-item) Re scale. They obtained a correlation of .22 between scores and ratings of responsibility for a sample of medical students. A correlation of .33 between Re-scale scores and ratings of positive character integration was reported for a sample of fourth-year graduate students.

In two studies, persons with higher Re-scale scores tended to have positions of leadership and responsibility. Knapp (1960) found that Marine Corps officers scored significantly higher on the MMPI Re scale than did enlisted men. Olmstead and Monachesi (1956) reported that fire captains scored higher on the MMPI Re scale than firefighters, but the difference was not statistically significant.

Duckworth and Duckworth (1975) suggested that the Re scale measures acceptance (high score) or rejection (low score) of a previously held value system, usually that of one's parents. For persons above the age of 25, high Re-scale scorers tend to accept their present value system and intend to continue using it, and low scorers may be questioning their current value system or rejecting their most recently held value system. For younger persons, high Re-scale scores indicate that they accept the value system of their parents, whereas low Re-scale scores indicate questioning or rejection of parental value systems. Duckworth and Duckworth (1975) also suggested that high Re-scale scorers, regardless of age, are more rigid in acceptance of existing values and are less willing to explore others' values. They also indicated that older persons tend to score higher than younger persons on the Re scale and that college students who are questioning parental values often receive quite low Re-scale scores.

Interpretation of High Re-Scale Scores

High Re-scale scorers tend to see themselves and are seen by others as willing to accept the consequences of their own behavior, as dependable and trustworthy, and as having integrity and a sense of responsibility to the group. They also are more likely than low Re-scale scorers to be in positions of leadership and responsibility. High Re-scale scorers are rigid in acceptance of existing values and are unwilling to explore others' values. Younger persons with high Re-scale scores tend to accept the values of their parents. Also, high scores may indicate persons who:

1. have deep concern over ethical and moral problems
2. have a strong sense of justice
3. set high standards for themselves
4. reject privilege and favor
5. place excessive emphasis on carrying their own share of burdens and duties
6. are self-confident
7. have trust and confidence in the world in general

In summary, high Re-scale scorers have incorporated societal and cultural values and are committed to behaving in a manner consistent with those values. In addition, they place high value on honesty and justice.

COLLEGE MALADJUSTMENT (Mt) SCALE

Scale Development

The College Maladjustment (Mt) scale was constructed to discriminate between emotionally adjusted and maladjusted college students (Kleinmuntz, 1961). Mt-scale items were selected from the MMPI item pool by comparing responses of forty adjusted male and female students and forty maladjusted male and female students. The adjusted students had contacted a university clinic to arrange for a routine mental health screening examination as part of teacher certification procedures, and none of them admitted to a history of psychiatric treatment. The maladjusted students had contacted the same clinic for help with emotional problems and had remained in psychotherapy for three or more sessions. Item-analytic procedures identified forty-three items that discriminated between the adjusted and maladjusted students. The MMPI-2 Mt scale has forty-one of the original forty-three items, with items scored such that higher scores on the scale are more indicative of greater maladjustment. Kleinmuntz (1961) found that scores on the forty-three-item scale administered separately corresponded quite well to Mt-scale scores derived from a standard MMPI administration.

Reliability and Validity

Internal-consistency coefficients (Alpha) for men and women in the MMPI-2 normative sample were .84 and .86, respectively (Butcher et al., 1989). When Kleinmuntz (1961) administered the Mt scale to college students twice with an interval of three days, a test-retest reliability coefficient of .88 was obtained. Moreland (1985b) reported test-retest coefficients (6-week interval) for male and female college students of .89 and .86, respectively. Test-retest coefficients (1-week interval) for subsamples of men and women in the MMPI-2 normative sample were .91 and .90, respectively (Butcher et al., 1989).

Kleinmuntz (1961) reported that college students who took the Mt scale when they entered college and later sought "emotional" counseling scored higher on the scale than did a similar group of students who sought out "vocational-academic" counseling. Using a Mt-scale cutoff of 15, Parker (1961) was able to classify correctly 74 percent of maladjusted students who completed the Mt scale at the time they sought counseling, but only 46 percent when the Mt scale was completed as part of a battery administered at the time of college admission. Parker's data and Kleinmuntz's own data led Kleinmuntz (1961) to conclude that the Mt scale is more accurate when it is used for identifying existing emotional problems than when it is used for predicting future emotional problems. Higher Mt-scale scores for maladjusted than for adjusted students were reported subsequently in several different settings (Kleinmuntz, 1963). Wilderman (1984) found that within a college counseling sample, higher Mt-scale scores were associated with more elevated MMPI profiles and with more severe psychopathology as indicated by therapist ratings than were relatively lower Mt-scale scores.

Female students tend to obtain higher Mt-scale raw scores than male students. Mt-scale scores of adjusted and maladjusted college students vary considerably from one college setting to another and among divisions within each college. This variation indicates that separate norms and cutoff scores should be established for each specific setting where the Mt scale is used.

Interpretation of High Mt-Scale Scores

Because of variations in Mt-scale scores among college settings, it is not possible to identify a cutoff score above which students should be considered to be maladjusted. However, among college students within a given setting, higher Mt-scale scores are more suggestive of maladjustment. Because the Mt scale has not been studied systematically in settings other than colleges and universities, its use is not recommended with subjects who are not college students. In addition to suggesting general maladjustment, high Mt-scale scores for college students may indicate persons who:

1. are ineffectual
2. are pessimistic
3. procrastinate
4. are anxious and worried
5. develop somatic symptoms during times of increased stress
6. feel that life is a strain much of the time

MASCULINE GENDER ROLE (GM) AND FEMININE GENDER ROLE (GF) SCALES

Scale Development

Peterson and Dahlstrom (1992) developed the Masculine Gender Role (GM) and Feminine Gender Role (GF) scales for the MMPI-2 as separate measures of the masculine and feminine components in the bipolar Masculinity-Femininity (Mf) scale of the MMPI. Items in the GM scale were those endorsed in the scored direction by a majority of men in the MMPI-2 normative sample and endorsed in that same direction by at least 10 percent fewer women in the MMPI-2 normative sample. Correspondingly, items in the GF scale were those endorsed in the scored direction by a majority of women in the MMPI-2 normative sample and endorsed in that same direction by at least 10 percent fewer men in the MMPI-2 normative sample. Only nine of forty-seven items in the GM scale and sixteen of the forty-six items in the GF scale also appear on the Mf scale.

Examination of the content of the items in the GM scale suggests that they deal primarily with the denial of fears, anxieties, and somatic symptoms. Some GM scale items have to do with interest in stereotypically masculine activities, such as reading adventure stories, and with denial of interests in stereotypically feminine occupations, such as nursing and library work. Other groups of GM-scale items have to do with denial of excessive emotionality and presentation of self as independent, decisive, and self-confident. The largest group of items in the GF scale has to do with the denial of asocial or antiso-cial acts, such as getting into trouble with the law or at school and excessive use of alcohol or other drugs. Many GF-scale items also have to do with liking stereotypically feminine activities, such as cooking and growing house plants, and with disliking stereotypically masculine activities, such as reading mechanics magazines and auto racing. A number of GF-scale items involve admissions of excessive sensitivity. There are also several items expressing early identification with a female figure and satisfaction with being female or wishing that one were not male.

Reliability and Validity

Internal-consistency estimates (Alpha) for the GM scale for men and women in the MMPI-2 normative sample were .67 and .75, respectively (Butcher et al., 1989). Test-retest reliability coefficients (1-week interval) for the GM scale for subsamples of men and women in the MMPI-2 normative sample were .73 and .89, respectively (Butcher et al., 1989).

The internal-consistency coefficient (Alpha) for the GF scale was .57 for both men and women in the MMPI-2 normative sample (Butcher et al., 1989). Test-retest reliability coefficients (1-week interval) for the GF scale were .86 and .78, respectively, for subsamples of men and women in the MMPI-2 normative sample (Butcher et al., 1989).

Peterson and Dahlstrom (1992) reported some preliminary data concern-ing behavioral correlates of the GM and GF scales. He found that for men the GM scale was related to high self-confidence, persistence, and wide interests, as well as to lack of fears or feelings of self-reference. For women the GM scale also was related to high self-confidence, as well as to honesty, to willingness to try new things, and to lack of worries or feelings of self-reference. For men the GF scale was related to religiosity, avoidance of swearing or cursing, and frank-ness in pointing out to others their personal faults. For men the GF scale also

was related to bossiness and poor control over one's temper, as well as susceptibility to the abuse of alcohol and nonprescription drugs. For women, the GF
scale also was related to religiosity and the use of nonprescription drugs.

Interpretation of GM- and GF-Scale Scores

Because very limited validity data are available concerning them, it is difficult
to generate descriptors that can be applied with confidence to high and low
scorers on the GM and GF scales. However, the preliminary data by Peterson
and Dahlstrom and the content of items in the GM and GF scales suggest that
both men and women who obtain high scores on the GM scale are characterized in positive ways. They tend to be self-confident and free from fears and
worries. For both men and women, higher scores on the GF scale are related
to religiosity and to the abuse of alcohol and nonprescription drugs. In addition, men scoring higher on the GF scale are seen as bossy and as having
poor control over their tempers.

  Peterson and Dahlstrom suggested that the conjoint interpretation of the
GM and GF scales can yield a gender-role typology similar to that used with
other instruments. Used in this manner, a high score on the GM scale and a
low score on the GF scale would indicate stereotypic masculinity; a high score
on the GF scale and a low score on the GM scale would indicate stereotypic
femininity; high scores on both the GM and GF scales would indicate androg-
yny; and low scores on both the GM and GF scales would indicate an undifferentiated orientation.

  Peterson and Dahlstrom suggested that high raw scores on the Mf scale are
indicative of stereotypically feminine subjects, and low raw scores on the Mf
scale are indicative of stereotypically masculine subjects. However, persons
falling in the middle range on the Mf scale can be either androgynous or
undifferentiated. He hypothesized that the conjoint interpretation of the GM
and GF scales can be helpful in making such distinctions concerning the pat-
terning of gender roles in persons in this middle range on the Mf scale. It
should be emphasized that the GM and GF scales have not been sufficiently
validated to permit their routine clinical use. They should be considered
experimental scales to be used for research purposes only.

POST-TRAUMATIC STRESS DISORDER SCALE: PK SCALE

Scale Development

The PK scale was developed by Keane, Malloy, and Fairbank (1984) to determine post-traumatic stress disorder (PTSD) by contrasting the MMPI item
responses of sixty male Vietnam combat veterans who had diagnoses of PTSD

based on structured interviews and a psychophysiological assessment procedure and sixty male veterans who had diagnoses other than PTSD. They identified forty-nine items that these two groups answered significantly differently. A raw cut-off score of 30 correctly classified 82 percent of subjects used in developing the scale and also 82 percent of cross-validation groups of PTSD and non-PTSD veterans. In the MMPI-2 version of the PK scale, three items that appeared twice in the original PK scale were eliminated and one item was slightly reworded.

The content of the PK-scale items is suggestive of great emotional turmoil. Some items deal with anxiety, worry, and sleep disturbance. Others are suggestive of guilt and depression. In their responses to certain items subjects are reporting the presence of unwanted and disturbing thoughts, and in others they are describing lack of emotional control. Feeling misunderstood and mistreated is also treated in item content.

Reliability and Validity

Internal-consistency coefficients for the PK scale for men and women in the MMPI-2 normative sample were .85 and .87, respectively. Test-retest reliability coefficients for subsamples of men and women in the MMPI-2 normative sample were .86 and .89, respectively.

Subsequent to the development of the PK scale by Keane et al. (1984), other investigations of its utility in diagnosing veteran PTSD patients have been published (Butler, Foy, Snodgrass, Hurwicz, & Goldfarb, 1988; Cannon, Bell, Andrews, & Finkelstein, 1987; Gayton, Burchstead, & Matthews, 1986; Hyer et al., 1986; Hyer, Woods, Summers, Boudewyns, & Harrison, 1990; Orr et al., 1990; Vanderploeg, Sisson, & Hickling, 1987; Watson, Kucala, & Manifold, 1986; Query, Megran, & McDonald, 1986). In virtually all of these studies, patients with PTSD diagnoses obtained significantly higher PK-scale scores than comparison groups (normals, substance abusers, and general psychiatric patients). However, classification rates using the PK scale have varied from study to study and generally have been somewhat lower than in the original Keane et al. (1984) study. Classification rates have been higher when more reliable diagnoses have been utilized. The raw cutoff score of 30, which was identified in the Keane et al. study, was not optimal in some other studies. Optimal cutoff scores have ranged from 8.5 to 30. Generally, the scale has been more effective in discriminating between PTSD patients and normals than between PTSD patients and patients with other diagnoses.

Several studies have reported positive correlations between PK-scale scores and symptoms of PTSD as determined by structured interviews (McFall, Smith, Roszell, Tarver, & Malas, 1990; Watson, Juba, Anderson, & Manifold, 1990). Berk et al. (1989) found that veterans who experienced greater noncombat traumas scored significantly higher on the PK scale. Watson, Kucala, Manifold, Vassar, and Juba (1988) did not find significant differences in PK-scale scores for veterans with delayed-onset PTSD and those with undelayed-onset PTSD.

Most of the PK-scale research to date has focused on combat-related PTSD. However, several studies have suggested that scores on the PK scale may also be related to civilian trauma. Koretzky and Peck (1990) reported that several small groups of civilian PTSD patients who had experienced life-threatening events (violent criminal victimization, industrial accidents, car or train accidents) obtained higher scores on the PK scale than a group of general psychiatric patients. The PK scale correctly classified 87 percent of subjects in one comparison and 88 percent in another. However, the civilian PTSD groups had lower mean PK-scale scores than those reported previously for combat-related PTSD groups. In addition, the optimal cutoff score for classifying these civilian groups was different from that typically reported for combat groups. Sloan (1988) studied thirty male survivors of the crash landing of an airplane immediately after the crash and several times later. He found that the mean PK-scale scores of these men were relatively high immediately after the crash and that they decreased markedly during a 12-month period after the crash. McCaffrey, Hickling, and Marrazo (1989) compared two small groups of civilians who had experienced traumatic events (typically motor vehicle accidents). The group that had been diagnosed with PTSD had higher PK-scale scores than the group with other diagnoses.

Fairbank, McCaffrey, and Keane (1985) indicated that caution should be used with the PK scale because it may be susceptible to faking by veterans who are motivated to appear to have PTSD in order to gain monetary compensation. The authors suggested that exaggeration be considered when the F-scale T score is greater than 88. However, Hyer, Fallon, Harrison, and Boudewyns (1987) found that many PTSD patients had very high F-scale scores and that their MMPIs would have been considered invalid according to the Fairbank et al. (1985) criterion.

Litz et al. (1992) compared the PK scales of the MMPI and the MMPI-2 by administering both versions to combat-related PTSD patients, general psychiatric patients, substance abusers, and normal men. The correlation between the two versions of the PK scale was .80. The mean raw scores for the MMPI and the MMPI-2 PK scales for the PTSD group were 43.2 and 36.2, respectively. The deletion of three items in the MMPI-2 version of the scale probably accounts for some of the difference. Litz et al. concluded that the MMPI and MMPI-2 PK scales were equally effective in discriminating PTSD patients from others. Although they did not report optimal PK-scale cutoff scores, they reasoned that, because of the differences in mean scores between the MMPI and MMPI-2 PK scales, different cutoff scores may be needed for the MMPI-2 scale. Litz et al. also pointed out that it is not appropriate to compare T scores for the original MMPI PK scale with T scores for the MMPI-2 PK scale, because the original T scores for the MMPI version of the scale were based on data from PTSD patients, whereas the T scores for the MMPI-2 PK scale are based on data from the MMPI-2 normative sample.

Lyons and Keane (1992) also pointed out that it may be appropriate to use different cutoff scores for the MMPI and MMPI-2 PK scales. They suggested that adding two raw-score points to the MMPI-2 PK-scale scores would make them comparable to MMPI PK-scale scores. In other words, a raw-score cutoff of 28 on the MMPI-2 PK scale would be equivalent to a raw-score cutoff of 30 on the MMPI PK scale. Because the three items deleted from the MMPI-2 PK scale are those that are repeated in the original scale, one might also assume that a subject would answer the items consistently if they are in the scale twice. Therefore, one could add one raw-score point to the MMPI-2 PK scale for each of these three items that is answered in the scored direction. Although both of these suggestions are reasonable and would probably not lead to vastly different decisions about subjects, more empirical research is needed to determine what MMPI-2 PK-scale cutoff scores are equivalent to MMPI PK-scale cutoff scores.

In summary, there appears to be considerable evidence that scores on the PK scale are related to PTSD diagnoses among veterans. Studies, such as that by Keane et al. (1984), that have used well-defined criteria to establish the diagnosis of PTSD, report higher classification rates than other studies that have used less reliable diagnostic procedures. Using the PK scale to classify veterans as PTSD or non-PTSD will produce more false-positive than false-negative errors. It is not clear to what extent scores on the PK scale are susceptible to faking by persons who are motivated to appear to have PTSD but who really do not have the disorder. The utility of the PK scale in identifying PTSD associated with noncombat stress remains to be demonstrated. As with other MMPI-2 scales, it is not responsible clinical practice to use a single scale to assign diagnostic labels.

Interpretation of High PK-Scale Scores

High scorers on the PK scale are likely to be manifesting many of the symptoms and behaviors typically associated with PTSD. When high PK-scale scores are encountered in persons who have experienced combat-related stress, the possibility of PTSD should be explored carefully. It is far less clear to what extent high PK-scale scores in other circumstances indicate the appropriateness of a PTSD diagnosis.

In addition to being associated with diagnoses of post-traumatic stress disorder, high scores on the PK scale indicate persons who:

1. are reporting intense emotional distress
2. report symptoms of anxiety and sleep disturbance
3. feel guilty and depressed
4. may be having unwanted and disturbing thoughts
5. fear loss of emotional and cognitive control
6. feel misunderstood and mistreated

POST-TRAUMATIC STRESS DISORDER SCALE: PS SCALE

The sixty items listed for the PS scale in the MMPI-2 manual do not really represent a formal scale. Rather, they are items that Schlenger and his associates (Schlenger & Kulka, 1989; Schlenger et al., 1989) at the Research Triangle Institute in North Carolina found to differentiate Vietnam veterans who had been diagnosed with post-traumatic stress disorders, Vietnam veterans with psychiatric diagnoses other than PTSD, and nonpatient Vietnam veterans. Forty-five of the items are also included in the PK scale described earlier. The additional fifteen items are those from the experimental form of the MMPI (Form AX) that were endorsed differentially by the three groups of veterans.

No research has yet been published concerning this group of items. Internal-consistency coefficients for men and women in the MMPI-2 standardization sample were .89 and .91, respectively. In a personal communication with this author, Schlenger reported a corrected split-half reliability coefficient of .94 for a scale based on the forty-five PS-scale items that are also found in the PK scale. Test-retest reliability coefficients for the PS scale for subsamples of men and women in the MMPI-2 normative sample were .92 and .88, respectively. Schlenger also reported that the 45-item scale correctly classified 81.6 percent of Vietnam veterans who had been diagnosed as PTSD based on structured interviews and other information and 87.7 percent of Vietnam veterans who did not have PTSD diagnoses.

In summary, the sixty items listed in the MMPI-2 manual for the PS scale do not really comprise a scale. Rather, they are items from the PK scale and other items from the MMPI-2 that were endorsed differentially by veterans with and without diagnoses of PTSD. The extent to which a reliable and valid scale can be developed from these items remains to be demonstrated. Schlenger and his associates plan to continue their research efforts with the items.

SUBTLE-OBVIOUS SUBSCALES

Wiener (1948) differentiated between MMPI items that were easy to detect as indicating emotional disturbance and items that were relatively difficult to detect as indicating emotional disturbance. The former were labeled as obvious items and the latter as subtle items. Wiener rationally developed obvious and subtle subscales for scales 2, 3, 4, 6, 8, and 9 of the original MMPI, hypothesizing that test subjects who were trying to fake bad on the MMPI would endorse many of the obvious and few of the subtle items in the clinical scales. Correspondingly, test subjects who were trying to fake good on the MMPI would endorse many of the subtle items and few of the obvious items in the clinical scales.

Although the subtle-obvious subscales are discussed in the MMPI-2 manual and the test publisher provides scoring keys and T-score transformation tables for the subscales, they will not be considered in detail in this chapter because of this author's belief that they are of little or no value in clinical practice. It seems likely that the subtle items came to be identified only by chance in the original item analyses. Had Hathaway and McKinley cross-validated their item analyses, these items probably would not have been included in the clinical scales.

Considerable research evidence indicates that nontest behaviors are most accurately predicted by the obvious rather than by the subtle items (Burkhart, Gynther, & Fromuth, 1980; Duff, 1965; Gynther, Burkhart, & Hovanitz, 1979; Snyter & Graham, 1984; Weed, Ben-Porath, & Butcher, 1990). In fact, including subtle items in the clinical scales may actually detract from the prediction of criterion variables.

The data concerning the utility of the subtle-obvious subscales in detecting deviant test-taking attitudes are somewhat more complex. Subjects who were known to be exaggerating psychopathology when responding to MMPI items tended to endorse many more obvious than subtle items. However, because all of the clinical scales except scale 9 have many more obvious than subtle items, persons who actually have considerable psychopathology also endorse more obvious than subtle items (Schretlen, 1988). Thus, the differential endorsement of the subtle and obvious items is not particularly useful in differentiating between exaggerated profiles and valid profiles indicating severe psychopathology. Several studies have demonstrated that the standard validity scales are better able than the subtle-obvious scales to detect malingering (Anthony, 1971; Berry et al., 1991; Dubinsky, Gamble, & Rogers, 1985; Grossman, Haywood, Ostrov, Wasyliw, & Cavanaugh, 1990; Schretlen, 1988; Timbrook, Graham, Keiller, & Watts, 1993).

When subjects are instructed to fake good on the MMPI or MMPI-2, they endorse fewer obvious items than under standard instructions. Interestingly, with the fake-good instructions subjects tend to endorse more of the subtle items than with standard instructions. This may be because the subtle items are seen by test subjects as representing socially desirable attitudes or actions (Burkhart, Christian, & Gynther, 1978). It is also possible that more subtle items are endorsed with fake-good instructions because most subtle items are keyed in the false direction, and persons who are faking good may have a naysaying response bias (Timbrook et al., 1993). Studies that have examined the utility of the subtle-obvious subscales in identifying fake-good response sets have not found them to be very useful (Dubinsky et al., 1985; Schretlen, 1988; Timbrook et al., 1993).

Several conclusions can be reached concerning the subtle-obvious subscales. First, the subtle items probably were included in the clinical scales because the item analyses were not cross-validated. Second, it is the obvious and not the subtle items that are most related to extratest behaviors. Third, although subjects who approach the MMPI-2 with motivation to minimize or to exaggerate psychopathology may endorse the subtle and obvious items differentially, the subtle-obvious subscales do not permit very accurate differentiation of faked and valid profiles. Fourth, the standard validity scales of MMPI-2 work as well as or better than the subtle-obvious subscales in identifying these deviant response sets. Because of complex relationships between subtle-obvious item endorsements, invalid response sets, and subject characteristics, this author does not use the subtle-obvious subscales in his own clinical work and does not recommend that others use them. Instead, it is recommended that scores and patterns of scores on the standard validity scales be used in making decisions about test invalidity. (See Chapter 3.)

8

Psychometric Considerations

Although the major purpose of this book is to help students and clinicians learn to use and interpret the MMPI-2 clinically, it also is very important for test users to understand the strengths and weaknesses of the MMPI-2 so they can evaluate the appropriateness of its use in various settings and for various purposes. This chapter provides a brief summary of information concerning the psychometric properties of the MMPI-2. Because of the continuity between the original MMPI and the MMPI-2, some information about both versions is included. Coverage is not exhaustive, and readers who require more information about topics covered here should consult other references such as the MMPI-2 manual (Butcher et al., 1989) and *An MMPI Handbook*, Volumes I and II (Dahlstrom et al., 1972, 1975).

The MMPI was the most widely used psychological test in the United States (Harrison et al., 1988; Keller & Piotrowski, 1989; Lubin et al., 1984; Piotrowski & Lubin, 1990). Far more research papers have been published about the MMPI than about any other psychological test (Graham & Lilly, 1984; Kramer & Conoley, 1992). Reviewers of the MMPI have been generally positive about its utility. Alker (1978) concluded that the MMPI can provide reliable indications of psychological treatments that will or will not work for specific patients. King (1978) concluded his review of the MMPI by stating: "The MMPI remains matchless as the objective instrument for the assessment of psychopathology . . . and still holds the place as the sine qua non in the psychologist's armamentarium of psychometric aids" (p. 938).

Although there have been some criticisms of the MMPI-2 (e.g., Adler, 1990; Duckworth, 1991b; Strassberg, 1991), reviews of the MMPI-2 in the *Eleventh Mental Measurements Yearbook* were quite positive. Archer (1992b) described the MMPI-2 as "a reasonable compromise of the old and the new; an appropriate balance between that which required change (norms) and that which required preservation (standard scales). It should prove to be a worthy successor to the MMPI" (p. 561). Nichols (1992) agreed with Archer, commenting that "what was broke was fixed, what was not broke was left alone" (p. 567) and adding that "the psychodiagnostician selecting a structured inventory for the first time will find that no competing assessment device for abnormal psychology has stronger credentials for clinical description and prediction" (p. 567).

STANDARDIZATION

Unlike many of the projective techniques, the MMPI was well standardized in terms of materials, administration, and scoring. Essentially the same items, scales, and profile sheets were used from its inception in the 1930s until the publication of the MMPI-2 in 1989. Although there were variations in the ways in which scores were interpreted, most users based interpretations on the sizable MMPI research literature. This standardization of materials and procedures ensured that data collected in diverse settings were comparable and led to the accumulation of a significant data base for interpreting results.

Although some changes were made in the MMPI-2, considerable effort was made to maintain continuity between the original test and the revised version. Although items were updated, a few deleted, and new ones added, the basic item pool is quite similar to that of the original MMPI. The MMPI-2 maintains the true-false response format of the original test. The basic validity and clinical scales remain essentially unchanged. Scores are arrayed on a profile sheet that bears strong resemblance to the original. Although uniform T scores are used for eight of the clinical scales, the resulting scores are very similar to the linear T scores of the original MMPI (Graham, Timbrook, Ben-Porath, & Butcher, 1991; Tellegen & Ben-Porath, 1992). Later in this chapter data will be reviewed suggesting that much of the research literature that has accumulated for the MMPI can be applied to the MMPI-2. Clearly, the continuity that exists between the two versions of the test makes much of what we have learned about the MMPI relevant to the MMPI-2.

SCALE CONSTRUCTION

As was discussed briefly in Chapter 1, the clinical scales of the MMPI were constructed according to empirical keying procedures. Items were selected for inclusion in a scale if patients diagnosed as having a particular clinical syndrome (e.g., hypochondriasis, depression) responded to the items differently from other patients and from normal subjects. The reader interested in details concerning original scale construction should consult a series of articles by Hathaway and his associates in *Basic Readings on the MMPI* (Dahlstrom & Dahlstrom, 1980). In order to ensure continuity between the original and revised instruments, the MMPI-2 maintains the basic clinical scales with only minor deletions and changes. (See Table 1.1 in Chapter 1.) Scale 5 had the most items deleted (four), and these dealt primarily with objectional sexual content. Ben-Porath and Butcher (1989a) studied the eighty-two rewritten items in the MMPI-2 and concluded that they are psychometrically equivalent to the original items. Correlations between raw scores on the original clinical scales and the clinical scales of the MMPI-2 are all above .98 (Graham, 1988).

Although the empirical keying approach was an improvement over the face-valid approach used in earlier personality inventories, the scale construction procedures were rather unsophisticated by current psychometric standards. The clinical samples were often very small. For example, only twenty criterion subjects were used to select items for scale 7. Although the test authors stressed that they tried to identify criterion groups composed of patients with only one kind of psychopathology, no data were presented concerning the reliability of the criterion placements. For most scales, cross-validation procedures were employed. The statistical analyses were not very sophisticated, and often only descriptive statistics were presented.

Because no attempt was made to ensure that items would appear on only one scale, there is considerable overlap between some of the scales. For example, thirteen of the thirty-nine items in scale 6 of the MMPI-2 also appear in scale 8. This item overlap contributes to high intercorrelations among the scales and limits the extent to which scores on a single scale contribute uniquely to prediction of appropriate criterion measures. The intercorrelations of the basic MMPI-2 scales for the normative sample are presented in Appendix L. It should be noted that Hathaway did not view item overlap and lack of homogeneity as problems, maintaining that external validity was the only relevant characteristic.

The empirical keying approach used with the MMPI did not emphasize item content and scale homogeneity. Hathaway and McKinley (1940) noted that the MMPI item content was heterogeneous, but they did not give attention to homogeneity of the individual scales. As a result, the internal consistency of the scales is not very high. More information about the internal consistency of MMPI-2 scales will be presented later in this chapter.

Various strategies have been used in developing supplementary scales for the MMPI and the MMPI-2. Certain scales (e.g., Ego Strength scale, MacAndrew Alcoholism scale) used empirical keying procedures similar to those employed in developing the clinical scales. Other scales (e.g., the Harris-Lingoes subscales) were constructed rationally with no attention given to external validity. Still other scales (e.g., the content scales) were constructed using a combined rational and statistical procedure that ensured greater item homogeneity. Supplementary scales also have been developed for the MMPI-2 utilizing these various methods.

NORMS

Test norms provide a summary of results obtained when the test is given to a representative sample of individuals. The sample is referred to as the normative or standardization sample. A person's score on a test typically has meaning only when it is compared with a normative sample.

The normal subjects used in constructing the original scales of the MMPI included 724 persons who were visiting friends or relatives at the University of Minnesota Hospitals. Only persons who reported that they were under the care of a physician were excluded from the sample. Other normal subjects used in various phases of scale development were 265 high school graduates who came to the University of Minnesota Testing Bureau for precollege guidance, 265 skilled workers involved with local Works Progress Administration projects, and 243 medical patients who did not report psychiatric problems.

Only the 724 hospital visitors were included in the sample that was used to determine T-score values for the original MMPI. All of the subjects in the standardization sample were Caucasians, and the typical person was about 35 years of age, married, residing in a small town or rural area, working in a skilled or semiskilled trade (or married to a man of this occupational level), and having about eight years of formal education (Dahlstrom et al., 1972). Hathaway and Briggs (1957) later refined this sample by eliminating persons with incomplete records or faulty background information. The refined sample was the one typically used for converting raw scores on supplementary MMPI scales to T scores.

Although some data collected on nonclinical subjects in a variety of research projects suggested that the original MMPI norms were still appropriate, concern was expressed that the MMPI norms had become outdated. Colligan, Osborne, and Offord (1980) collected contemporary data from normal subjects in the same geographic area where the original MMPI norms had been collected. These investigators found that on some MMPI scales contemporary subjects endorsed more items in the scored direction than did Hathaway's normal subjects. The Colligan et al. data were of limited utility because their sample was limited geographically and demographically. In addition, their data were presented as normalized T scores. Although Colligan et al. argued that normalized scores were appropriate, Hsu (1984) offered convincing arguments to the contrary, and Graham and Lilly (1986) demonstrated that the use of normalized T scores led to underdiagnosis of psychopathology in psychiatric patients.

The MMPI-2 normative sample is larger and more representative than that of the original MMPI (Butcher et al., 1989). The 1,138 men and 1,462 women in the MMPI-2 normative sample were selected from diverse geographic areas of the United States, and their demographic characteristics closely parallel 1980 census data. They were community residents who were solicited randomly for participation in the restandardization study. Approximately three percent of the men and six percent of the women indicated that they were in treatment for mental health problems at the time of their participation in the study.

The sample includes representatives of minority groups. For men, 82 percent were Caucasians, 11.1 percent were African Americans, 3.1 percent were Hispanics, 3.3 percent were Native Americans, and .5 percent were Asian

Americans. For women, 81 percent were Caucasians, 12.9 percent were African Americans, 2.6 percent were Hispanics, 2.7 percent were Native Americans, and .9 percent were Asian Americans. Although the MMPI-2 normative sample does not match census data exactly, it is far more ethnically diverse than the original MMPI standardization sample. Although the educational level of subjects in the MMPI-2 normative sample (mean = 14.72 years) is somewhat higher than that of the general population in the 1980 census data, it probably is representative of persons to whom the MMPI-2 is likely to be administered. Persons with little or no formal education are represented in the census data but are not likely to take the MMPI-2. Butcher (1990b) reported that persons of different educational levels obtain different mean scores on some MMPI-2 scales, but the differences are small and probably not clinically important.

Normative subjects ranged in age from 18 to 85 years. In a separate project, normative data were collected from large, diverse samples of adolescent subjects. Based on those data, a decision was made to publish a separate version of the test (MMPI-A) for adolescent subjects. (See Chapter 12.)

Analyses of MMPI-2 scores were conducted for various subgroupings of normative subjects. These analyses indicated that separate norms were not needed for subjects of differing ages, geographic areas, or ethnicity. There were important differences between the men and women, so separate norms were developed for these two groups.

In summary, whereas the normative sample for the original MMPI was small and not very representative of the general population of the United States, the normative sample for MMPI-2 is large and representative of the population on major demographic variables. This represents a highly significant improvement in the instrument.

T-SCORE TRANSFORMATIONS

As indicated in Chapter 2, raw scores on the various MMPI-2 scales are converted to T scores to facilitate interpretation. With the original MMPI, raw scores were converted to linear T scores having a mean of 50 and a standard deviation of 10. Because raw scores for the MMPI and MMPI-2 scales are not normally distributed, linear T scores, which maintain the same distributions as the raw scores on which they are based, do not have exactly the same meaning for every scale. For example, a T score of 70 on any particular clinical scale of the original MMPI did not have the same percentile value as a T score of 70 on another MMPI clinical scale. Although the percentiles were not greatly different from one scale to another, the differences led to problems in profile interpretation.

For the eight basic clinical scales (excluding scales 5 and 0), the MMPI-2 utilizes a different kind of T-score transformation from that used with the original MMPI. This transformation, called a uniform T score, assures that a T score of a given level (e.g., 65) has the same percentile value for all scales (Butcher et al., 1989). A composite (or average) distribution of the non-K-corrected raw scores of the normative sample on the eight basic clinical scales was derived. The distribution of each of the eight clinical scales was adjusted so that it would match the composite distribution. This procedure resulted in uniform T scores that are percentile equivalent and whose distributions are closely matched in terms of skewness and kurtosis (Tellegen & Ben-Porath, 1992). The change in the distribution of any particular scale is not great, so the profile retains most of its familiar characteristics. Percentile equivalents for various uniform T scores are reported in Appendix M.

Although it has been suggested by some that the use of uniform T scores significantly affects scale elevations and configurations, this is not the case. The MMPI-2 manual (Butcher et al., 1989) presented data concerning profile elevation and linear versus uniform T scores. Raw scores of psychiatric patients were transformed to both linear and uniform T scores using the MMPI-2 normative data, and the numbers of patients with elevated scores were calculated. There were only negligible differences between linear and uniform T scores. Graham, Timbrook, Ben-Porath, and Butcher (1991) compared congruence of MMPI and MMPI-2 code types using both linear and uniform T-score transformations. Again, the effect of using uniform T scores on code type congruence was negligible.

Uniform T scores were not derived for scales 5 and 0 or for the validity scales, because the distributions of scores on these scales differ considerably from those of the eight clinical scales. For these scales, linear T-score transformations, comparable to those used with the original MMPI, were derived. The norms for most MMPI-2 supplementary scales are expressed as linear T scores. The exceptions are the MMPI-2 content scales and the Marital Distress Scale (MDS), for which uniform T scores were derived using the same composite distribution used for the eight clinical scales. This permits direct comparison between the basic clinical scales and these supplementary scales.

TEMPORAL STABILITY

That scores on tests of ability, interest, and aptitude should have high temporal stability is quite accepted by most psychologists. What should be expected from tests of personality and psychopathology in this regard is not as clear. Although personality test scores should not be influenced by sources of error variance, such as room temperature, lack of sleep, and the like, it must be recognized that many personality attributes change over relatively short periods of time. Dahlstrom (1972) pointed out that some of the inferences made from personality test data involve current emotional status, whereas others deal with personality structure. Scales assessing personality structure should have high temporal stability, but those designed to measure current emotional status should be sensitive to rather short-term fluctuations. This is particularly true in clinical settings in which individuals are experiencing heightened levels of distress that might be expected to decrease over time and with treatment.

From the MMPI's inception there has been an awareness of the importance of scale stability. Dahlstrom et al. (1972), Graham (1977), and Schwartz (1977) summarized temporal stability data for the individual validity and clinical scales of the MMPI. Table 8.1 reports ranges and typical values of test-retest correlations of the original MMPI scales for various samples and varying test-retest intervals. For normal subjects the test-retest coefficients for relatively short intervals were relatively high and comparable to coefficients for other personality tests. For longer intervals the coefficients were considerably lower. The data for psychiatric patients were very similar to those for normals. For criminal samples the short-term coefficients were a bit lower than for the normal and psychiatric samples. Schwartz (1977) concluded that the temporal stability of the original MMPI scales was not related systematically to gender of subjects or MMPI form used. Further, no MMPI scale appeared to be consistently more stable than other scales. A meta-analysis conducted by Parker, Hanson, and Hunsley (1988) reported an average stability coefficient for MMPI scales across a variety of samples to be .74. This value was only slightly lower than the value of .82 that was reported for the Wechsler Adult Intelligence Scale. In summary, short-term temporal stability of the original MMPI scales compared favorably with that of scores from other psychological tests.

The MMPI-2 manual (Butcher et al., 1989) reports test-retest reliability coefficients for the basic MMPI-2 validity and clinical scales for 82 men and 111 women in the normative sample. The retest interval for these subjects was approximately one week. Table 8.2 summarizes these coefficients. Butcher, Graham, Dahlstrom, and Bowman (1990) reported test-retest stability of MMPI-2 scales for a sample of 42 male and 79 female college students. They concluded that the test-retest correlations for the college students are comparable to those reported for the MMPI-2 normative sample. Test-retest reliabil-

Table 8.1. Summary of Test-Retest Reliability Coefficients for the Original MMPI Scales

| | Test-Retest Interval | | | | | |
| | One Day or Less | | One to Two Weeks | | One Year or More | |
Samples	Actual Range	Typical Range	Actual Range	Typical Range	Actual Range	Typical Range
Normal	.49–.96	.80–.85	.29–.92	.70–.80	.13–.73	.35–.45
Psychiatric	.61–.94	.80–.85	.43–.86	.80–.85	.22–.72	.50–.60
Criminal	.40–.86	.70–.80	.21–.84	.60–.70	—	—

Source: Schwartz, G.F. (1977). *An investigation of the stability of single scale and two-point MMPI code types for psychiatric patients.* Unpublished doctoral dissertation, Kent State University, Kent, Ohio. Reproduced by permission.

ity coefficients for the MMPI-2 supplementary scales are reported in the chapters of this book in which the development of the scales is described. In summary, the short-term temporal stability of the MMPI-2 scales for normal subjects appears to be at least as high as or higher than that of the original MMPI scales and compares quite favorably with that of other psychological tests.

Because MMPI interpretive strategies have emphasized configural aspects of profiles, it is important to consider the temporal stability of such configurations. Many clinicians assume that because the individual clinical scales have reasonably good temporal stability, the configurations based on those scales also have good temporal stability. Only limited data are available concerning the stability of scale configurations for the MMPI-2. Because of the continuity between the original and revised instruments, one can assume that the stability of MMPI-2 configurations is very similar to the stability of configurations on the original MMPI.

Table 8.3 summarizes the results of studies that have reported stability of high-, two-, and three-point code types for the original MMPI. Although the kinds of subjects and test-retest intervals have differed across studies, the results have been consistent. About one half of subjects have had the same high-point code type on two administrations; about one fourth to one third have had the same two-point code type; and about one fourth have had the same three-point code type. It should be recognized that the stability indicated by these studies is lower than we would expect if the analyses had been limited to well-defined code types. (See Chapter 5 for discussion of profile definition.)

Table 8.2. Test-Retest Coefficients for MMPI-2 Scales for Men and Women in the Normative Sample (1-Week Interval)

Scales	Men (n = 82)	Women (n = 111)
L	.77	.81
F	.78	.69
K	.84	.81
Hs	.85	.85
D	.75	.77
Hy	.72	.76
Pd	.81	.79
Mf	.82	.73
Pa	.67	.58
Pt	.89	.88
Sc	.87	.80
Ma	.83	.68
Si	.92	.91

Source: Butcher, J.N., Dahlstrom, W.G., Graham, J.R., Tellegen, A., & Kaemmer, B. (1989). *Minnesota Multiphasic Personality Inventory-2 (MMPI-2): Manual for administration and scoring.* Minneapolis: University of Minnesota Press. Copyright © 1989 by the University of Minnesota. Reproduced by permission.

Table 8.3. Percentages of High-Point, Two-Point, and Three-Point MMPI Codes Remaining the Same on Retest[a]

Source	Sample	n	Gender	Test-retest	High-point	Two-point	Three-point
Ben-Porath & Butcher (1989b)	College student	86	M	1–2 weeks	54	35	—
Ben-Porath & Butcher (1989b)	College student	102	F	1–2 weeks	54	44	—
Chojnacki & Walsh (1992)	College student	86	M	1–2 weeks	64	42	43
Chojnacki & Walsh (1992)	College student	94	F	1–2 weeks	57	45	45
Graham (1977)	College student	43	M	1 week	51	35	23
Graham (1977)	College student	36	F	1 week	50	31	28
Fashingbauer (1974)	College student/ Psychiatric	61	M/F	1 day	63	41	23
Lichtenstein & & Bryan (1966)	Volunteer/ Psychiatric	82	M/F	1–2 days	50	—	—
Kincannon (1968)	Psychiatric	60	M/F	1–2 days	61	—	—
Pauker (1966)	Psychiatric	107	F	13–176 days	44	25	—
Sivanich (1960)[b]	Psychiatric	202	F	2–2230 days	48	20	—
Uecker (1969)	Organic	30	M	1 week	—	23	—
Lauber & Dahlstrom (1953)	Delinquent	19	F	?	90	95	—

[a] For two-point and three-point code types, scales are used interchangeably.

[b] Used only four two-point code types (4–6, 4–2, 6–8, 2–7).

Source: Graham, J.R., Smith, R.L., & Schwartz, G.F. (1986). Stability of MMPI configurations for psychiatric inpatients. *Journal of Consulting and Clinical Psychology, 54,* 375–380. Copyright © 1986 by the American Psychological Association. Adapted by permission of the publisher.

Graham et al. (1986) concluded that no particular MMPI high- or two-point code type was significantly more stable than other code types. They also reported that configurations tended to be more stable when the scales in the code types were more elevated initially and when there was a greater difference between these scales and other scales in the profile. When code types of subjects changed from test to retest administrations, the second code type often was in the same diagnostic grouping (neurotic, psychotic, characterological) as was the first code type. From one half to two thirds of subjects had code types in the same diagnostic grouping on test and retest. When the code types were from different diagnostic groupings on test and retest, the most frequent change was from psychotic on the initial test to characterological on the retest. The implication of these data is that many of the inferences that would be made would be the same even though the patients did not have exactly the same two-point code type on the two occasions.

What about subjects whose code types changed from one major diagnostic grouping on the initial test to another major grouping on the retest? Were these changes due to the unreliability of the MMPI, or did they reflect significant changes in the status of the patients? Graham et al. (1986) addressed this issue to a limited extent. They studied psychiatric inpatients who produced psychotic two-point code types at the time of the initial testing and nonpsychotic code types at the time of retesting. They compared these patients with patients who had psychotic code types for both administrations or for neither administration. External psychiatric ratings of psychotic behaviors were available for patients. Patients who changed from psychotic to nonpsychotic showed concomitant changes in psychiatric ratings. Patients who had nonpsychotic code types on both administrations were given relatively low psychosis ratings on both occasions. Complicating the results of the study was the finding that patients who had psychotic code types on both test and retest were given lower psychosis ratings at retest than at the time of the initial test. The psychiatrists who completed the ratings also were case managers for the patients they rated. It may be that they were reluctant to indicate that patients that they were treating, and perhaps were about ready to discharge from the hospital, had not shown decreases in psychotic behaviors.

Graham, Timbrook, Ben-Porath, and Butcher (1991) reported temporal stability of one-, two-, and three-point code types for the MMPI-2 normative sample based on both the MMPI and MMPI-2 norms. As reported in Table 8.4, the stability of the MMPI and MMPI-2 code types was very similar. Because the sample of normative subjects who took the test twice was relatively small, the Graham et al. analysis could not be restricted to well-defined code types. Thus, all of the values presented in Table 8.4 are somewhat lower than we would expect if only well-defined code types had been used in the analyses.

Several conclusions can be reached about the temporal stability of MMPI-2 scores. Individual MMPI-2 scales seem to be as reliable temporally as the scales of the original MMPI and as other personality measures. Code types are likely to be more stable when their scales are more elevated and when they are well defined. Although configurations probably are not as stable as they are assumed to be by test users, many subjects are likely to produce the same code types on different administrations of the test. When the code types

Table 8.4. MMPI and MMPI-2 Code-Type Congruence Rates for Test-Retest Subjects in the Normative Sample (82 Men and 111 Women)

	One-Point	Two-Point	Three-Point
MMPI	52%	28%	19%
MMPI-2	49%	26%	15%

Source: Graham, J.R., Timbrook, R.E., Ben-Porath, Y.S., & Butcher, J.N. (1991). Code-type congruence between MMPI and MMPI-2: Separating fact from artifact. *Journal of Personality Assessment, 57,* 205–215. Copyright © 1991 by Lawrence Erlbaum Associates, Inc. Reproduced by permission.

change from one administration to another, they are likely to remain in the same diagnostic grouping. When the code types change dramatically, there are likely to be concomitant behavioral changes.

INTERNAL CONSISTENCY

Because of the empirical keying procedures used in constructing the basic validity and clinical scales, little or no attention was given by the test authors to internal consistency of most of these scales. Dahlstrom et al. (1975) summarized MMPI internal-consistency data for a variety of samples. Estimates of internal consistency varied considerably (from -.05 to +.96), with typical values ranging from .60 to .90. Scales 3, 5, and 9 appeared to be the least consistent ones, whereas scales 1, 7, and 8 appeared to be the most internally consistent. A meta-analysis of MMPI studies conducted by Parker et al. (1988) determined an average internal-consistency coefficient of .87 across a number of samples.

The MMPI-2 manual (Butcher et al., 1989) reports internal-consistency values for the standard validity and clinical scales of the MMPI-2. These values are summarized in Table 8.5. The coefficients are similar to the typical values

Table 8.5. Internal-Consistency Coefficients (Alpha) for the MMPI-2 Validity and Clinical Scales for Men and Women in the Normative Sample

Scale	Men	Women
	(n = 1138)	(n = 1462)
L	.62	.57
F	.64	.63
K	.74	.72
Hs	.77	.81
D	.59	.64
Hy	.58	.56
Pd	.60	.62
Mf	.58	.37
Pa	.34	.39
Pt	.85	.87
Sc	.85	.86
Ma	.58	.61
Si	.82	.84

previously reported for the original MMPI scales. Scales 1, 7, 8, and 0 appear to be the most internally consistent scales, whereas scales 5, 6, and 9 appear to be the least internally consistent scales. Internal-consistency data for the MMPI-2 supplementary scales are presented in the chapters of this book in which the development of the scales is described.

Factor analyses of items within each standard scale of the MMPI have indicated that most of the scales are not unidimensional (Ben-Porath, Hostetler, Butcher, and Graham, 1989; Comrey, 1957a, 1957b, 1957c, 1958a, 1958b, 1958c, 1958d, 1958e; Comrey & Margraff, 1958; Graham et al., 1971). The one exception seems to be scale 1, where most of the variance is associated with a single dimension: concern about health and bodily functioning. Because little attention was given to internal consistency when the original MMPI scales were constructed, it is not surprising that the scales are not as internally consistent as other personality scales that were developed according to internal-consistency procedures. To date only one study has been published concerning the factor structure of an MMPI-2 scale. In their development of subscales for scale 0, Ben-Porath, Hostetler, Butcher, and Graham (1989) conducted item-level factor analyses and identified three rather distinct factors.

SCALE-LEVEL FACTOR STRUCTURE

Two consistent dimensions have emerged whenever scores on the basic MMPI validity and clinical scales have been factor analyzed (Block, 1965; Eichman, 1961, 1962; Welsh, 1956). Scales 7 and 8 had high positive loadings on Factor I, and the K scale had a high negative loading on this factor. Welsh and Eichman both labeled this factor "Anxiety," whereas Block scored it in the opposite direction and called it "Ego Resiliency." Welsh developed the Anxiety (A) scale to assess this dimension. This scale seems to assess a general maladjustment dimension.

Scales 1, 2, and 3 had high positive loadings on Factor II, and scale 9 had a moderately high negative loading on this factor. Welsh and Eichman labeled this dimension "Repression," and Block called it "Ego Control." Welsh developed the Repression (R) scale to assess this dimension. This scale seems to assess denial, rationalization, lack of insight, and overcontrol of needs and impulses.

The MMPI-2 manual (Butcher et al., 1989) reports results of factor analyses of the MMPI-2 scales for subjects in the normative sample. The results of these analyses are quite consistent with previous studies reported in the literature for the original MMPI. One major factor seems to be related to general maladjustment and psychotic mentation, whereas another major factor seems to be more related to neurotic characteristics. Weaker factors seem to be related to gender-role identification and social introversion. The factor structure was somewhat different for men and women, suggesting that certain MMPI-2 patterns may have somewhat different interpretive meanings for each gender.

ITEM-LEVEL FACTOR STRUCTURE

Some investigators factor analyzed or cluster analyzed responses to the entire original MMPI item pool (Barker, Fowler, & Peterson, 1971; Chu, 1966; Johnson, Null, Butcher, & Johnson, 1984; Lushene, 1967; Stein, 1968; Tryon, 1966; Tryon & Bailey, 1965). Most of these early studies were limited by small sample sizes or analyses based on subsets of the total MMPI item pool. Only the study of Johnson et al. (1984) utilized a very large sample (more than 11,000 subjects) and analyzed the entire item pool in a single computational pass.

Using replication procedures, Johnson et al. identified the following twenty-one factors in the MMPI item pool: Neuroticism—General Anxiety and Worry; Psychoticism—Peculiar Thinking; Cynicism—Normal Paranoia; Denial of Somatic Problems; Social Extroversion; Stereotypic Femininity; Aggressive Hostility; Psychotic Paranoia; Depression; Delinquency; Inner Directedness; Assertiveness; Stereotypic Masculinity; Neurasthenic Somatization; Phobias; Family Attachment; Well-being—Health; Intellectual Interests; Rebellious Fundamentalism; Sexual Adjustment; and Dreaming. The authors noted the similarity of these factors to the content dimensions represented in the Wiggins content scales and in the original content categories presented by Hathaway and McKinley. They also commented that the item pool seemed to be measuring more aspects of personality than merely emotional stability. No factor analyses of the MMPI-2 item pool have been published to date. However, because of the considerable overlap between items in the original MMPI and the MMPI-2, one would expect to find some of the same factors represented in the MMPI-2 item response data. However, because of item deletions and additions, some of the weaker factors present in the original MMPI item data might not emerge from analysis of MMPI-2 item data, and other factors not present in the original MMPI item data probably would be identified with MMPI-2 item data.

Studies of the factor structure of item responses to the original MMPI offered interesting insights into the psychometric properties of the MMPI and suggested the basis for new and potentially useful scales. Barker et al. (1971) and Stein (1968) developed scales to assess their factor dimensions and demonstrated them to be as reliable as the standard MMPI scales and as good as or better than the standard scales in discriminating among diagnostic groups. However, the limited amount of data concerning the utility of scales based on factor analyses precluded including them in the MMPI-2. Whether reliable and valid scales can be developed through factor analyses of item responses to MMPI-2 remains to be determined.

RESPONSE SETS AND STYLES

Over the years some critics (e.g., Edwards, 1957, 1964; Edwards & Clark, 1987; Edwards & Edwards, 1992; Messick & Jackson, 1961) have argued that the MMPI scales were of limited utility because most of the variance in their scores could be attributed to response sets or styles. Messick and Jackson argued that subjects who obtained high scores on the MMPI scales did so only because of an acquiescence response style (i.e., a tendency to agree passively with inventory statements). In support of their argument, Messick and Jackson pointed out that the standard MMPI scales were not balanced for proportion of items keyed as true or false. Further, it was shown that scores on Welsh's Anxiety scale (a measure of one major source of variance in MMPI responses) correlated positively with an acquiescence measure.

Edwards maintained that scores on the standard scales of the MMPI were grossly confounded with a social desirability response set. Subjects who obtained higher scores on the clinical scales were hypothesized to be persons who were willing to admit to socially undesirable behaviors, whether or not these behaviors really were characteristic of them. Major support for this position came from data indicating that scores on the standard MMPI scales and on Welsh's Anxiety scale had high negative correlations with a social desirability scale.

Although a number of persons argued against the acquiescence and social desirability interpretations of the MMPI scales, Block (1965) most thoroughly reviewed the arguments in support of acquiescence and social desirability and pointed out some statistical and methodological problems. He also presented new evidence that clearly rebutted the arguments. Block modified the standard MMPI scales, balancing the number of true and false items within each scale. Contrary to the prediction of Messick and Jackson, the factor structure of the MMPI with these modified scales was essentially the same as with the standard scales. Block also developed a measure of Welsh's Anxiety scale that was free of social desirability influences. Correlations between this modified Anxiety scale and standard MMPI scales were essentially the same as for Welsh's original scale. Finally, Block demonstrated that the MMPI scales had reliable correlates with important extratest behaviors even when the effects of social desirability and acquiescence were removed.

In summary, the criticism directed at the MMPI by critics such as Messick, Jackson, and Edwards was severe. However, the MMPI withstood their challenges. Because of the continuity between the original MMPI and the MMPI-2, the data presented by Block concerning response sets probably can be generalized to the MMPI-2 as well.

VALIDITY

Because of the continuity between the MMPI and the MMPI-2, validity studies of both versions are relevant to an evaluation of the validity of the MMPI-2. This section will review MMPI validity studies, consider the extent to which these studies are relevant to the MMPI-2, and summarize the MMPI-2 validity studies that have been reported to date.

Validity of the Original MMPI

In Volume II of *An MMPI Handbook*, Dahlstrom et al. (1975) cited over six thousand studies involving the MMPI. In trying to reach conclusions about the validity of the original MMPI, it seems helpful to group research studies into three general categories. First, studies have compared the MMPI profiles of relevant criterion groups. Most of these studies have identified significant differences on one or more of the MMPI scales among groups formed on the basis of diagnosis, severity of disturbance, treatment regimes, and numerous other criteria. Lanyon (1968) published average or typical profiles for many of these criterion groups. Efforts to develop classification rules to discriminate among groups of subjects also investigated the validity of the MMPI (Goldberg, 1965; Henrichs, 1964; Meehl & Dahlstrom, 1960; Peterson, 1954; Taulbee & Sisson, 1957).

A meta-analysis of 403 control and psychiatric samples (Zalewski and Gottesman, 1991) indicated that the MMPI was effective in discriminating between psychiatric and control groups, neurotic and psychotic groups, and depression and anxiety disorder groups. In a different meta-analysis, Parker et al. (1988) found that the average validity coefficient for MMPI studies conducted between 1970 and 1981 was .46. Although this average coefficient was somewhat lower than for the Wechsler Adult Intelligence Scale ($r = .62$), Parker et al. concluded that the MMPI had acceptable validity. Atkinson (1986) applied meta-analytic procedures to a sample of MMPI studies conducted between 1960 and 1980. The results indicated that MMPI scores accounted for approximately 12 percent of the variance in criterion measures across many samples. It was noted that the MMPI was more valid when studies were conceptually based rather than undirected and when more reliable criterion measures were employed.

A second category of studies included efforts to identify reliable behavioral correlates of MMPI scales and configurations. Extratest correlates of high or low scores on individual clinical scales have been identified for adolescents (Archer, Gordon, Giannetti, & Singles, 1988; Hathaway & Monachesi, 1963), normal college students (Black, 1953; Graham & McCord, 1985), student nurses (Hovey, 1953), normal Air Force officers (Block & Bailey, 1955; Gough, McKee, & Yandell, 1955), medical patients (Guthrie, 1949), and psychiatric patients (Boerger, Graham, & Lilly, 1974; Hedlund, 1977). Correlates for configurations of two or more MMPI scales were reported for normal adults (Hathaway & Meehl, 1952), normal college students (Black, 1953), medical patients (Guthrie, 1949), and psychiatric patients (Boerger et al., 1974; Gilberstadt & Duker, 1965; Gynther et al., 1973; Lewandowski & Graham, 1972; Marks et al., 1974; Meehl, 1951). The results of the numerous empirical studies with adults are summarized in Chapters 4 and 5 of this book, and empirical studies of adolescents are summarized in Chapter 12. Clinicians drew heavily on the results of these studies in making inferences about MMPI scores and configurations. These data suggested that there are reliable extratest correlates for MMPI scores and configurations, but they also indicated that exactly the same correlates may not always be found for subjects of differing demographic backgrounds.

A third category of studies considered the MMPI scores and the person interpreting them as an integral unit and examined the accuracy of inferences based on the MMPI. In an early study of this type, Little and Shneidman (1959) asked expert test interpreters to provide diagnoses, ratings, and descriptions of subjects based on data from the MMPI, Rorschach, Thematic Apperception Test (TAT), and Make a Picture Story Test (MAPS). The accuracy of these judgments was determined by comparing them with judgments based on extensive case history data. The average correlations between judges' descriptions of subjects based on the MMPI, Rorschach, TAT, and MAPS were .28, .16, .17, and .11, respectively. Kostlan (1954) reported data suggesting that the MMPI leads to the most accurate inferences when it is used in conjunction with social case history data. Sines (1959) reported a mean correlation of .38 between judgments based on MMPI and interview data and criterion ratings provided by patients' therapists. In a study limited to MMPI data, Graham (1967) reported correlations between MMPI-based descriptions and criterion descriptions of .31, .37, and .29 for judges of high, medium, and low experience, respectively. Other studies (Graham, 1971a, 1971b; Henrichs, 1990) have indicated that judges can learn to make more accurate inferences from MMPI data if they are given feedback concerning their performance.

Reviewing clinical judgment research, of which the above cited studies are a part, Goldberg (1968) concluded that, in general, clinical judgments tend to be rather unreliable, only minimally related to the confidence and the amount of experience of test judges, relatively unaffected by the amount of information available to judges, and rather low in validity on an absolute basis. In a more recent review of the clinical judgment literature, Garb (1984) was more optimistic, concluding that adding MMPI data to demographic information led to significant increases in validity of inferences. Graham and Lilly (1984) pointed out that, compared with Rorschach and other projective techniques, personality descriptions based on the MMPI have been relatively more accurate. Also, descriptions based on MMPI data have been more accurate than descriptions based on judges' stereotypes of typical patients. When MMPI data have been used in conjunction with social history and/or interview data, the resulting descriptions have been more valid than when the MMPI data have been used alone.

Comparability of MMPI and MMPI-2 Raw Scores

As clinicians have switched from the MMPI to the MMPI-2, they understandably have wondered to what extent their clients' scores and configurations of scores on the MMPI-2 are similar to the scores and configurations of scores that they would have obtained if the MMPI had been used. Graham (1988) reported data suggesting that raw scores on the standard validity and clinical scales of the MMPI-2 and of the original MMPI are remarkably similar. MMPI and MMPI-2 raw scores were correlated for normal and psychiatric subjects, and all correlations were greater than .98. Ben-Porath and Butcher (1989b) compared MMPI and MMPI-2 raw scores for a sample of college students. They found that, except for the F scale for women, correlations between the MMPI and MMPI-2 validity and clinical scale raw scores were not significantly different from scores on the MMPI taken on two occasions with a 1-week interval between testings. A significant difference for the F scale for women was attributed to an atypically high correlation for this scale for two administrations of the original MMPI. Also using college student subjects, Chojnacki and Walsh (1992) found correlations between the MMPI and MMPI-2 raw scores to be slightly lower than those reported by Ben-Porath and Butcher. Chojnacki and Walsh attributed the difference to the fact that they administered both the MMPI and MMPI-2, whereas Ben-Porath and Butcher abstracted MMPI and MMPI-2 scores from a single administration of the adult experimental (AX) form.

Correlations, such as those reported above, indicate that the order of raw scores on the MMPI and the MMPI-2 is very similar. However, correlational data do not address the issue of similarities and differences in the magnitude of raw scores. In other words, do persons obtain the same levels of raw scores on the two versions of the test? As part of the revision process, the MMPI restandardization committee rescored data from the original MMPI standardization sample to obtain raw scores for the validity and clinical scales as they appear in the MMPI-2. These raw scores were then compared with raw scores for the MMPI-2 standardization sample. For most scales the contemporary normative sample endorsed more items in the scored direction than did the original MMPI normative sample. The difference in mean scores for the MMPI and MMPI-2 normative samples typically was between one and two points. Men and women in the MMPI-2 normative sample scored somewhat lower than their counterparts in the MMPI normative sample on the L scale. Men in the MMPI-2 normative sample scored slightly lower than men in the MMPI normative sample on scale 1, and women in the MMPI-2 normative sample scored slightly lower than men in the MMPI normative sample on scale 7.

The somewhat higher raw scores for the MMPI-2 normative sample could indicate that the sample is more pathological than the original sample, but this seems unlikely given that the differences were consistent across most clinical scales. A more likely explanation has to do with the instructions given to subjects. The original MMPI normative subjects were permitted, and even encouraged, to omit items if they felt unable to answer them. By contrast, the MMPI-2 normative subjects were given the instructions that have been in use since the original MMPI was published. These instructions encourage subjects to try to respond to every statement in the test booklet. Thus, the MMPI-2 normative subjects omitted many fewer items than the MMPI normative subjects. In the process of answering more items, the MMPI-2 normative subjects endorsed more items in the scored direction.

Comparability of MMPI and MMPI-2 T Scores

Because of the differences in raw scores between the MMPI and MMPI-2 normative samples, a particular raw score does not yield the same T score for both tests. Because raw scores tend to be higher for the MMPI-2 than for the MMPI normative sample, T scores tend to be lower on the MMPI-2 than on the MMPI. These findings led to a recommendation in the MMPI-2 manual (Butcher et al., 1989) that less restrictive criteria that recognize potential significance of elevation in the 65–69 T-score range be used in interpreting MMPI-2 profiles.

Munley (1991) used information presented in the MMPI-2 manual to determine MMPI T scores that are equivalent to MMPI-2 T scores. Strassberg (1991) did the same kinds of calculations and presented tables indicating how much different MMPI and MMPI-2 T scores are at various raw-score levels. Both investigators concluded that the relationship between MMPI and MMPI-2 T scores is complex. Although the average T-score difference across all comparisons is close to the five points mentioned in the MMPI-2 manual, some scales differ by as much as 10–15 points and for some scales the MMPI-2 T scores are actually lower than the corresponding MMPI T scores. Relationships differ across gender and across specific ranges within the scales. Thus, one cannot assume that every person who produces a specific MMPI-2 profile would have produced the same MMPI profile at a level five T-score points higher.

The MMPI-2 manual (Butcher et al., 1989) reported data concerning elevated scores for psychiatric patients using the MMPI and MMPI-2 norms. As would be expected, there were fewer patients with elevated T scores on the clinical scales when the MMPI-2 norms were used. Consistent with information presented earlier in this chapter, the differences were found for both linear and uniform T-score transformations of MMPI-2 raw scores.

MMPI and MMPI-2 T scores of psychiatric outpatients and inpatients were compared by Harrell, Honaker, and Parnell (1992). For all clinical scales, the

MMPI-2 T scores were lower than the MMPI T scores. Differences were largest for scales 2 (8.90 points), 4 (10.23 points), and 8 (8.12 points) and smallest for scales 1 (3.64 points) and 0 (3.90 points). Correlations between MMPI and MMPI-2 T scores were not significantly different from correlations based on two administrations of either the MMPI or the MMPI-2. Lachar et al. (1991) reported differences between MMPI and MMPI-2 T scores similar to those found by Harrell et al.

Comparability of MMPI and MMPI-2 Configurations

If all of the MMPI-2 scales differed from their corresponding MMPI scales in the same direction and by the same amount, MMPI and MMPI-2 code types and other configurations would be the same for both versions of the test. That differences are not equal for all scales leads to lack of congruence in configural aspects of the MMPI and MMPI-2 profiles. The MMPI-2 manual (Butcher et al., 1989) reported data concerning the congruence of MMPI and MMPI-2 code types for a sample of 232 male and 191 female psychiatric patients. Approximately two thirds of the patients had the same two-point code type for both versions of the test. Graham, Timbrook, Ben-Porath, and Butcher (1991) refined the analyses of the psychiatric patient data by including only well-defined code types (i.e., those in which the lowest scale in the code type was at least five T-score points higher than the next-highest clinical scale in the profile). They found that congruence of two-point code types increased markedly. Similar increases in congruence rates were also noted for one- and three-point code types. Graham et al. reported corresponding analyses for the MMPI-2 normative sample and noted even higher congruence rates for well-defined code types. Table 8.6 reports congruence rates for one-, two-, and three-point code types for normative subjects and psychiatric patients at several levels of code type definition. Dahlstrom (1992) reported data concerning comparability of specific MMPI and MMPI-2 two-point code types for the MMPI-2 normative sample and concluded that some code types are quite comparable and others are not. However, Dahlstrom's findings are difficult to interpret because he chose not to take code type definition into account in his analyses.

Ben-Porath and Butcher (1989b) examined the congruence of MMPI and MMPI-2 high-point and two-point code types for a sample of college students. Because they did not use well-defined code types, congruence rates were somewhat lower than in the Graham, Timbrook, Ben-Porath, and Butcher (1991) study. Ben-Porath and Butcher reported that 58.7 percent of their male and 58.7 percent of their female subjects had the same high-point code type, and 35.9 percent of their male subjects and 30.9 percent of their female subjects had the same two-point code type. They concluded that there was as much congruence between MMPI and MMPI-2 code types as between code types resulting from two administrations of the MMPI.

Table 8.6. Congruence of High-Point and Two-Point Code Types for Male and Female Psychiatric Patients as a Function of Code-Type Definition

		Percentage Agreement MMPI-2 and MMPI			
		Normative Sample		Psychiatric Sample	
Code Type	Definition[a]	Men ($n = 232$)	Women ($n = 191$)	Men ($n = 232$)	Women ($n = 191$)
High-Point	0	75.9	74.4	73.3	77.0
	5	94.8	96.0	83.4	90.6
	10	98.8	100.0	91.6	98.0
Two-Point	0	64.5	62.3	62.5	67.5
	5	94.8	93.4	81.6	94.3
	10	97.6	100.0	82.4	96.6
Three-Point	0	60.9	56.8	56.5	64.4
	5	93.8	93.6	75.3	90.9
	10	100.0	100.0	91.3	100.0

[a]T-score points between lowest scale in the code type and next-highest clinical scale.

Source: Graham, J.R., Timbrook, R.E., Ben-Porath, Y.S., Butcher, J.N. (1991). Code-type congruence between the MMPI and MMPI-2: Separating fact from artifact. *Journal of Personality Assessment, 57,* 205–215. Copyright © 1991 by Lawrence Erlbaum Associates, Inc. Reproduced by permission of the publisher.

Chojnacki and Walsh (1992) extended the Ben-Porath and Butcher study by including three-point code types and utilizing somewhat different procedures for examining congruence between MMPI and MMPI-2 code types. When the standard eight clinical scales were used, 52.2 percent of the male subjects and 58.3 percent of the female subjects had the same high-point code type; 34.8 percent of the male subjects and 26.0 percent of the female subjects had the same two-point code type; and 42.4 percent of the male subjects and 25.0 percent of the female subjects had the same three-point code type. Chojnacki and Walsh also examined the congruence of well-defined code types. A subject was considered to have a well-defined MMPI code type if he or she had a one-, two-, or three-point code type that was at least five T-score points above the next-highest scale. If a subject had more than one well-defined code type, the simplest code type was used. On the MMPI-2, 55.7 percent of the male subjects and 61.5 percent of the female subjects had the same code type. Chojnacki and Walsh concluded that overall the MMPI-2 seems to produce scores and configural patterns that are similar to the MMPI, but they pointed out that score changes on scales L, 5, and 8 warrant further investigation.

Lachar et al. (1991), using a sample of 100 psychiatric inpatients, found that 60 percent had the same MMPI and MMPI-2 two-point code types. Harrell et al. (1992) determined concordance rates for MMPI and MMPI-2 high-point and two-point code types for their sample of psychiatric outpatients and inpatients. For both defined and undefined code types, concordance was not significantly different from concordance of code types based on two administrations of either the MMPI or the MMPI-2.

Munley and Zarantonello (1990) translated the mean MMPI code types reported previously by Gilberstadt and Duker (1965) and Marks et al. (1974) into MMPI-2 T scores. They found that the MMPI and estimated MMPI-2 T-score profiles correlated highly with each other.

Clavelle (1992) asked clinicians to indicate how similar MMPI and MMPI-2 profile pairs were to each other and how similar or different their interpretations would be for each pair of profiles. Clinicians indicated that 92–96 percent of their diagnoses and 89–93 percent of their interpretations would be essentially the same or only slightly different from one version of the test to the other. Poorly defined MMPI-2 code types were more likely to be viewed as somewhat different or quite different from the MMPI. It should be emphasized that Clavelle asked clinicians to indicate how they would interpret the profiles, but actual interpretations were not made and compared.

In summary, existing data indicate that MMPI and MMPI-2 code types are quite congruent for normative subjects, college students, and psychiatric inpatients and outpatients. Congruence rates are higher for well-defined code types than for code types that are not well defined. Several studies have indicated that the congruence of MMPI and MMPI-2 code types, while not very high on an absolute level, is not significantly lower than congruence of code types based on two administrations of either the MMPI or the MMPI-2.

Some differences between scores on the MMPI and the MMPI-2 are expected, and, as pointed out by Ben-Porath and Graham (1991), such differences are necessary if the MMPI-2 is to be an improvement over the MMPI. If both versions yielded exactly the same scores, there would be little to justify or recommend the revised instrument. Critics of the MMPI-2 (e.g., Caldwell, 1991; Duckworth, 1991a, 1991b) seem to conclude that in cases where the MMPI and MMPI-2 results are not comparable, something is wrong with the MMPI-2. This assumption may not be justified. Graham, Timbrook, Ben-Porath, and Butcher (1991) identified persons from the MMPI-2 normative sample for whom well-defined MMPI and MMPI-2 two-point code types were not congruent. They compared descriptions based on each code type for each person and determined their accuracy by comparing the descriptions with information provided by persons who knew the subjects very well. Results indicated that the descriptions based on the MMPI-2 were at least as accurate as the MMPI-based descriptions, and perhaps even more accurate. Clearly, additional research is needed to address this important issue.

Another important consideration is the extent to which specific MMPI-2 code types have extratest correlates similar to corresponding code types on the MMPI. For example, are the correlates of an MMPI-2 4–8 two-point code type similar to the correlates of an MMPI 4–8 two-point code type? Only very limited data are available to address this point. Moreland and Walsh (1991) determined MMPI and MMPI-2 two-point code types for a sample of psychiatric inpatients. Psychiatric symptom ratings were identified that differentiated patients with a particular code type from patients with other code types. Correlates associated with frequently occurring MMPI-2 code types were quite similar to those associated with corresponding MMPI code types. These preliminary analyses support the notion that MMPI-2 code types can be interpreted similarly to MMPI code types.

Validity of the MMPI-2

Although the congruence between scores on the original MMPI and the MMPI-2 offers information concerning the validity of the MMPI-2, it is important that the validity of the MMPI-2 also be established by comparing its scores with relevant extratest measures. Graham (1988) reported some preliminary data concerning external correlates of MMPI-2 scores. Using data from 822 couples who participated in the restandardization project together, scores on the basic clinical scales of the MMPI-2 (abstracted from Form AX) were correlated with behavioral ratings of subjects provided by their partners. Table 8.7 lists the behavioral ratings that were most highly correlated with the clinical scales. The pattern of correlations suggests both convergent and discriminant validity for the clinical scales of MMPI-2. Most of the correlates are quite consistent with those previously reported for the original MMPI. However, it should be noted that there were important differences in correlates between men and women. For example, on scale 4, higher-scoring men were more likely to get into trouble with the law and use nonprescription drugs than were lower-scoring men on this scale. The relationship did not hold true for women. Also, higher-scoring women on scale 6 were more likely than lower-scoring women on the scale to be moody and lacking in emotional control. The relationship did not hold true for men. These gender differences suggest that additional research is needed to clarify the extent to which different interpretations should be considered for men and women.

Graham (1988) also reported behavioral correlate data for psychiatric patients. Scores on the MMPI-2 clinical scales (abstracted from Form AX) were correlated with ratings of symptoms that were completed by psychiatrists and psychologists who had observed and interviewed the patients. Table 8.8 lists the symptoms that were most highly correlated with the clinical scales of the MMPI-2. As with the data for the normal subjects, the data from this study suggest that there are reliable extratest correlates for MMPI-2 scores and that these extratest correlates are quite similar to those previously reported for the original MMPI. Again, as with normal subjects, different correlates were found for men and women for some of the scales. For example, male patients scoring higher on scale 7 were more likely to manifest hallucinatory behavior and unusual thought content than males scoring lower on this scale, but the relationship was not found for female patients.

Moreland and Walsh (1991) reported correlates of some frequently occurring two-point code types for an inpatient psychiatric sample. The correlates were quite similar to those previously reported for MMPI code types. Dwyer

Table 8.7. External Correlates for the MMPI-2 Clinical Scales for 822 Men and 822 Women in the MMPI-2 Normative Sample

Scale	Rating-Scale Item	Correlation Coefficients Men	Correlation Coefficients Women
1—Hs	Worries about health	27	23
	Lacks energy	22	18
	Headaches, stomach trouble	22	27
	Trouble sleeping	20	21
2—D	Lacks energy	29	24
	Lacks interest	23	23
	Self-confident	−22	−24
	Worries, frets	22	23
	Sad, blue	21	23
3—Hy	Trouble sleeping	18	20
	Worries about health	17	15
4—Pd	Angry, yells	09	25
	Cooperative	−11	−24
	Irritable, grouchy	10	23
	Trouble with the law	21	13
	Uses nonprescription drugs	16	13
5—Mf	NONE		
6—Pa	Sad, blue	11	21
	Moody	12	19
	Bad dreams	08	19
	Cries easily	11	19
	Lacks emotional control	03	15
7—Pt	Bad dreams	14	23
	Many fears	21	21
	Self-confident	−18	−21
	Indecisive	17	18
8—Sc	Many fears	19	18
	Bad dreams	17	16
9—Ma	Uses nonprescription drugs	22	14
	Wears unusual clothing	10	22
	Bossy	18	06
	Talks too much	15	21
0—Si	Shy	28	28
	Self-confident	−28	−28
	Avoids people	24	28
	Lacks interest	23	20

Note: Because of the large sample sizes, correlations greater than .06 are statistically significant. Only items with correlations of .15 or greater for at least one gender are reported.

Source: Graham, J.R. (1988, August). *Establishing validity of the revised form of the MMPI.* Symposium Presentation at the 96th Annual Convention of the American Psychological Association, Atlanta, Georgia.

Table 8.8. Symptom Descriptors for the MMPI-2 Clinical Scales for 232 Male and 191 Female Psychiatric Patients

Scale	Symptom Descriptor	Correlation Coefficients	
		Men	Women
1—Hs	Somatic concern	18	14
	Hallucinatory behavior	14	17
	Grandiosity	−13	−16
	Unusual thought content	13	02
2—D	Depressive mood	30	26
	Grandiosity	−25	−36
	Guilt feelings	24	25
	Hallucinatory behavior	23	27
3—Hy	Grandiosity	−19	−25
	Somatic concern	18	15
	Depressive mood	17	12
4—Pd	Guilt feelings	17	02
	Grandiosity	−16	−21
	Emotional withdrawal	03	20
	Depressive mood	11	21
5—Mf	Motor retardation	−01	19
6—Pa	Suspiciousness	02	20
	Emotional withdrawal	−02	19
	Unusual thought content	18	05
	Motor retardation	−15	−03
	Anxiety	−10	19
7—Pt	Hallucinatory behavior	22	09
	Grandiosity	−22	−27
	Guilt feelings	21	15
	Unusual thought content	21	00
	Depressive mood	18	20
8—Sc	Suspiciousness	04	25
	Unusual thought content	22	02
9—Ma	Depressive mood	−17	−05
	Hostility	09	16
	Conceptual disorganization	15	18
0—Si	Grandiosity	−23	−25
	Hallucinatory behavior	23	17
	Depressive mood	22	15
	Unusual thought content	20	14
	Guilt feelings	15	21

Note: With these sample sizes correlations greater than .11 are statistically significant. Only descriptors with correlations equal to or greater than .15 for at least one gender are listed.

Source: Graham, J.R. (1988, August). *Establishing validity of the revised form of the MMPI.* Symposium Presentation at the 96th Annual Convention of the American Psychological Association, Atlanta, Georgia.

et al. (1992) found that both the clinical and content scales of the MMPI-2 were related significantly to symptom ratings for a sample of psychiatric inpatients. The relationships between the MMPI-2 clinical scales and the symptom ratings were very consistent with those previously reported for the MMPI.

In summary, although research with the MMPI-2 is limited at this time, existing data suggest that the MMPI-2 has validity. First, scores and configurations of scores on the MMPI-2 and the original MMPI are quite congruent. Second, initial research has indicated that there are reliable extratest correlates for the MMPI-2 clinical scales and code types and that these correlates are consistent with previously reported correlates for the original MMPI. However, as Dahlstrom (1992) concluded, it is extremely important for additional research to be conducted to establish the external correlates of scores and patterns of scores on the MMPI-2. Several research projects of this kind are being conducted, and their results should add significantly to our understanding of the validity of the MMPI-2.

9

Use with Special Groups

The original MMPI was developed for use with adult psychiatric patients. Its norms were based on Caucasian adult subjects living in the cities and towns surrounding the University of Minnesota. Considerable caution was indicated in using the instrument with subjects whose demographic characteristics were different from those of the standardization sample or in other than traditional psychiatric settings. Although the MMPI-2 norms are more representative of the population of the United States than were those for the original MMPI, not much information is available yet concerning the use of the MMPI-2 with special groups. This chapter reviews information concerning the use of the original MMPI with some special groups. Because of the continuity between the MMPI and the MMPI-2, much of this information is likely to apply also to the MMPI-2. In addition, some preliminary information concerning the use of the MMPI-2 with special groups is summarized.

ADOLESCENTS

Although the original MMPI was intended for use with adults, it also became popular as an instrument for assessing adolescents (e.g., Hathaway & Monachesi, 1963). Many concerns and problems were expressed about such use of the test (e.g., Archer, 1987; Williams, 1986). In 1992, a version of the MMPI developed specifically for use with adolescents was published (Butcher et al., 1991). The MMPI-2 should not be used with test subjects younger than 18 years of age. Instead, the MMPI-A, which is described in some detail in Chapter 12 of this book, should be used.

OLDER ADULTS

The relationship between age and MMPI scores was noted quite early (e.g., Brozek, 1955). Although research results have not been totally consistent, most MMPI studies have found that older persons obtain somewhat higher scores on scales 1, 2, 3, and 0 and lower scores on scales 4 and 9 (e.g., Butcher et al., 1991; Gynther, 1979; Leon, Gillum, Gillum, & Gouze, 1979; Lezak, 1987). It has been suggested that the differences between scores of older and younger persons should not be interpreted as indicating that the older persons are more psychologically disturbed. Rather, they seem to reflect realistic increases in concern about health and decreases in activity and energy levels.

Taylor, Strassberg, and Turner (1989) studied groups of elderly community residents and elderly psychiatric patients. They found that MMPI T scores based on the standard norms were significantly higher than average for both groups. However, the scores significantly discriminated between the two groups. In addition, scores of the elderly psychiatric patients were related in expected ways to information obtained from a structured psychiatric interview. They concluded that the MMPI, as generally applied, is valid for use with older adults.

Swenson (1961) suggested that the use of the standard MMPI norms results in inaccurate inferences about older adults. Using data from thirty-one men and sixty-four women, Swenson constructed T-score norms and recommended their use with older adults. Colligan, Osborne, Swenson, and Offord (1983) reported age-specific data for larger groups of persons. The Colligan et al. data were later reanalyzed by Colligan and Offord (1992) to provide estimates of how much difference one should expect between the general norms and persons in specific age groups. Consistent with other studies, elderly subjects (60–99 years of age) scored slightly higher (1–4 T-score points) on scales 1, 2, and 0 and slightly lower (1–6 T-score points) on scales 4 and 9. It should be noted that these data were cross-sectional in nature, making it difficult to determine if scores change with increasing age or if the observed differences are due to cohort effects.

In deciding whether to use age-specific norms for the MMPI (or MMPI-2) one should consider the purposes for which the assessment is being conducted. If one wants to compare a specific older adult with other persons of similar age, age-specific norms probably are appropriate. However, the use of age-specific norms can mask important differences between older and younger persons. For example, it could be important to know that a specific older adult has greater somatic concern than the average adult (regardless of age). The use of norms that compare older adults only with other older adults would not reveal this somatic concern. We do not know from existing research findings if T scores of older adults based on age-specific norms result in more accurate inferences about those persons. We need research that compares the accuracy of inferences based on the general test norms with inferences based on age-specific norms.

Most studies of the relationship between MMPI scores and age have used cross-sectional designs. Large samples of persons are subdivided into age groups and their MMPI scores are compared. Such designs do not take into consideration that differences could be attributable to cohort effects. It may be that MMPI scores do not change as persons age. Rather, differences in age groups may be reflecting the different times at which individuals were growing up. However, a longitudinal study by Leon et al. (1979) found differences between older and younger persons similar to those resulting from cross-sectional studies. Koeppl, Bolla-Wilson, and Bleecker (1989) failed to replicate age-related differences reported in earlier studies and suggested that the earlier results could have been due to geographic, societal, and population factors.

Several conclusions can be reached about the relationship between age and scores on the original MMPI. Older persons have tended to obtain higher scores on scales 1, 2, 3, and 0 and lower scores on scales 4 and 9. It is not clear to what extent these differences can be attributed to the cross-sectional designs used in most studies and to what extent they indicate that MMPI scores change as individuals age. The differences in MMPI scores between older adults and adults in general probably do not reflect greater psychopathology in the elderly. Rather, they may indicate realistic changes in health concerns, energy levels, and other attitudes that often, but not always, occur as persons age.

The MMPI-2 normative sample included persons ranging in age from 18 to 84 years (mean = 41.71) (Butcher et al., 1989). However, older adults (70 years of age or older) were underrepresented. Although the MMPI-2 manual does not report scores separately by age groups, the committee that developed the MMPI-2 conducted appropriate analyses and concluded that age-specific norms were not needed.

Butcher et al. (1991) compared MMPI-2 scores of men in the normative sample with those of a group of healthy elderly men from the Normative Aging Study (NAS). Ages of men in the NAS ranged from 40 to 90 years (mean = 61.27). Scores of the two groups of men were very similar. Of the 567 MMPI-2 items, only 14 differed by more than 20 percent in endorsement percentages between the two groups. Interestingly, the item with the greatest difference in endorsement was "I have enjoyed using marijuana," supporting earlier speculation that differences between older and younger subjects may reflect cohort effects. The content of the remaining items that were endorsed differently by the groups suggests that the older men were indicating less stress and turmoil in their lives and greater satisfaction and contentment.

Within each of their samples, Butcher et al. (1991) compared scores of men of different ages. The differences between age groups in each sample were consistent with previously reported data for the original MMPI. However, the differences were small and probably not clinically important. Butcher et al. concluded that the differences may represent the single or combined effects of cohort factors and age-related changes in physical health status rather than age-related changes in psychopathology per se. They also concluded that special, age-related norms for the MMPI-2 are not needed for older men.

Considering both MMPI and MMPI-2 data, it seems appropriate to conclude that there are small differences between scores of older adults and adults in general. The differences tend to be small (less than five T-score points) and not clinically important. They probably reflect changes in concerns, attitudes, and behaviors that often, but not always, are associated with aging. The use of age-specific norms for older adults does not seem appropriate. However, clinicians should take examinee age into account in interpreting MMPI-2 scores. Future research should utilize longitudinal designs and determine the accuracy of scores based on standard and age-specific norms.

ETHNIC MINORITIES

Almost all of the psychiatric patients and nonpatients used in the development of the original MMPI were Caucasians. The MMPI norms were based on the responses of Caucasian adult normals in Minnesota. Therefore, there was understandable concern that the MMPI might not be appropriate for assessing persons from minority groups. Dahlstrom, Lachar, and Dahlstrom (1986) and Greene (1987) presented detailed reviews of the literature concerning the use of the original MMPI with minority subjects.

African Americans

The basic concern has been that, because African Americans were not included in the scale development or normative sample, the MMPI might be biased against such persons. Early studies reporting that African Americans obtained higher scores than Caucasians on some of the MMPI clinical scales reinforced concerns about test bias (e.g., Ball, 1960; Butcher, Ball, & Ray, 1964). A general finding was that African Americans tended to score higher (approximately five T-score points) than Caucasians on scales F, 8, and 9. Later studies found that differences between African Americans and Caucasians were small and not clinically meaningful or did not exist when groups were matched for age, education, and other demographic characteristics (Dahlstrom et al., 1986; Penk, Robinowitz, Roberts, Dolan, & Atkins, 1981).

Several other factors also were found to be related to MMPI differences between African Americans and Caucasians. When invalid profiles were excluded the from analyses, differences lessened dramatically (Costello, Tiffany, & Gier, 1972). Several studies also indicated that persons who were more identified with the African-American culture differed more from Caucasians on the MMPI (Costello, 1977; Harrison & Kass, 1968).

Finding significant test score differences between majority and minority group members does not necessarily mean that a test is biased. As Pritchard and Rosenblatt (1980) pointed out, for a test to be biased the accuracy of inferences or predictions based on test scores must be different for majority and minority group members. Several kinds of MMPI studies bear on this issue.

Some studies have compared MMPI scores of African Americans and Caucasians who had been assigned the same psychiatric diagnoses. For example, Cowan, Watkins, and Davis (1975) and Davis (1975) found that matched groups of African-American and Caucasian schizophrenics did not have significantly different MMPI scores. Similarly, Johnson and Brems (1990) compared the MMPI scores of twenty-two African-American and twenty-two Caucasian psychiatric inpatients who had been matched for age, gender, and psychiatric diagnoses. They found no statistically or clinically significant differences in scores between the African-American and Caucasian patients.

Other studies have compared extratest correlates of MMPI scales and code types for African Americans and Caucasians. The results of these studies have been mixed. Gynther et al. (1973) were able to demonstrate reliable correlates of very high F-scale scores for Caucasians but not for African Americans. However, Smith and Graham (1981) found no significant differences in F-scale correlates for African-American and Caucasian psychiatric patients. Elion and Megargee (1975) concluded that scale 4 is sensitive to antisocial behavior patterns for African Americans and Caucasians. Clark and Miller (1971) reported that the cardinal features of the 8–6 code type were similar for African Americans and Caucasians. Several other studies found differences in code type correlates for African Americans and Caucasians (Gynther et al., 1973; Strauss, Gynther, & Wallhermfechtel, 1974).

Another approach to studying potential test bias has been to determine if MMPI differences between African Americans and Caucasians are associated with important extratest differences. Butcher, Braswell, and Raney (1983) found that MMPI differences between African-American and Caucasian psychiatric inpatients were meaningfully associated with actual symptomatic differences. Dahlstrom et al. (1986) reported analyses suggesting that MMPI scores were not differentially related to psychopathology for African-American and Caucasian psychiatric patients. Dahlstrom et al. suggested that the higher MMPI scores of African Americans, especially young men, might reflect the various coping and defense mechanisms to which some minority group subjects may resort in their efforts to deal with the special circumstances they too often encounter in America today. These investigators made reasonable recommendations concerning the interpretation of the MMPIs of African-American subjects. They felt that the best procedure was to accept the pattern of MMPI scores that results from the use of the standard norms and, when profiles are markedly deviant, take special pains to explore in detail the subjects' life circumstances to understand as fully as possible the nature and degree of their problems and demands and the adequacy of their efforts to deal with them.

As was discussed earlier, the normative sample for the MMPI-2 includes African Americans in approximately the same proportion as in the 1980 census. This is certainly an improvement over the original MMPI and increases the likelihood that the MMPI-2 will be useful in assessing African Americans. However, inclusion of African Americans in the normative sample does not necessarily mean that the test is not biased. We must still determine if significant MMPI-2 differences exist between African Americans and Caucasians. If such differences are found, we must then find out if the differences are associated with relevant extratest characteristics. Further, it will be important to determine if African Americans and Caucasians with the same MMPI-2 scores or code types should be described in similar ways.

Appendix H of the MMPI-2 manual (Butcher et al., 1989) reports MMPI-2 summary data separately for the various ethnic groups included in the normative sample. African Americans scored slightly higher than Caucasians on most scales. However, T-score differences of more than five points occurred only for scale 4 for women (six T-score points). It is important to recognize that the African-American and Caucasian groups were not matched for age, socioeconomic status, or other demographic variables. As stated earlier, MMPI differences between African Americans and Caucasians were much smaller when such variables were taken into account. The MMPI-2 restandardization committee interpreted these summary data as indicating that separate norms were not needed for minority groups. No data were presented concerning whether these small differences between African Americans and Caucasians in the normative samples were associated with extratest differences.

Shondrick, Ben-Porath, and Stafford (1992) reported MMPI-2 data for 106 Caucasian and 37 African-American men who were undergoing forensic evaluations. Scores of the two groups on the validity and clinical scales were remarkably similar. The groups differed significantly only on scale 9, with African Americans scoring approximately seven T-score points higher than the Caucasians. Because no data were presented concerning extratest characteristics of the Caucasians and African Americans, we do not know if the scale 9 differences are indicative of important behavioral differences.

Very little is known about differences between African Americans and Caucasians on supplementary scales of the MMPI or MMPI-2. Greene (1991) summarized several MMPI studies and concluded that fewer differences between African Americans and Caucasians have been found for the supplementary than for the standard validity and clinical scales. Shondrick et al. (1992) found that African-American and Caucasian men undergoing forensic evaluations differed significantly on only two of the MMPI-2 content scales. African Americans scored significantly higher on the Cynicism and Antisocial Practices scales. Again, no data were presented concerning the extent to which the African Americans and Caucasians differed on relevant extratest characteristics.

As indicated in Chapter 7, several studies have suggested caution in using the MacAndrew Alcoholism scale (MAC) with African-American subjects (Graham & Mayo, 1985; Walters et al., 1984; Walters et al., 1983). These studies indicate that African-American alcoholics tend to obtain relatively high scores on the MAC, but classification rates for African Americans are not very good because nonalcoholic African-American psychiatric patients also tend to obtain rather high MAC scores. All of these studies used military personnel or veterans, so the extent to which the findings can be generalized to other African Americans is unclear. However, until additional information is available, caution should be exercised when the MAC-R scale is used with African Americans.

Although more research is needed concerning MMPI-2 differences between African Americans and Caucasians, some tentative conclusions may be reached. Because of the continuity between the MMPI and MMPI-2, it seems likely that the MMPI findings in this area will be replicated with the MMPI-2. Based on MMPI studies, it seems that differences between African Americans and Caucasians are small when groups are matched on variables such as age and socioeconomic status. When differences are found, they tend to be associated with relevant extratest characteristics and should not be attributed to test bias. Only small MMPI-2 differences have been reported between African Americans and Caucasians in the normative sample, and these differences could be accounted for by socioeconomic differences between the African Americans and Caucasians. Although research data are not yet available concerning the extent to which MMPI-2 scores should be interpreted similarly for African Americans and Caucasians, data from the original MMPI suggest that similar interpretations probably are appropriate.

Hispanics

Although Velasquez (1992) published a bibliography of eighty-six MMPI studies involving Hispanics, Greene (1987) identified only ten empirical studies comparing Hispanic and Caucasian subjects on the MMPI. Although there were some significant differences between Hispanic and Caucasian subjects in these studies, Greene concluded that there was no pattern to them. The data did not support Greene's earlier contention (Greene, 1980) that Hispanics frequently score higher on the L scale and lower on scale 5. It also appeared that there were fewer differences between Hispanics and Caucasians on the MMPI than between African Americans and Caucasians.

Campos (1989) conducted a meta-analysis of sixteen studies that compared Hispanics and Caucasians and concluded that the only consistent finding was that Hispanics scored higher (approximately four T-score points) than Caucasians on the L scale. Campos also concluded that the limited data available suggested that the MMPI predicts job performance of peace officers equally well for Hispanic and Caucasian groups.

Several studies have reported that there are marked similarities between MMPI characteristics of Hispanic and non-Hispanic male sex offenders and Hispanic and non-Hispanic schizophrenics (Velasquez and Callahan, 1990b; Velasquez, Callahan, & Carrillo, 1989). Velasquez and Callahan (1990a) found that Hispanic male veterans seeking treatment for alcoholism scored significantly lower on scales 4, 5, and 0 than did Caucasian veterans seeking treatment for alcoholism. It should be noted that none of these studies of clinical groups controlled for variables such as age, socioeconomic status, or level of acculturation.

Little information is available concerning Hispanic-Caucasian differences on the supplementary scales of the MMPI. McCreary and Padilla (1977) found no meaningful differences between Hispanic and Caucasian prisoners on most supplementary scales. However, Hispanic prisoners scored higher than Caucasian prisoners on the Overcontrolled-Hostility scale. Page and Bozlee (1982) found no differences between Hispanic and Caucasian substance abusers on the MacAndrew Alcoholism scale. Dolan, Roberts, Penk, Robinowitz, and Atkins (1983) found no meaningful differences between Hispanic and Caucasian heroin addicts on the Wiggins content scales. Although Montgomery, Arnold, and Oroco (1990) found that Mexican-American college students scored slightly higher than Caucasian college students on most MMPI scales, these differences diminished when scale validity, age, and level of acculturation were controlled.

The data concerning MMPI differences between Hispanics and Caucasians are difficult to interpret for several reasons. First, there have not been many studies concerning MMPI characteristics of Hispanics. Second, there is considerable heterogeneity among persons designated as Hispanic. Typically studies have not indicated whether Hispanic subjects were Mexican-American, Cuban, or Puerto Rican. Finally, most studies have not controlled for differences between Hispanic and Caucasian groups on variables such as socioeconomic status or level of acculturation. However, it appears that MMPI differences between Hispanic and Caucasian groups have been small and, in most cases, not clinically significant. When variables such as socioeconomic status have been controlled, differences have been even smaller.

To date the only MMPI-2 data available concerning Hispanics are reported in Appendix H of the MMPI-2 manual (Butcher et al., 1989). Summary data are reported separately for Caucasian and Hispanic subjects in the normative sample. Careful examination of these data reveals that Hispanic men in the normative sample scored slightly higher than Caucasian men on most scales. However, none of the differences between Hispanic and Caucasian men was greater than five T-score points. Hispanic women in the normative sample scored higher than Caucasian women on all scales except L, K, and 0. On scales L and K the Hispanic women scored lower than the Caucasians, and on scales 5 and 0 there were no differences between the two groups. For scales F, 1, 4, 7, 8, and 9, the differences were greater than five T-score points.

It appears that the differences between Hispanic and Caucasian men in the normative sample are small and not clinically meaningful. The larger differences between Hispanic and Caucasian women in the normative sample are of more concern. However, it should be noted that the ethnic groups in the normative sample were not matched for variables such as age or education. Additional analyses controlling for these variables are indicated.

Adequate data do not exist to permit conclusions concerning how Hispanic-Caucasian differences should be interpreted. We need to know if, when such differences occur, they are associated with important extratest differences between the groups. For example, are Hispanic women who score higher than Caucasian women on scale 4 more likely to have the antisocial characteristics typically associated with higher scores on this scale? Based on the data that have been published concerning the meaning of MMPI differences between African Americans and Caucasians, we can speculate that Hispanic-Caucasian differences will be associated with important extratest differences and not be attributable to test bias.

Native Americans

Greene (1987) identified only seven studies that compared MMPI scores of Native Americans and Caucasians. Although the Native Americans tended to score higher than Caucasians on certain clinical scales, there was no clear pattern to these differences across the studies. In an early study, Arthur (1944) compared MMPI scores of a small group of adolescent and young-adult Native Americans with scores of Caucasian college students. She concluded that there were more similarities than differences between the groups and that the MMPI scores realistically assessed the circumstances of the Native-American and Caucasian subjects. Herreid and Herreid (1966) studied groups of native and non-native Alaskan college students. They found that the natives obtained somewhat higher MMPI scores than the non-natives, although only one difference (Mf for women) was as large as five T-score points. Data were not available to examine relationships between differences in MMPI scores and differences in extratest characteristics.

Several studies have compared MMPI scores of Native-American alcoholics with scores of other groups of alcoholics. Kline, Rozynko, Flint, and Roberts (1973) examined a small group of male Native-American alcoholics. They reported that, compared with groups of Caucasian alcoholics reported previously in the literature, the Native Americans had more deviant MMPI scores. However, there was no attempt to determine if these more deviant scores were associated with more deviant extratest characteristics. Uecker, Boutiller, and Richardson (1980) compared Native-American and Caucasian alcoholics and concluded that the groups were very similar in terms of MMPI scores. However, the Caucasians scored significantly higher than the Native Americans on scales 4 and 5. Page and Bozlee (1982) compared very small groups of Native-American and Caucasian male alcoholics and concluded that the groups were more similar than different. It is interesting to note that no significant differences between Native-American and Caucasian alcoholics were reported for the MacAndrew Alcoholism scale in the two studies that considered that scale (Page & Bozlee, 1982; Uecker et al., 1980).

Only two studies involving Native-American psychiatric patients have been reported. Pollack and Shore (1980) compared several different diagnostic groups of patients from Pacific-Northwest tribes and concluded that the

MMPI scores were very similar across diagnostic groups. They saw their data as demonstrating a significant cultural influence on the MMPI results of these groups of Native Americans. Butcher et al. (1983) included a small group of Native Americans in their study of psychiatric inpatients. They found that the MMPI scores of the Native-American patients were less deviant than those of Caucasian or African-American patients.

Taken together, the results of these MMPI studies of Native Americans suggest that there are few important differences between Native-American and Caucasian subjects on the MMPI. What has been lacking in all of the studies described above has been the inclusion of extratest data that could be used to determine if MMPI differences are associated with other important characteristics of test subjects.

There are thirty-eight Native-American men and thirty-nine Native-American women in the MMPI-2 normative sample. Appendix H of the MMPI-2 manual (Butcher et al., 1989) summarizes scores for these two groups of subjects. Native Americans scored higher than Caucasians on most scales. The Native-American men scored more than five T-score points higher than the Caucasian men on scales F and 4. The Native-American women scored more than five T-score points higher than the Caucasian women on scales F, 1, 4, 5, 7, and 8. Although the Native-American groups are small and not necessarily similar to the Caucasian groups in terms of age, socioeconomic status, or other demographic characteristics, the differences between Native-American and Caucasian groups are potentially important. However, there have not yet been analyses reported that would indicate if the MMPI-2 differences are associated with important extratest characteristics. Clearly, analyses of this kind are needed before we will know how to interpret the MMPI-2 differences between the groups.

Asian Americans

Greene (1987) reported only three studies comparing MMPI performance of Asian-American and Caucasian subjects. Sue and Sue (1974) found that male Asian-American (Chinese, Japanese, and Korean) college student counselees obtained higher scores than Caucasian counselees on most MMPI scales. Because mean scores were not reported, it is not possible to determine how large the differences were. It also was noted that the Asian-American and Caucasian groups had similar patterns of scores on the MMPI. The authors concluded that differences in MMPI scores reflected the cultural backgrounds and conflicts of Asian Americans. Because extratest measures were not available, it was not possible to determine if differences in MMPI scores were associated with differences in psychopathology.

Marsella, Sanborn, Kameoka, Shizuru, and Brennan (1975) reported that Asian-American (Chinese, Japanese) college students residing in Hawaii had higher scores on scale 2 of the MMPI than Caucasian college students. Interestingly, the Asian-American students also had higher scores on the Beck Depression Inventory, suggesting that the higher scores on scale 2 were reflecting differences in symptoms of depression.

Tsushima and Onorato (1982) compared MMPI scores of Asian-American (Chinese and Japanese) and Caucasian patients diagnosed as having somatization disorders or organic brain syndromes. They concluded that there were no important race-related differences in MMPI scores if diagnostic classifications were held constant.

Because very few Asian Americans were included in the MMPI-2 normative sample (6 men, 13 women), it is not appropriate to reach conclusions about differences between Asian Americans and Caucasians on the MMPI-2. However, the data in Appendix H of the MMPI-2 manual suggest that there were few differences between the Asian Americans and Caucasians included in the normative sample and that the Asian Americans did not score consistently higher than the Caucasians.

Except for the data in Appendix H of the MMPI-2 manual (Butcher et al., 1989), no data have been published concerning MMPI-2 differences between Caucasians, African Americans, Hispanics, Native Americans, and Asian Americans. Additional research is needed concerning MMPI-2 differences between these groups. It will be especially important to determine if MMPI-2 differences between these groups are associated with differences in important extratest characteristics of subjects. Based on studies involving the original MMPI, it seems likely that such an association will be found.

MEDICAL PATIENTS

The original MMPI was used frequently by psychologists in medical settings (Piotrowski & Lubin, 1990). Although no data have been published yet concerning how frequently the MMPI-2 is used in medical settings, it seems likely that it will be used even more frequently than the original instrument in such settings. Until more data are available concerning the use of the MMPI-2 in medical settings, it will be necessary to rely primarily on information concerning the use of the original MMPI in these settings.

Osborne (1979) and Henrichs (1981) presented overviews of the use of the MMPI in medical settings. Swenson, Pearson, and Osborne (1973) reported item, scale, and pattern data for 50,000 medical patients at the Mayo Clinic. Swenson, Rome, Pearson, and Brannick (1965) reported data indicating that most medical patients (89%) readily agreed to take the MMPI and completed and returned the test booklet. Swenson et al. (1965) also surveyed 158 physicians who had used the MMPI routinely for at least four months. For all fourteen items on the questionnaire, 70 to 85 percent of the physicians indicated that the MMPI was useful with their patients. It will not be possible in this chapter to review even briefly the voluminous research literature concerning the relationship between MMPI data and characteristics of medical patients. Rather, an attempt will be made to indicate the general purposes for which the MMPI-2 can be used in medical settings.

Screening for Psychopathology

One important use of the MMPI with medical patients has been to screen for serious psychopathology that might not have been reported or that has been minimized by patients. The indicators of serious psychopathology discussed elsewhere in this book (e.g., F-scale level and overall profile elevation) should be considered when examining the profiles of medical patients.

Many clinicians assume that medical problems are going to be very emotionally distressing to patients and that this distress will be reflected in highly deviant scores on the MMPI-2. It is important to develop some expectations concerning typical MMPI-2 scores and profiles produced by medical patients. Swenson et al. (1973) reported summary MMPI data for approximately 25,000 male and 25,000 female patients at the Mayo Clinic. The mean profiles for both male and female patients fell within normal limits. The validity scales suggested a slightly defensive test-taking attitude. T scores on scales 1, 2, and 3 were near 60. Apparently, the medical problems of these patients were not psychologically distressing enough to lead to grossly elevated scores on the MMPI scales. Based on what we know about the continuity between the MMPI and the MMPI-2, we expect that MMPI-2 profiles of medical patients will be within normal limits with T scores on scales 1, 2, and 3 between 55 and 60 and scores on the rest of the clinical scales near 50.

Screening for Substance Abuse Problems

The MMPI-2 can be useful in alerting clinicians to the possibility of substance abuse problems among medical patients. Some patients develop physical symptoms because of chronic substance abuse; some patients develop substance abuse problems because of their medical problems; and some patients have substance abuse problems that are not directly related to their physical symptoms. Regardless of the reasons for the relationship between substance abuse problems and medical problems, early awareness of these problems facilitates treatment planning.

Research with the original MMPI did not reveal a single pattern of MMPI scales associated with substance abuse problems (e.g., Graham & Strenger, 1988). However, there is convincing evidence that scale 4 is likely to be elevated among groups of persons who abuse substances. Scale 4 typically is not significantly elevated among groups of medical patients. Thus, when a person is presenting primarily physical symptoms in a medical setting and has significant scale 4 elevation, the possibility of abuse of substances should be considered.

Fordyce (1979) suggested that chronic-pain patients can easily become addicted to narcotics, barbiturates, or muscle relaxants. He reported that such persons often obtain elevations on scales 2 and 9. When both of these scales are elevated above T = 65, the possibility of addiction to prescription medications should be considered.

The 24/42 two-point code type often is found among male alcoholics in treatment, and this same code type and the 46/64 code type often are found for female alcoholics in treatment. Neither of these code types is common among medical patients who do not abuse alcohol. Thus, when these code types are encountered in medical patients (particularly if the scores are greater than T = 65), the possibility of alcohol abuse should be explored carefully.

The MacAndrew Alcoholism Scale—Revised (MAC-R) was described in Chapter 7 of this book. Although the original MAC was developed by comparing the item responses of male alcoholic outpatients and male psychiatric outpatients, subsequent research indicated that the scale was useful for identifying substance abuse problems of various kinds for men and women in a variety of settings (Graham & Strenger, 1988). If significant elevation is found on the MAC-R for patients in a medical setting, careful consideration should be given to the possibility that these persons are abusing substances. As discussed in Chapter 7, MAC-R raw scores of 28 or above suggest substance abuse. Scores between 24 and 27 are somewhat suggestive of such abuse. Scores below 24 suggest that substance abuse problems are not very likely. It should be emphasized that research concerning the MAC and the MAC-R has emphasized abuse of alcohol and other nonprescribed drugs. Little is known about the extent to which the MAC-R is sensitive to the abuse of prescribed drugs, such as those used by chronic-pain patients.

Two other scales, which were discussed in Chapter 7, can provide information about possible substance abuse problems in medical patients. Patients who obtain T scores greater than 60 on the Addiction Acknowledgment Scale (AAS) (Weed et al., 1992) are openly acknowledging substance abuse problems, and additional assessment in this area is indicated. T scores greater than 60 on the Addiction Potential Scale (APS) (Weed et al., 1992) also should alert clinicians that additional information concerning possible substance abuse should be obtained.

Organic versus Functional Etiology

The original MMPI often was used to try to determine if the somatic symptoms presented by patients were organic or functional in origin. It is never appropriate to use the MMPI-2 alone to diagnose an organic condition or to rule out such a condition. The most that the MMPI-2 can do is to give some information concerning the underlying personality characteristics of the patient. This information can be used, along with other available information, to make inferences concerning the compatibility of the personality characteristics and a functional explanation of symptoms.

Osborne (1979) and Keller and Butcher (1991) summarized studies with the original MMPI that tried to determine if patients' symptoms were functional or organic. The symptoms studied have included low back pain, sexual impotence, neurologic complaints, and others. In general, the research indicated that patients with symptoms that were exclusively or primarily psychological in origin tended to score higher on scales 1, 2, and 3 than patients with similar symptoms that were organic in origin. Particularly common among groups of patients with symptoms of psychological origin was the 13/31 two-point code type. When this code type was found, and scales 1 and 3 were elevated above T = 65 and were considerably higher than scale 2, the likelihood of functional origin increased. However, Osborne and Keller and Butcher pointed out that the differences between the functional and organic groups have not been large enough to permit predictions in individual cases.

Several supplementary scales have been developed specifically to try to determine if physical symptoms are of functional or organic origin. Two of the most commonly used of these scales are the Low Back Pain (Lb) scale (Hanvik, 1949, 1951) and the Caudality (Ca) scale (Williams, 1952). Because of inadequate research data to support the usefulness of these scales, they were not included in the MMPI-2.

Establishing Homogeneous Subtypes

A number of investigators have used the MMPI to establish homogeneous subtypes within a particular medical disorder. Most of this research has been in relation to chronic pain. The goals underlying the subtyping approach have been to identify etiological factors that are unique to particular subtypes and to determine specific treatment interventions that are appropriate for the subtypes.

Sometimes the subtyping has been based on MMPI code types or other configural aspects of test performance. Keller and Butcher (1991) reviewed this literature and concluded that MMPI researchers consistently had identified two subtypes. One subtype was characterized by the "conversion-V" pattern (scales 1 and 3 as highest scales and both considerably higher than scale 2, which is third-highest in the profile). The second subtype was characterized by elevated scores on scales 1, 2, and 3, but without significant differences among these three scales. There were some suggestions that the 1–3–2 pattern was associated with poorer treatment outcome. However, Keller and Butcher concluded that using this approach to predict treatment outcome was not very effective.

Other investigators have used more sophisticated procedures (e.g., cluster analysis) to identify homogenous subgroups among pain patients (e.g., Costello, Hulsey, Schoenfeld, & Ramamurthy, 1987; Costello, Schoenfeld, Ramamurthy, & Hobbs-Hardee, 1989). Four cluster types have emerged consistently (the conversion-V pattern, the neurotic triad pattern, the normal-limits profile, and the general elevation profile). Other, less stable cluster types have been identified by some investigators.

Regardless of how subtypes are determined, the clinical utility of this approach depends on demonstrating empirically that there are important extratest differences associated with the cluster types. As Keller and Butcher (1991) pointed out, most studies have generated interpretive schemata based on the supposed significance of various scale configurations, but data to support these schemata are very limited. Clearly, if the subtyping approach is to be useful with the MMPI-2, future research must concentrate on establishing empirically that persons in the various subtypes differ in important ways (e.g., in their responsiveness to treatment interventions of various kinds).

Psychological Effects of Medical Conditions

The MMPI-2 can be used to understand how persons with medical problems are affected psychologically by them. When used for this purpose, the clinician should consider indicators of emotional disturbance that have been discussed previously in this book. For example, high scores on scale 2 or on the Depression (DEP) content scale suggest that medical patients are experiencing dysphoria/depression, and elevated scores on scale 7, on Welsh's Anxiety scale, or on the Anxiety (ANX) content scale indicate that medical patients are anxious, worried, and tense. Research with the MMPI has indicated that patients with disorders such as head injury (e.g., Diamond, Barth, & Zillmer, 1988; Nockleby & Deaton, 1987), stroke (e.g., Gass & Lawhorn, 1991), and cancer (Chang, Nesbit, Youngren, & Robison, 1988) have MMPI scores suggestive of significant psychological distress and maladjustment. Gass and Lawhorn also reported that some of the deviance in MMPI scores of stroke patients had to do with bona fide stroke symptoms. When a correction factor was applied, there was a significant reduction in MMPI scores of the stroke patients. The Gass and Lawhorn finding suggests that clinicians should consider the possibility that the elevation in MMPI-2 scores of medical patients can be attributed, at least in part, to symptoms associated with their disorders and are not due entirely to psychological maladjustment.

Empirically derived correlates of various MMPI code types for patients of an internist were given by Guthrie (1949). Henrichs (1981) provided a useful summary table describing characteristics associated with MMPI profile patterns of medical patients. Although these correlates have not yet been replicated for the MMPI-2, the continuity between the MMPI and the MMPI-2 leads us to expect that they will apply.

Response to Medical Treatment

The MMPI-2 also can provide important information concerning how medical patients are likely to respond psychologically to medical interventions. Several examples can be given to illustrate this potential use. Henrichs and Waters (1972) investigated the extent to which MMPIs administered preoperatively to cardiac patients could predict emotional or behavioral reactions to surgery. Based on prior literature concerning cardiac patients, five types were conceptualized, and rules for classifying MMPI profiles into these types were developed. The rules were able to classify 97 percent of MMPIs given to patients preoperatively. Postoperative course was recorded for all patients to determine if they had behavioral or emotional problems. The occurrence of such postoperative problems was significantly different for the different MMPI types. Only 6 percent of patients who were in the well-adjusted MMPI type preoperatively had postoperative emotional or behavioral problems. By contrast, 44 percent of persons having seriously disturbed preoperative MMPIs had postoperative behavioral or emotional problems. Henrichs and Waters pointed out that different types of preoperative interventions could be developed to address anticipated postoperative problems.

Sobel and Worden (1979) demonstrated that the MMPI was useful in predicting psychosocial adjustment of patients who had been diagnosed as having cancer. The MMPI typically was administered following the diagnosis of cancer, and patients were studied for 6 months following this testing. Patients were classified as having high distress or low distress at follow-up on the basis of multiple measures of emotional turmoil, physical symptoms, and effectiveness of coping with the demands of daily living. Using discriminant analyses of MMPI scores, 75 percent of patients were correctly classified as having high distress or low distress. The authors pointed out that interventions could be developed to assist cancer patients who are considered to be at high risk for psychosocial problems.

A general finding in the literature has been that persons who are well adjusted emotionally before they develop serious medical problems and/or before they are treated for such problems seem to handle the illness-related stress better than persons who are emotionally less well adjusted. Additionally, better-adjusted persons seem to have better postoperative courses than do less well adjusted persons. However, not all studies have found positive relationships between MMPI results and response to medical interventions (e.g., Guck, Meilman, & Skultety, 1987). Although the MMPI-2 seems to have a great deal of potential in predicting response to medical interventions, there is a need for additional research to establish more clearly the extent to which the MMPI-2 can and should be used for this purpose. It should be noted that the Negative Treatment Indicators (TRT) content scale, which was discussed in Chapter 6, may be useful in this regard.

Psychological Effects of Medical Treatment

Several investigators have used the MMPI to examine psychological status following medical interventions. For example, Clark and Klonoff (1988) identified five clusters of MMPI profiles of patients awaiting coronary bypass surgery. The different clusters were associated with different degrees of anxiety and depression before surgery, and these differences were also present at the same levels after surgery. Kirkcaldy and Kobylinska (1988) compared breast cancer patients after they had undergone treatment (mastectomy or chemotherapy). They found that both groups of patients had more deviant MMPI scores than a control group of healthy nurses. Unfortunately, no pre-treatment MMPI data were reported, so it is not possible to determine if the distress suggested by the MMPI scores of the patients resulted from the treatment or was present prior to the interventions.

Predisposing Factors

Numerous research projects have sought to identify factors that place persons at risk for serious medical problems such as coronary heart disease or cancer. The MMPI has been utilized in some of these projects. In a number of early studies the MMPI was administered to persons who already had a serious medical disorder, and the results were compared with those of healthy persons. These studies were of little value because it was not possible to determine if the psychological characteristics assessed by the MMPI had predisposed the persons to the disorders or had resulted from the disorders.

More sophisticated prospective studies have also been reported. In these studies large groups of persons have been assessed using a variety of procedures, including the MMPI, and followed for many years to determine which of them developed serious medical disorders.

Although the MMPI clinical scales were analyzed in some of these prospective studies (e.g., Persky, Kempthorne-Rawson, and Shekelle, 1987), most have focused on the Hostility (Ho) scale (Cook & Medley, 1954) of the MMPI. This scale originally was developed to predict performance of classroom teachers. Subsequently, it became obvious that the scale assessed lack of confidence in other people and hostility. Although the Ho scale is not among those described in the MMPI-2 manual, it is possible to score the scale because all of its fifty items are included in the MMPI-2.

Several studies have reported that persons with lower Ho-scale scores are less likely than persons with higher scores to develop coronary heart disease (CHD). Williams et al. (1980) studied men and women who had diagnoses of CHD. Angiography indicated that those with higher Ho-scale scores were more likely to have significant occlusion of at least one artery. Although these results are very interesting, they are of limited utility because the study was cross-sectional in nature.

Barefoot, Dahlstrom, and Williams (1983) conducted a 25-year follow-up study of 255 physicians who had completed the MMPI while in medical school. They found that higher Ho-scale scorers were more likely to have been diagnosed as having CHD and that mortality from all causes was greater for persons with higher Ho-scale scores. Shekelle, Gale, Ostfeld, and Paul (1983) evaluated 1,877 male employees of the Western Electric Company 10 and 20 years after an initial assessment that included the MMPI. They found that, even after other risk factors such as blood pressure and cholesterol levels were factored out, persons with higher Ho-scale scores were more likely to receive CHD diagnoses during the follow-up period.

Colligan and Offord (1988) cautioned that higher Ho-scale scores are not necessarily associated with a greater risk for CHD. They studied groups of contemporary normal subjects, general medical patients, persons undergoing treatment for alcoholism, and persons hospitalized for psychiatric treatment and found that many persons in these groups had Ho-scale scores that would suggest that they were at risk for CHD. Of course, this was not a prospective study, so we do not know how many of the persons with higher Ho-scale scores would, in fact, develop CHD in the future.

Several prospective studies have failed to replicate the earlier relationship between Ho-scale scores and CHD. Hearn, Murray, and Luepker (1989) conducted a 33-year follow-up study of men who had taken the MMPI when they entered the University of Minnesota. Higher Ho-scale scores did not predict CHD mortality, CHD morbidity, or total mortality either before or after adjustment for baseline risk factors. Leon, Finn, Murray, and Bailey (1988) studied 280 men who had been healthy when they completed the MMPI in 1947. In a 30-year follow-up there was not a significant relationship between Ho-scale scores and CHD. Similarly, McCranie, Watkins, Brandsma, and Sisson (1986) completed a follow-up study of persons who had taken the MMPI when in medical school. Ho-scale scores were not significant predictors of CHD incidence or total mortality in this sample.

In summary, there are data suggesting that the MMPI, especially the Ho scale, may be helpful in identifying psychological risk factors in coronary heart disease. However, research results have not been completely consistent. Studies have differed in terms of subject age, geographic location, follow-up period, and methods used to assess CHD. Additional research is needed to clarify the extent to which the Ho scale or other scales of the MMPI-2 can be used clinically to assess psychological risk for CHD and other serious medical disorders. Future research in this area should take into account other MMPI-2 scales, such as the Anger, Cynicism, and Type A content scales.

MMPI-2 and Medical Disorders

To date only two studies have been published concerning the relationship between MMPI-2 performance and medical disorders. Keller and Butcher (1991) studied 590 chronic-pain patients. The mean profile of these patients was characterized by the conversion-V pattern that had been reported previously for the MMPI. As one might expect, the pain patients also obtained high scores on the Health Concerns (HEA) content scale. Three MMPI-2 clusters, similar to those previously reported for the MMPIs of pain patients, were identified for male and female patients. They differed primarily in terms of elevation rather than configuration. There were few differences between clusters in patient characteristics as determined from clinical records. Keller and Butcher concluded that the MMPI-2 is useful with chronic-pain patients primarily as a way to assess general level of distress and disability. Scale 2 was especially useful in this regard.

Gass (1992) studied a group of 110 patients with cerebrovascular disease (CVD). Although mean MMPI-2 scores were not reported, Gass concluded that scores on scales 1, 2, 3, and 8 were inflated because of endorsement of items having to do with bona fide symptoms of CVD. He developed a correction factor and recommended that it be applied to the MMPI-2 scores of such CVD patients.

CORRECTIONAL SUBJECTS

Dahlstrom et al. (1975) reviewed the major ways in which the MMPI has been used with correctional subjects and presented lists of references concerning such use. Dahlstrom et al. indicated that the MMPI profiles of prisoners have seemed to be remarkably homogeneous. Numerous studies have documented that scale 4 usually is the most elevated scale in the mean profiles of prisoners. The 4–2 and 4–9 code types have been identified as the most frequently occurring ones for prisoners.

A major use of the MMPI in correctional settings is to classify prisoners. Accurate classification permits correctional administrators to make more efficient use of limited resources and to avoid providing resources for offenders who do not require them. Early work by Panton and his associates in North Carolina (e.g., Panton, 1958) and by Fox and his associates in California (reported by Dahlstrom et al., 1975) indicated moderate success in using the MMPI to classify prisoners in meaningful ways when they entered the correctional system.

A comprehensive and useful system of classifying criminal offenders based on the MMPI was developed by Megargee and his associates (Megargee, Bohn, Meyer, & Sink, 1979). These investigators used hierarchical profile analysis to identify clusters among the MMPIs of offenders. Classification rules were developed for placing offenders into the groups defined by the cluster analyses. Following some use of the system with other samples of offenders, the original classification rules were refined and expanded.

The resulting classification system involved ten types of offender MMPIs and explicit rules for classifying offenders into the type to which they were most similar. Megargee (1979) reported that in a variety of correctional settings, mechanical application of the rules classified about two thirds of offender profiles. Most of the remaining one third of the profiles could be classified by clinicians using published guidelines and additional data. Overall, 85 to 95 percent of offender profiles could be classified using the ten profile types. Although the derivational work for the Megargee system took place in a federal facility and involved primarily youthful male offenders, subsequent data indicated that the classification system also worked well in other settings (e.g., state prisons; Nichols, 1980) and with other kinds of offenders (e.g., women; Schaffer, Pettigrew, Blouin, & Edwards, 1983).

If a classification system for offenders is to be useful, the groups identified using the system must differ from each other on important extratest variables. Many studies have been conducted concerning extratest characteristics associated with the various Megargee types.[1] Most studies have found significant differences between the Megargee types on demographic characteristics, criminal behavior patterns, rated personality characteristics, measures of institutional adjustment (including frequency of disciplinary infractions, incidence of reports to sick call, interpersonal relations, and work performance), and recidivism rates. (See Zager, 1988, for a review of these and other studies involving the Megargee system.) Table 9.1 summarizes MMPI and extratest characteristics for the ten Megargee types. Readers who want more detailed information about the development and use of the Megargee classification system should consult Megargee's 1979 book.

With the revision of the MMPI it became necessary to determine to what extent Megargee's MMPI-based classification rules are still appropriate for the MMPI-2. Megargee has conducted impressive research relevant to this issue (Megargee, personal communication, August 11, 1992). Although the results of this research have not yet been published, the results will be summarized briefly here. More details are forthcoming in journal articles and possibly a monograph.[2]

A preliminary study determined that only 51.6 percent of youthful male offenders would be classified into the same categories when the original classification rules were applied to estimated MMPI-2 scores. In a second analysis, 1,164 youthful male offenders who had been classified earlier on the basis of MMPI scores were reclassified based on estimated MMPI-2 scores. Of the 1,075 subjects who could be classified on both measures, 59.9 percent received identical classifications. In a third analysis, MMPI-2 responses of 422 federal prisoners were rescored to yield estimated MMPI scores. The classification rules were applied to both sets of scores. Of those subjects who could be classified for both sets of scores, 65.2 percent received the same classification.

Megargee then devised a new set of rules that would enable the classification of MMPI-2 profiles that would approximate classification that would have resulted if the MMPI had been used. A final set of revised rules and procedures was applied to the scores of the 422 prisoners. Of the 380 that could be

[1]A comprehensive list of references concerning Megargee's typology can by obtained from Edwin I. Megargee, Ph.D., Department of Psychology, Florida State University, Tallahassee, FL 32306.

[2]Research concerning the use of the Megargee classification system with the MMPI-2 was supported by a grant to Dr. Edwin I. Megargee from the National Institute of Justice (No. 89-IJ-CX-0026). Thanks are extended to Dr. Megargee for sharing the findings of his research prior to their publication.

Table 9.1. Summary of Megargee types

Type	MMPI Characteristics	Behavioral Characteristics
Item	Scales generally unelevated.	Stable, well adjusted, with minimal problems and lacking conflicts with authorities.
Easy	Moderate elevations; scales. 4 and 3 often elevated.	Bright, stable; good adjustment, personal resources, and interpersonal relationships; underachievers.
Baker	Moderate elevations; scales 4 and 2 often elevated.	Inadequate, anxious, constricted, dogmatic; tendency to abuse alcohol.
Able	Moderate elevations; scales 4 and 9 typically elevated.	Charming, impulsive, manipulative; achievement-oriented; adjust well to incarceration.
George	Moderate elevations; scales 1, 2, and 3 elevated.	Hardworking, submissive, anxious; have learned criminal values; often take advantage of educational and vocational programs.
Delta	Moderate to high elevation on scale 4; other scales lower.	Amoral, hedonistic, egocentric, manipulative, and bright; impulsive sensation seekers; poor relations with peers and authorities.
Jupiter	Moderate to high elevations typically on scales 8, 9, and 7.	Often overcome deprived backgrounds to do better than expected in prison and upon release.
Foxtrot	High elevations; scales 8, 9, and 4 often highest.	Tough, streetwise, cynical, antisocial; deficits in most areas; extensive criminal histories; poor prison adjustment.
Charlie	High elevations; peaks typically on scales 8, 6, and 4.	Hostile, misanthropic, alienated, aggressive, antisocial; extensive histories of poor adjustment, criminal convictions, and mixed substance abuse.
How	Many very high scores.	Unstable, agitated, disturbed mental health cases; extensive needs; function ineffectively in major areas.

Source: Zager, L.D. (1988). The MMPI-based criminal classification system: A review, current status, and future directions. *Criminal Justice and Behavior, 15,* 39–57. Copyright © 1988 American Association for Correctional Psychology. Reprinted by permission of Sage Publications, Inc.

classified on the basis of both MMPI and MMPI-2 scores, identical classifications were obtained for 80 percent.

Megargee has recommended that the new rules be used instead of the original rules when the MMPI-2s of male offenders are to be classified. He is conducting additional research to determine if the new rules should also be used for classifying female prisoners using MMPI-2 scores. Megargee emphasizes that future research should focus on establishing whether the empirically-determined correlates of the ten types observed using the original MMPI equally characterize types classified according to the new rules applied to the MMPI-2.

SUBJECTS IN NONCLINICAL SETTINGS

The MMPI was developed in a psychiatric hospital setting, and most of the research done with the instrument has been with subjects in clinical settings. However, the use of the MMPI with nonclinical subjects increased dramatically in the years before its revision (Graham & McCord, 1985). The MMPI has been used in two basic ways in relation to selection of employees or students (Butcher, 1979, 1985). First, the test has been used to screen for psychopathology among applicants. Second, the test has been used to try to predict quality of job performance by matching individuals with certain personal characteristics to jobs or positions that are believed to require such characteristics.

Using the MMPI-2 to screen for psychopathology among applicants is most justified when individuals are being considered for employment in occupations involving susceptibility to occupational stress, personal risk, and personal responsibility. Such sensitive occupations include those of air traffic controller, airline pilot, police officer, fire fighter, and nuclear power plant operator. Routine use of the MMPI-2 for personnel selection is not recommended. For many jobs the primary requirements are appropriate training and ability, and personality factors may be unimportant or irrelevant.

Research data suggest that the MMPI could be used effectively to screen for psychopathology in normal groups. Lachar (1974c) demonstrated that the MMPI could predict serious psychopathology leading to dropouts among Air Force cadets. Strupp and Bloxom (1975) found that men with certain MMPI code types were likely to have difficulty with personal adjustment, graduating from college, finding a job, and deciding on a career. Richard, Wakefield, and Lewak (1990) reported that congruence of responses to MMPI items was a reliable predictor of marital satisfaction.

There also are data suggesting that the MMPI could be used to predict effective hotline workers (Evans, 1977), competent clergy (Jansen & Garvey, 1973), and successful businesspeople (Harrell & Harrell, 1973). The success of police applicant selection is well documented (e.g., Bernstein, 1980; Beutler, Storm, Kirkish, Scogin, & Gaines, 1985; Costello, Schoenfeld & Kobos,

1982; Hartman, 1987; Inwald, 1988). The MMPI also has been used success-fully in selecting physicians' assistants (Crovitz, Huse, & Lewis, 1973), medical assistants (Stone, Basset, Brosseau, Demers, & Stiening, 1972), psychiatric residents (Garetz & Anderson, 1973), clinical psychology graduate students (Butcher, 1979), nurses (Kelly, 1974), fire fighters (Avery, Mussio, & Payne, 1972), probation officers (Solway, Hays, & Zieben, 1976), and nuclear power plant personnel (Dunnette, Bownas, & Bosshardt, 1981). In virtually all of these studies the most effective way to use the MMPI has been to exclude persons with very elevated scores on one or more of the clinical scales.

No data have been published yet concerning the use of the MMPI-2 to screen for psychopathology in applicants or to predict effective functioning in various occupations or training programs. However, given the continuity between the original MMPI and the MMPI-2, one would expect that the revised instrument will be at least as effective as the original one for these purposes. The Negative Work Attitudes (WRK) content scale may prove to be especially useful in this regard. Butcher, Graham, Williams, and Ben-Porath (1990) reported some preliminary data concerning WRK-scale scores of men who could be assumed to have differing levels of work performance. Airline pilots, who were assumed to have highly successful work skills, scored well below the mean for the MMPI-2 normative sample, and active-duty military personnel, who volunteered to participate and were as a group not experiencing occupational problems, scored at about the mean for the MMPI-2 normative sample. Alcoholics and hospitalized psychiatric patients, who typically have very poor work histories, scored well above the mean for the normative sample. Although these group data are encouraging, research is needed in which scores on the WRK scale are compared with measures of actual job performance.

There is not much information available concerning use of the MMPI to match persons with certain personality characteristics with jobs requiring those characteristics. There are several problems with using the MMPI-2 in this manner. First, there is only limited information concerning the personality characteristics of normal persons with particular MMPI scores and profiles. Graham and McCord (1985) suggested a strategy for generating inferences about personality characteristics of normal persons who did not have extremely elevated MMPI scores. For such persons clinicians were advised to use the descriptive information generated for clinical subjects but to eliminate those inferences that deal with serious psychopathology. For example, a normal person with a 6–8 two-point code type and no T scores above 70 might have been described as suspicious and distrustful of others, as deficient in social skills, and as avoiding deep emotional ties. However, a normal person with this kind of profile would not have been seen as having confused thinking, delusions, or hallucinations. Although Graham and McCord presented data indicating that this approach is justified, they stressed the need for additional research before the approach could be recommended for use on a routine basis.

Another problem in using the MMPI-2 to match persons with certain personality characteristics with jobs requiring those characteristics is that for most jobs there is not a clear understanding of the kinds of personality characteristics associated with successful performance. Future research in this area should emphasize thorough job analyses prior to attempts to use the MMPI-2 to identify persons suitable for certain jobs.

National Computer Systems markets a computerized MMPI-2 interpretive report intended for use in personnel selection situations (The Minnesota Report: Personnel Selection System) (Butcher, 1989c). This report provides profiles of scores on the standard validity and clinical scales and several supplementary scales and lists of scores on many other supplementary scales. Based on a set of MMPI-2 decision rules, summary evaluations in the form of ratings on several dimensions relevant to how employees function are provided: openness to evaluation; ability to relate to other people; potential for drug or alcohol addiction; tolerance for stress; and overall level of adjustment.

Although no studies have yet been published concerning the validity of the evaluations provided in the MMPI-2 Personnel Report, Butcher (1989c) summarized several studies utilizing the MMPI version of the report. Butcher (1988) compared ratings of overall level of adjustment of airline pilot applicants based on the decision rules used in the Personnel Report with corresponding ratings by experienced clinicians. There was striking agreement between the computer rules and the experienced clinicians. Muller and Bruno (1988) compared evaluations of police applicants based on the Personnel Report decision rules with evaluations based on other data (interview, background check, polygraph). The MMPI decision rules were quite effective in identifying persons who subsequently were rejected on the basis of other data as psychologically unsuited for the job of police officer. Although these data concerning the validity of evaluations based on the Personnel Report are encouraging, research is needed in which evaluations based on the MMPI-2 are compared with reliable measures of actual job performance.

There are several other important considerations in using the MMPI-2 for personnel selection purposes. It has been suggested that requiring applicants for jobs to complete tests like the MMPI-2 is an invasion of privacy (Brayfield, 1965). Many persons feel that items dealing with sex, religious beliefs, or bowel and bladder functioning are inappropriate for job applicants because there is not likely to be any relationship between such items and job performance (Butcher & Tellegen, 1966), and recent court decisions, for example, *Soroka* v. *Dayton Hudson Corporation* (Cox, 1991) have supported these beliefs. The elimination of many objectional items from the MMPI-2 item pool was intended to address these concerns. It can be argued that some invasion of privacy is justified when evaluating applicants for psychologically sensitive or stress-vulnerable occupations. Applicants are less likely to object to taking the MMPI-2 if they are given explanations of how the test works and why they are being asked to take it.

Because job applicants understandably are motivated to present themselves in the best possible light, they are likely to produce defensive profiles (Butcher, 1979). If a profile is considered to be invalid because of the defensiveness, it should not be interpreted and the applicant should be evaluated by other means (e.g., interview). Butcher (1989a) suggested guidelines for interpreting defensive profiles that are not considered to be invalid. If there are T scores above 65 on any of the clinical scales of MMPI-2, the elevations probably accurately reflect important problems, because they were obtained when subjects were presenting a very favorable view of themselves. Since they were trying to present an overly favorable view, T scores in the 60–65 range on the clinical scales should also be interpreted as indicating significant problems. If all clinical scale T scores are below 60, such a profile will not provide much useful information. It is not possible to determine if such a profile indicates a person who is not functioning very well and is being defensive or if it indicates a person who is functioning within normal limits.

Not much research has yet been reported concerning the use of the MMPI-2 with nonclinical subjects. In Chapter 8, data were presented suggesting that there were reliable extratest correlates of MMPI-2 scores for persons in the MMPI-2 normative sample. For the most part, the correlates are consistent with data previously reported for the MMPI.

Also using data from the MMPI-2 normative subjects, Keiller and Graham (1993) investigated the extent to which normal persons having high, medium, or low MMPI-2 scores differed according to ratings provided by persons who were well acquainted with the subjects. Results indicated that there were important differences in personality and behavior associated with levels of scores on the MMPI-2 clinical scales. For the most part, high scorers were characterized as having more symptoms and negative characteristics than medium and low scorers. Low scorers differed from medium scorers in important ways. For the most part, low scorers were rated as better adjusted and as having fewer negative characteristics than medium scorers. Keiller and Graham stressed the need for additional studies with nonclinical subjects.

Hjemboe and Butcher (1991) evaluated the extent to which the MMPI-2 could be used to assess marital distress. Using 841 couples from the MMPI-2 normative sample and 150 couples in marital therapy, MMPI-2 scores were compared with scores on the Dyadic Adjustment Scale (DAS), a standardized measure of marital satisfaction. Because scores on scale 4 and the Family Problems (FAM) content scale were strongly related to DAS scores, Hjemboe and Butcher suggested using scores on these two scales as indexes of marital distress.

10

An Interpretive Strategy

In 1956 Paul Meehl made a strong plea for a "good cookbook" for psychological test interpretation. Meehl's proposed cookbook was to include detailed rules for categorizing test responses and empirically-determined extratest correlates for each category of test responses. The rules could be applied automatically by a nonprofessional worker (or by a computer) and the interpretive statements selected for a particular type of protocol from a larger library of statements. Although efforts have been made to construct such a cookbook (e.g., Gilberstadt & Duker, 1965; Marks & Seeman, 1963; Marks et al., 1974), the current status of psychological test interpretation is far from the automatic process Meehl envisioned. All tests, including the MMPI-2, provide opportunities for standardized observation of the current behavior of examinees. On the basis of these test behaviors, inferences are made about other extratest behaviors. The clinician serves both as information processor and clinical judge in the assessment process. The major purpose of this chapter is to suggest one approach (but by no means the only one) to using MMPI-2 data to make meaningful inferences about examinees.

The MMPI-2 should be used solely to generate hypotheses or inferences about an examinee, for the interpretive data presented earlier will not apply completely and unfailingly to each and every person with a specified MMPI-2 protocol. In interpreting MMPI-2s, one must deal in probabilities. A particular extratest characteristic is more likely than another to hold true for a person with a particular type of MMPI-2 protocol, but one can never be completely certain that it will. The inferences generated from an individual's MMPI-2 protocol should thus be validated against other test and nontest information available about that individual.

The MMPI-2 will be most valuable as an assessment tool when it is used in conjunction with other psychological tests, interview and observational data, and appropriate background information. Although blind interpretation of the MMPI-2 certainly is possible and in fact is the procedure involved in computerized interpretations, such interpretations should be used only to generate hypotheses, inasmuch as more accurate person-specific inferences are likely to occur when the MMPI-2 is viewed in the context of all information available about an individual. This position is consistent with research findings by investigators such as Kostlan (1954) and Sines (1959).

Two kinds of interpretive inferences can be made on the basis of MMPI-2 data. First, some characteristics of an examinee with a particular kind of MMPI-2 protocol are those that, with better than chance probability, differentiate that examinee from other persons in a particular setting (e.g., hospital or clinic). For example, one might infer from a hospitalized patient's MMPI-2 profile that he or she is likely to abuse alcohol or other substances. Because most patients do not abuse substances, this inference clearly differentiates this particular patient from most other patients. A second kind of inference is one that involves a characteristic common to many individuals in a particular setting. For example, the inference that a hospitalized psychiatric patient does not know how to handle stress in an effective manner is one that probably is true for most patients in that setting. Although the differential, patient-specific inferences tend to be more useful than the general ones, the latter are important in understanding an individual case, particularly for clinicians and others involved in the treatment process who might not have a clear understanding of what behaviors are shared by most persons in a particular setting.

Whereas Meehl envisioned the assessment process as dealing exclusively with nontest behaviors that are directly and empirically tied to specific aspects of test performance, the current status of the assessment field is such that only limited relationships of this sort have been identified. Often it is possible and necessary to make higher-order inferences about examinees based on a conceptualization of their personalities. For example, currently no clear data indicate that a particular kind of MMPI-2 profile is predictive of a future suicide attempt. However, if we have inferred from an individual's MMPI-2 that he or she is extremely depressed, agitated, and emotionally uncomfortable, is impulsive, and shows poor judgment much of the time, the higher-order inference that such a person has a higher risk of suicide than patients in general is a logical one. Although it is legitimate to rely on such higher-order inferences in interpreting the MMPI-2, one should have greater confidence in the inferences that are more directly related to MMPI-2 scores and configurations.

A GENERAL STRATEGY

In his own clinical work the author utilizes an approach to MMPI-2 interpretation that involves trying to answer the following questions about each MMPI-2 protocol:

1. What was the test-taking attitude of the examinee, and how should this attitude be taken into account in interpreting the protocol?
2. What is the general level of adjustment of the examinee?
3. What kinds of behaviors (e.g., symptoms, attitudes, defenses) can be inferred about or expected from the examinee?

4. What etiology or set of psychological dynamics underlies these behaviors?
5. What are the most appropriate diagnostic labels for the person who produced the protocol?
6. What are the implications for treatment of the examinee?

Test-Taking Attitude

The ideal examinee is one who approaches the task of completing the MMPI-2 in a serious and cooperative manner. This individual reads each MMPI-2 item and responds to the item in an honest, direct manner. When such an ideal situation is realized, the examiner can feel confident that the test responses are a representative sample of the examinee's behavior and can proceed with the interpretation of the protocol. However, as suggested in Chapter 3, for various reasons examinees may approach the test-taking task with an attitude that deviates from the ideal situation described above. Specification of test-taking attitude for each individual examinee is important because such differential attitudes must be taken into account in generating inferences from the MMPI-2 protocol. In addition, such attitudes may be predictive of similar approaches to other nontest aspects of the examinee's life situation.

Qualitative aspects of an examinee's test behavior often serve to augment inferences based on the more quantitative scores and indices. One such factor is the amount of time required to complete the MMPI-2. As stated in Chapter 2, the typical examinee takes between 1 and 1 ½ hours to complete the test. Excessively long testing times may be indicative of indecisiveness, psychomotor retardation, confusion, or passive resistance to the testing procedures. Extremely short times suggest that either examinees were quite impulsive in responding to the test items or they did not read and consider their content.

Examinees occasionally become very tense, agitated, or otherwise upset in the MMPI-2 test-taking situation. Such behavior may be predictive of similar responses to other stressful situations. Some examinees, who are obsessive or indecisive, write qualifications to their true-false responses in the margins of the answer sheet.

Although the qualitative features of test performance discussed above can offer important information about an examinee, the validity scales are the primary objective sources of inferences about test-taking attitude. The Cannot Say (?) score indicates the number of items omitted by the examinee. A large number of omitted items may indicate indecisiveness, ambivalence, or an attempt to avoid admitting negative things about oneself without deliberately lying. Examinees who answer all or most of the items are not availing themselves of this simplistic way of attempting to present a positive picture of themselves.

Above-average scores on the L scale indicate that examinees may have used a rather naive and global denial of problems and shortcomings in an attempt to present themselves favorably. When L-scale scores are moderately high, the protocol can be interpreted but adjustments must be make to take into account the examinee's defensiveness. (See Chapter 3.) Extremely high scores on the L scale indicate that the protocol is invalid and should not be interpreted. (See Chapter 3.)

Scores on the F scale reflect the extent to which an examinee's responses to a finite pool of items compare to those of the standardization sample, with higher F scores reflecting greater deviance. Scores that are considerably higher than average suggest that examinees are admitting to many clearly deviant behaviors and attitudes. There are several reasons for such admissions. (See Chapter 3.) Examinees might have responded to the test items randomly or with a deliberate intention of appearing very emotionally disturbed. Another possibility is that the examinees are emotionally disabled and are using the MMPI-2 as a vehicle to express a cry for help. F-scale scores that are considerably below average indicate that the examinees are admitting fewer than an average number of deviant attitudes and behaviors. They may be overly defensive and trying to create unrealistically positive pictures of themselves. F-scale scores in the average range indicate that the examinees have been neither hypercritical of themselves nor overly denying in responding to the test items.

Whereas F-scale scores provide information about examinees' attitudes in responding to approximately the first 370 items of the test, the Back-Page Infrequency (Fb) scale provides similar information for items that occur later in the test booklet. High and low Fb scores should be interpreted very similarly to those of the standard F scale.

The K scale can serve as another index of defensiveness. Above-average K-scale scores indicate that examinees have been overly defensive, whereas below-average scores indicate lack of defensiveness and a highly self-critical attitude. (See Chapter 3.) Average-range scores on the K scale suggest that examinees have been neither overly defensive nor overly self-critical in endorsing the MMPI-2 items.

The Variable Response Inconsistency (VRIN) and True Response Inconsistency (TRIN) scales offer additional information concerning the possibility of response sets that can invalidate the protocol. (See Chapter 3.) Significantly elevated scores on VRIN indicate that the examinee probably responded to the items without reading and considering their content. When the TRIN scale is elevated (in either the true or false direction), the possibility of a yea-saying or nay-saying response set must be considered.

As discussed in Chapter 3, the configuration of the validity scales is important for understanding examinees' test-taking attitudes. In general, persons who are approaching the test with the intention of presenting themselves in an overly favorable way have L- and K-scale scores greater than the F-scale score, producing a V-shape in the validity scale portion of the profile. On the other hand, persons who are using the test to be overly self-critical and/or to exaggerate their problems produce an inverted V-shape in the validity scales (i.e., the L-scale and K-scale scores will be significantly lower than the F-scale score).

In summary, a first step in interpreting an MMPI-2 profile is to make some judgments concerning the test-taking attitude of the examinee. If the decision is made that the test was approached in a manner that invalidates the protocol (e.g., inconsistent responding, faking good, faking bad), no additional interpretation of the profile is in order. If there are less extreme response sets operating (e.g., defensiveness or exaggeration), it may be possible to make some tentative interpretations of the profile, but adjustments in interpretations must be made to take into account these response sets.

Adjustment Level

There are two important components to psychological adjustment level. First, how emotionally comfortable or uncomfortable are individuals? Second, how well do they carry out the responsibilities of their life situations regardless of how conflicted they might be? For most people these two components are very much related. Persons who are psychologically comfortable tend to function well and vice versa. However, for certain individuals (e.g., some neurotics) a great deal of discomfort and turmoil can be present, but adequate functioning continues. For other persons (e.g., chronic schizophrenics), quite serious impairment in coping with responsibilities can be found without accompanying emotional discomfort. The MMPI-2 potentially can permit inferences about both of these aspects of adjustment level.

The F scale serves as one index of degree of psychopathology. If one rules out the possibility of deviant response sets or styles that can invalidate the protocol (e.g., an angry person who decides to answer true to all of the deviant items), high F-scale scores suggest intense emotional turmoil and/or serious impairment in functioning. For example, most acutely psychotic subjects tend to obtain high scores on the F-scale. However, some neurotic individuals and persons undergoing severe situational stress also achieve high F-scale scores. On the other hand, some clearly psychotic individuals, particularly those for whom the disorder has been present for quite some time, do not achieve very high F-scale scores.

A second simple but meaningful index of adjustment has to do with the overall elevation of the clinical scales. In general, as more of the clinical scales are elevated (and as the degree of elevation increases), the probability is greater that serious psychopathology and impaired functioning are present. To obtain a crude, quantitative index of the degree of this psychopathology, some clinicians find it useful to compute a mean T score for eight clinical scales (excluding scales 5 and 0). Another index of psychopathology is the number of clinical scales with T scores above 65. Higher mean scores and more scores above 65 are indicative of greater psychopathology.

The slope of the profile also yields important inferences about adjustment level. If the clinical scales are elevated and a positive slope (left side low, right side high) is present, the likelihood of severe psychopathology, and perhaps even psychosis, should be considered. A negative slope (left side high, right side low) is more indicative of a neurotic individual or one who is internally conflicted and miserable but who still is able to function fairly well.

Scores on several of the standard clinical and supplementary scales also can serve as indexes of level of adjustment. Welsh's Anxiety (A) and Barron's Ego Strength (Es) scales are measures of general maladjustment. High scorers on the A scale and low scorers on the Es scale tend to be rather disturbed emotionally. The A scale seems to be more sensitive to subjective emotional turmoil than inability to cope behaviorally. The Es scale indicates an individual's ability to cope with the stresses and problems of everyday life, with high scorers generally better able to cope than low scorers. Scale 2 (Depression) is a good indicator of dissatisfaction with one's life situation. As scores on scale 2 become higher, greater dissatisfaction is suggested. Scale 7 (Psychasthenia) is perhaps the single best measure of feelings of anxiety and agitation. High scale 7 scorers usually are overwhelmed by anxiety, tension, fear, and apprehension.

It may be helpful to check responses to the Koss-Butcher or Lachar-Wrobel critical items. These items deal with some blatantly psychotic behaviors and attitudes, sexual deviation, excessive use of alcohol, homicidal and/or suicidal impulses, and other manifestations of serious maladjustment. The critical-item lists are reproduced in Appendix I of this book. As was discussed in Chapter 6, care should be taken not to overinterpret these individual item responses.

Characteristic Traits and Behaviors

At this point in the interpretive process, the clinician's goal is to describe the examinee's symptoms, traits, behaviors, attitudes, defenses, and so forth, in enough detail to allow an overall understanding of the person. Although not every protocol permits inferences about all of the points listed below, in general this author tries to make statements or inferences about each of the following:

1. symptoms
2. major needs (e.g., dependency, achievement, autonomy)
3. perceptions of the environment, particularly of significant other people in the examinee's life situation
4. reactions to stress (e.g., coping strategies, defenses)
5. self-concept
6. sexual identification
7. emotional control
8. interpersonal relationships
9. psychological resources

Inferences about these various aspects of behavior and personality are based primarily on analysis of two- and three-point configurations and on scores on individual validity and clinical scales. (See Chapters 3, 4, and 5.) In addition, the supplementary scales, discussed in Chapters 6 and 7, often add important information about examinees. One way for the beginner with the MMPI-2 to use the standard and supplementary scales is to consider each configuration of scores and each high score in turn, consult the appropriate sections in earlier chapters of this book, and write down the appropriate hypotheses or inferences for each scale or configuration.

Initially there may appear to be inconsistencies among the various inferences that have been generated. The clinician should first consider the possibility that the apparent inconsistencies are accurately reflecting different facets of the examinee's personality and behavior. Consider, for example, a profile with high scores (T > 65) on both scales 2 and 4. The scale 2 score suggests sensitivity to the needs and feelings of others, whereas the scale 4 score suggests insensitivity to the needs and feelings of others. It is possible that the same individual may show both characteristics at different times. In fact, research with the 24/42 code type indicates that persons with this code type tend to alternate between periods of great sensitivity to others and periods of gross insensitivity to others.

Sometimes inconsistencies in inferences cannot be reconciled as easily as in the above example. In these instances the clinician must decide in which inferences to have the most confidence. In general, greater confidence should be placed in inferences that occur for several scales or configurations than in those that occur for only a single scale. Inferences based on very high scores should receive greater emphasis than those based on moderately high scores. Inferences stemming from scale configurations typically will be more accurate than those based on a single scale. In most instances more confidence should be placed in inferences coming from the standard validity and clinical scales than from the supplementary scales, because there is more research underlying the standard scale inferences.

As mentioned earlier in this chapter, certain inferences about an examinee do not result directly from scores on a scale or from configurations of scales. Rather, they are higher-order inferences generated from a basic understanding of the examinee. For example, there are no data indicating that scores on particular MMPI-2 scales are predictive of suicide attempts. However, it would be reasonable to be concerned about such attempts in a person whose MMPI-2 scores led to inferences of serious depression, agitation, impulsivity, and similar characteristics.

Dynamics and Etiology

In most assessment situations it is desirable to go beyond a description of an individual's behavior and to make inferences about the dynamics underlying the behavior or about the etiology of a particular problem or condition. For some MMPI-2 scales and configurations, the interpretive information presented in previous chapters of this book includes statements about these underlying factors. In addition, it is possible and/or necessary to make higher-order inferences about dynamics. For example, if one infers from an examinee's MMPI-2 protocol that he or she is afraid of becoming emotionally involved with other people because of fear of being hurt or exploited, one might then speculate that the person has been hurt and/or exploited in earlier emotional relationships. Furthermore, if one interprets an MMPI-2 protocol as indicating strong resentment of authority, it is reasonable to infer that the resentment has its origins in parent-child relationships. The higher-order inferences often are based on MMPI-2 data combined with other information about examinees and on the clinician's basic understanding of behavior, personality, and psychopathology.

Diagnostic Impressions

Although the usefulness of psychiatric diagnoses per se has been questioned by many clinicians, referral sources often request information about diagnosis. In addition, it often is necessary to assign diagnostic labels for purposes such as insurance claims, disability status, or competency status. Many of the interpretive sections in earlier chapters of this book present diagnostic information for the individual clinical scales and for two- and three-point code types. In addition, it can be useful to consider the slope of the profile. As stated earlier, a negative slope (left side high, right side low) is suggestive of nonpsychotic disorders, whereas a positive slope (left side low, right side high) is suggestive of psychotic disorders.

Goldberg (1965) derived a linear regression equation for discriminating psychotic and neurotic profiles. To compute the Goldberg index, one simply inserts T-score values into the following formula: $L + Pa + Sc - Hy - Pt$. Goldberg found for his samples that a cutoff of 45 on his index provided the best discrimination between psychotic and neurotic profiles. However, it is recommended that a specific cutoff score be derived in each setting where the index is used. It is important to note that the index is useful only when the clinician is relatively sure that the person being considered is either psychotic or neurotic. When the index is applied to the scores of normal persons or those with personality disorder diagnoses, most of them are considered to be psychotic. It also should be noted that the index was based on data from the original MMPI. Data have not yet been published concerning the use of the index with the MMPI-2.

In using the MMPI-2 for diagnostic purposes, the clinician must consider that most research studies concerning diagnostic inferences were based on the original MMPI and the original standardization sample. However, because of the continuity between the original MMPI and the MMPI-2, it is expected that data based on the original instrument will be applicable to the revised one. It also is important to know that most of the studies of the relationship between MMPI data and diagnoses were conducted prior to the publication of the DSM-III and DSM-III-R. In presenting diagnostic inferences for particular scores and configurations of scores in this book, the earlier diagnostic labels were translated into the more contemporary ones.

Treatment Implications

A primary goal in most assessments is to be able to make meaningful recommendations about treatment. Sometimes, when demand for treatment exceeds the resources available, the decision simply is whether or not to accept a particular person for treatment. Such a decision may involve clinical judgment about how much the person needs treatment as well as how likely the person is to respond favorably to available treatment procedures. When differential treatment procedures are available, the assessment may be useful in deciding which procedures are likely to be most appropriate. When the decision has been made before the assessment that a person will receive a particular treatment procedure, the assessment can be helpful in providing information about problem areas to be considered in treatment and by alerting the therapist (or others involved in treatment) to assets and liabilities that could facilitate or hinder progress in therapy. The MMPI-2 can provide information relevant to all of these aspects of treatment. Butcher (1990a) has provided a detailed account of the use of MMPI-2 in treatment planning.

Many of the inferences generated from scores on individual validity and clinical scales and from code types will have direct relevance to treatment considerations. Of special importance is the pattern of scores on the standard validity scales. A defensive validity scale pattern (the L and K scales considerably higher than the F scale) suggests that the examinee is not admitting to problems or symptoms and is not likely to be very receptive to therapeutic intervention. By contrast, an examinee whose F scale is considerably higher than the L and K scales is likely to be admitting to problems, symptoms, and emotional distress. This person is likely to be motivated to enter and remain in treatment.

Several of the MMPI-2 supplementary scales can provide useful information concerning treatment. The Ego Strength (Es) scale, which was discussed in Chapter 7, was designed to predict response to psychotherapy. If one is dealing with a neurotic individual, a high Es score is likely to mean that such a person will benefit from traditional, individual psychotherapy. With other kinds of persons and/or treatment procedures the relationship between Es scores and treatment outcome is less clear, but in general, higher Es scores can be interpreted as suggestive of greater psychological resources that can be used in treatment. It should be noted that persons who approach the MMPI-2 in a defensive manner tend to achieve relatively high Es scores that are not indicative of a positive prognosis for treatment. Also, persons who try to exaggerate problems and symptoms tend to achieve very low Es scores, which do not necessarily indicate a negative prognosis.

The Negative Treatment Indicators (TRT) scale may be helpful in predicting response to psychological treatment. Although no research concerning the relationship between scores on the TRT scale and response to treatment has yet been published, examination of the content of scale items suggests that higher scorers on this scale perceive themselves as unwilling or unable to change their life situations, are pessimistic about the possibility of positive change, are uncomfortable discussing problems with others, and are likely to be rigid and noncompliant in therapy.

As was discussed above in relation to characteristic traits and behaviors, many of the inferences about treatment will not come directly from scores on specific MMPI-2 scales or configurations of scales. Rather, they are higher-order inferences based on other inferences that already have been made about the examinee. For example, if one has inferred from the MMPI-2 that an examinee is in a great deal of emotional turmoil, it can further be inferred that this person is likely to be motivated enough to change in psychotherapy. On the other hand, if one has inferred that a person is very reluctant to accept responsibility for her or his own behavior and blames others for problems and shortcomings, the prognosis for traditional psychotherapy is very poor. A person who has been described by the MMPI-2 as very suggestible is apt to respond more favorably to direct advice giving than to insight-oriented therapy. A person described as psychopathic (based on elevations on scales 4 and 9), who enters therapy rather than going to jail, is likely to terminate therapy prematurely. Obviously, there are many other examples of higher-order inferences related to treatment.

SPECIFIC REFERRAL QUESTIONS

The interpretive strategy that has just been discussed is intended to help the clinician generate as many meaningful inferences as possible about subjects' MMPI-2s. It should be recognized that very often persons are given the MMPI-2 so that some very specific questions can be answered about them or so that very specific predictions can be made about their behavior. Unless these specific questions or predictions are dealt with directly, it is unlikely that the MMPI-2 interpretation will be regarded as useful.

Specific issues for which the MMPI-2 may be useful include, but are not limited to, detecting acting-out behaviors, psychotic disorders, or substance abuse problems; differentiating functional from organic somatic disorders; and predicting response to treatment. Sometimes data are produced that are directly relevant to the issue of concern. For example, scores on the substance abuse scales (MAC-R, AAS, APS) permit inferences about alcohol or other drug problems. At other times, conclusions about issues of concern are dependent on clinical integration of many different kinds of MMPI-2 data and second-order inferences.

AN ILLUSTRATIVE CASE

To illustrate the strategy discussed above, an actual case will now be considered and a step-by-step analysis of the MMPI-2 protocol will be presented. As a practice exercise, readers may interpret the profiles (Figures 10.1–10.4) and then compare their interpretations with the one presented below.

Background Information

Jeff is a 23-year-old Caucasian male. He has never been married and lives with his older sister and her family. He graduated from high school with somewhat below-average grades. He did not get into trouble in school. He did not participate in sports or other school activities, and he did not have any close friends. He has never had a serious relationship with a woman. Since completing high school he has had several jobs in fast-food restaurants and service stations, and he currently is unemployed.

The MMPI-2 was administered to Jeff when he requested services at a local community mental health center. He was cooperative and completed the MMPI-2 in about an hour and a half. Jeff had not previously been involved with mental health services. He was referred to the mental health center by emergency room staff at a local hospital. Apparently, during a period of excessive drinking he took a large number of aspirin. He was treated at the emergency room and released. Jeff admitted that he had been feeling increasingly upset and depressed lately. His sister had encouraged him to seek professional help.

Test-Taking Attitude

Jeff completed the MMPI-2 in about an average amount of time for a psychiatric outpatient, indicating that he was neither excessively indecisive nor impulsive in responding to the items. He omitted no items, suggesting that he was cooperative and did not use this rather simple way of avoiding unfavorable self-statements. His T score of 48 on the L scale is only slightly below average, so we may infer that he was not blatantly defensive and denying in his approach to the test. Jeff's T score of 55 on the F scale suggests that he was admitting more than an average number of deviant attitudes and behaviors. The F-scale score is not high enough to suggest random responding or a

Figure 10.1. MMPI-2 profile for basic scales for practice case (Jeff). Reproduced by permission of University of Minnesota Press.

Figure 10.2. MMPI-2 profile for supplementary scales for practice case (Jeff). Reproduced by permission of University of Minnesota Press

Name _Jeff_

Address _____

Occupation _Unemployed_   Date Tested _/ /_

Education _12_ Age _23_ Marital Status _Single_

Referred By _____

Scorer's Initials _JRG_

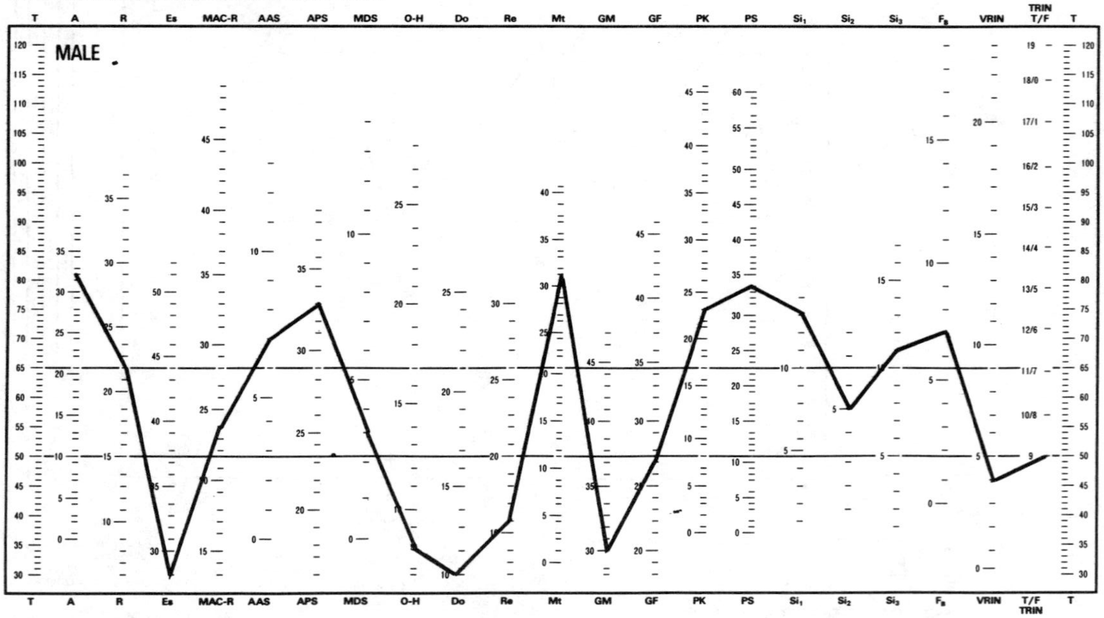

MALE

LEGEND

A	Anxiety
R	Repression
Es	Ego Strength
MAC-R	MacAndrew Alc.
AAS	Addiction Admission
APS	Addiction Potential
MDS	Marital Distress
F_b	Back F
VRIN	Variable Resp. Incon.
TRIN	True Resp. Incon.
O-H	Overcon. Host.
Do	Dominance
Re	Social Responsibility
Mt	College Maladjustment
GM	Gender Role–Masculine
GF	Gender Role–Feminine
PK	Post Traum.–Keane
PS	Post Traum.–Schlenger

	T	A	R	Es	MAC-R	AAS	APS	MDS	O-H	Do	Re	Mt	GM	GF	PK	PS	Si₁	Si₂	Si₃	F₆	VRIN	TRIN T/F
Raw Score		32	22	25	24	7	33	3	8	8	16	31	30	27	23	34	13	5	11	7	4	9

PRINTED WITH AGRI-BASE INK

NATIONAL COMPUTER SYSTEMS
PROFESSIONAL ASSESSMENT SERVICES
24004

230

Name __Jeff__

Address _____

Occupation __Unemployed__ Date Tested __/ /__

Education __12__ Age __23__ Marital Status __Single__

Referred By _____

Scorer's Initials __JRG__

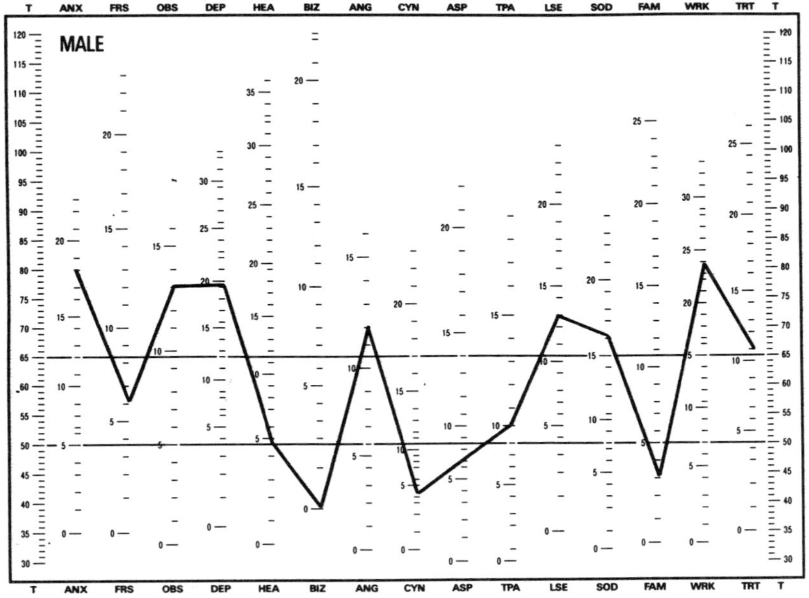

MMPI-2™
S.R. Hathaway and J.C. McKinley
Minnesota Multiphasic Personality Inventory-2™

Profile for Content Scales
Butcher, Graham, Williams and Ben-Porath (1989)

Minnesota Multiphasic Personality Inventory-2
Copyright © by THE REGENTS OF THE UNIVERSITY OF MINNESOTA
1942, 1943 (renewed 1970), 1989. This Profile Form 1989.
All rights reserved. Distributed exclusively by NATIONAL COMPUTER SYSTEMS, INC.
under license from The University of Minnesota.
"MMPI-2" and "Minnesota Multiphasic Personality Inventory-2" are trademarks owned by
The University of Minnesota. Printed in the United States of America.

MALE

	ANX	FRS	OBS	DEP	HEA	BIZ	ANG	CYN	ASP	TPA	LSE	SOD	FAM	WRK	TRT
Raw Score	18	6	13	19	5	0	12	4	7	10	13	16	3	24	11

Figure 10.3. MMPI-2 profile for content scales for practice case (Jeff). Reproduced by permission of University of Minnesota Press.

231

Figure 10.4. MMPI-2 profile for Harris-Lingoes subscales for practice case (Jeff). Reproduced by permission of University of Minnesota Press.

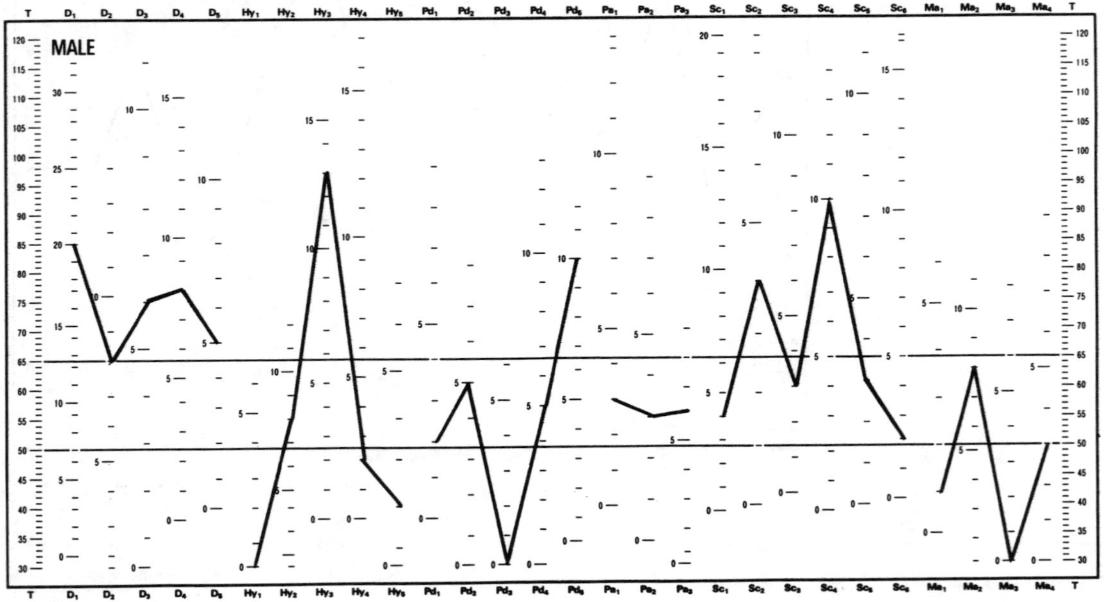

fake-bad response set. The VRIN raw score of 4 (T = 46) does not suggest random responding. The TRIN raw score of 9 (T = 50) is not suggestive of all-true or all-false responding. His raw score of 7 (T = 71) on the Back-Page Infrequency (Fb) scale suggests that he responded consistently to items appearing later in the test booklet.

Jeff's T score of 41 on the K scale indicates that he was not defensive. In fact, he was rather self-critical in responding to the items. Such low K-scale scores are rather common among persons voluntarily seeking mental health services. Jeff's configuration of scores on the validity scales resembles the inverted V-shape discussed above.

In summary, Jeff seems to have approached the MMPI-2 in an honest and open manner, admitting to some symptoms and problem behaviors. There are no indications that he approached the MMPI-2 in an invalid manner. Therefore, interpretation of the clinical, content, and supplementary scales can be undertaken.

Adjustment Level

Jeff's T score of 55 on the F scale does not indicate serious psychopathology. His mean T score on the eight clinical scales (excluding scales 5 and 0) is approximately 64, suggesting some psychological problems. This impression is reinforced by the fact that three of the eight clinical scale T scores are greater than 65. The slope of the profile is slightly negative (left side high, right side low), also suggesting that he is experiencing emotional turmoil but probably continues to function on a day-to-day basis, although at a reduced level of efficiency. The high scores on scales 2 (T = 81) and 7 (T = 77) and on the Anxiety (A) scale (T = 81) and the very low score on the Ego Strength (Es) scale (T = 30) support the inferences of significant problems and considerable emotional turmoil. Examination of Jeff's endorsement of the Koss-Butcher and Lachar-Wrobel critical items reveals that he admitted to a wide variety of deviant attitudes and behaviors, including anxiety, depression, substance abuse, some antisocial activities, and problems with concentration, attention, and anger control. (See the computerized interpretation in Chapter 11.)

In summary, Jeff's scores indicate that he has serious emotional problems. Although the level of his F-scale score does not indicate serious psychopathology, other scores and responses present a consistent picture of significant maladjustment.

Characteristic Traits and Behaviors

At this point in the interpretation we want to generate as many inferences about Jeff as we possibly can from his MMPI-2 scores. A first step in trying to generate inferences is to determine the code types for Jeff's profile and to consult Chapter 5 to generate inferences for them. It is best to start with the most complex code type for which interpretive information is available. When scales 5 and 0 are excluded, Jeff's three highest clinical scale scores are on 7, 2, and 6. Because there is not a five T-score point difference between scales 2 and 6, a defined three-point code type does not exist. Thus, no inferences are made on the basis of the configuration of these three highest scales.

27/72 Code Type

Jeff's two highest clinical scales are 2 and 7, and there is a nine T-score point difference between scale 7 and the next-highest clinical scale (6). Thus, Jeff has a defined two-point code type. This is a commonly occurring code type among mental health outpatients, and considerable information is available concerning its correlates. The two scales in the code type are quite elevated and the code type is well defined. Therefore, we are confident that Jeff is likely to have the same code type if retested in the near future and that inferences based on the code type are likely to fit him well. Because the scales in the code type are quite elevated, we can make inferences both about symptoms and personality characteristics commonly associated with the code type.

Reference to Chapter 5 indicates that numerous inferences about Jeff can be made on the basis of the 27/72 code type. Although he may not be feeling extremely sad or blue, he is reporting symptoms of clinical depression, including weight loss, slow personal tempo, and slowed thought processes. He seems to be extremely pessimistic about the world in general and more specifically about the likelihood of overcoming his problems, and he tends to brood about his problems much of the time.

Jeff also is reporting symptoms of anxiety. He is likely to feel anxious, nervous, tense, high-strung, and jumpy. He worries excessively and is vulnerable to real and imagined threat. He tends to anticipate problems before they occur and to overreact to minor stress. Vague somatic symptoms and complaints of fatigue, tiredness, and exhaustion may be reported.

Jeff seems to have a strong need for achievement and for recognition of his accomplishments. He has high expectations for himself, and he feels guilty when he falls short of his goals. He tends to be indecisive and to harbor feelings of inadequacy, insecurity, and inferiority. He is intropunitive, blaming himself for the problems in his life. He tends to be rigid in his thinking and problem solving, and he is meticulous and perfectionistic in his daily activities. He may be excessively religious and extremely moralistic.

Jeff tends to be rather docile and passive-dependent in his relationships with other people. In fact, he often finds it difficult to be even appropriately assertive. He has the capacity for forming deep emotional ties, and in times of stress he may become overly dependent and clinging. He is not aggressive or belligerent, and he tends to elicit nurturance and helping behavior from other people.

Other Configural Aspects

Several other configural aspects of Jeff's profile can be examined. The relationship among scales 1, 2, and 3 tells us something about the use of denial and repression. Because Jeff's scale 2 score is much higher than scores on scales 1 and 3, we would infer that his defenses are not working well and that he may be reporting a variety of symptoms and complaints. Because scales 3 and 4 are not much different from each other and neither is very elevated, we would not place much emphasis on their configuration. However, as discussed in Chapter 5, when scale 3 is higher than scale 4, there is less likelihood of problems with impulse control than when scale 4 is higher than scale 3. The relative scores on scales 7 and 8 can be helpful in making inferences about thought disorder. Because Jeff's scale 7 score is much higher than his scale 8 score, we would infer that he is not likely to have a thought disorder.

Validity Scales

L (T = 48). A T score of 48 on the L scale is about average and would not lead to any inferences beyond those made previously concerning test-taking attitude.

F (T = 55). A T score of 55 on the F scale is only slightly above average. We would not make inferences other than those previously made concerning the endorsement of some deviant items.

K (T = 41). A T score of 41 is a below-average K-scale score. It suggests that Jeff is self-critical and self-dissatisfied. He has little insight into his own motives and behaviors and is not very effective in dealing with the problems of daily life. He is socially conforming and overly compliant with authority. He has a slow personal tempo and may appear to be inhibited, retiring, and shallow. He feels socially awkward and at times may be blunt and harsh in social situations. He is cynical and skeptical and may be suspicious about the motives of other people.

Clinical Scales

Scale 1 (T = 57). A T score of 57 on scale 1 is in the average range, so no inferences would be made.

Scale 2 (T = 81). A T score of 81 on scale 2 is a very high score and is suggestive of clinical depression. Jeff is likely to feel blue, unhappy, and dysphoric. He may cry, refuse to speak, and show psychomotor retardation. He engages in self-depreciation and has guilt feelings. He is quite pessimistic about the future and may talk about committing suicide. He may report bad dreams, physical complaints, weakness, fatigue, and loss of energy. He is likely to be worried, agitated, tense, irritable, and high-strung. He may have a sense of dread that something bad is going to happen to him. He lacks self-confidence and feels useless and unable to function. He acts helpless and gives up easily. He feels like a failure in many things that he undertakes. He is a shy, introverted person whose lifestyle can be characterized by withdrawal and lack of involvement with other people. He has a restricted range of activities and may have withdrawn from activities in which he previously participated. He seems to feel that other people do not care about him, and his feelings are easily hurt. He is seen by others as aloof and distant. He is very cautious and conventional and not creative in problem situations. He is likely to be indecisive and to feel overwhelmed by major life decisions. He tends to be overcontrolled, denying his impulses, avoiding unpleasantness, and making concessions to avoid confrontations.

Scale 3 (T = 64). A T score of 64 on scale 3 is in the average range, so no inferences would be made.

Scale 4 (T = 62). A T score of 62 on scale 4 is in the average range, so no inferences would be made.

Scale 5 (T = 48). A T score of 48 on scale 5 is in the average range. It would lead us to infer that Jeff's interests and attitudes are similar to that of most men.

Scale 6 (T = 68). Although a T score of 68 represents a significant elevation on scale 6, the score is not high enough to lead to inferences concerning psychotic symptoms and behavior. Rather, we would expect Jeff to be excessively sensitive and overly responsive to the opinions of others. He feels that he is getting a raw deal from life. He is suspicious and guarded in relationships. He tends to rationalize and blame others for difficulties and may be hostile, resentful, and argumentative in relationships. He may appear to overemphasize rationality and to be moralistic and rigid in opinions and attitudes.

Scale 7 (T = 77). A scale 7 score of 77 suggests that Jeff is experiencing considerable psychological turmoil and discomfort. He is likely to feel anxious, tense, agitated, worried, fearful, apprehensive, high-strung, and jumpy. He may report difficulties in concentrating. He is quite introspective and may be obsessive in thinking and compulsive and ritualistic in behavior. Because he may not always feel in control of his thought processes, he may fear that he is losing his mind. He may report somatic symptoms of various kinds. Jeff is likely to be plagued by self-doubts. He has high standards of performance for himself and others and is neat, organized, meticulous, perfectionistic, persistent, and reliable. However, he lacks ingenuity in his approach to problems. He is likely to feel depressed and guilty when he falls short of his goals. Others are likely to see him as dull, formal, rigid, and moralistic. He has difficulty making decisions. He distorts the importance of problems and overreacts to stressful situations. Jeff tends to be shy, does not interact well socially, and is described as hard to get to know. He may also be described by others as sentimental, peaceable, soft-hearted, sensitive, kind, dependent, unassertive, and immature. He worries about popularity and social acceptance.

Scale 8 (T = 56). A T score of 56 on scale 8 is in the average range, so no inferences would be made about Jeff based on it.

Scale 9 (T = 45). A T score of 45 is a somewhat below-average score on scale 9. No inferences would be made based on this score.

Scale 0 (T = 73). A T score of 73 on scale 0 suggests that Jeff is socially introverted, shy, timid, reserved, and retiring. He is not likely to participate in many social activities and when he does he probably feels very insecure and uncomfortable. He is especially likely to feel uncomfortable around members of the opposite sex. He may feel more comfortable alone or with a few close friends. Jeff lacks self-confidence and is likely to be described by others as cold, distant, and hard to get to know. He is quite troubled by lack of involvement with other people. He tends to be overcontrolled and is not likely to display feelings openly. He is submissive and compliant in interpersonal relationships and is overly accepting of authority. He is described as having a slow personal tempo. He also is described as cautious, conventional, and unoriginal in his approach to problems, and he gives up easily when things are not going well. He is somewhat rigid and inflexible in attitudes and opinions, and he may have great difficulty making even minor decisions. He enjoys work and gets pleasure from productive personal achievement. He tends to worry, to be irritable, and to feel anxious. Others describe him as moody, and he may experience episodes of depression characterized by lack of energy and lack of interest in the activities of daily living.

Harris-Lingoes Subscales

As stated in Chapter 6, the Harris-Lingoes subscales can be helpful in understanding high scores on the clinical scales. We would next examine the Harris-Lingoes subscales for any clinical scales with high scores (T > 65). Although one should be very cautious in interpreting high scores on Harris-Lingoes subscales whose parent scale scores are not high, several of Jeff's subscale scores are so high that they should be noted even though scores on the parent scales are not very high.

Scale 2 subscales. Jeff obtained T scores equal to or greater than 65 on all five of the scale 2 subscales. This suggests that he is reporting a wide variety of symptoms of depression. His T score on the D1 (Subjective Depression) was much higher than on the other scale 2 subscales, indicating that he feels unhappy, blue, or depressed much of the time. He is not very interested in what goes on around him, and he lacks energy for coping with the problems of his everyday life. He feels nervous and tense much of the time, has difficulty in concentrating and attending, and has a poor appetite and trouble sleeping. He broods and cries frequently. He lacks self-confidence, feels inferior and useless much of the time, and is easily hurt by criticism. He feels uneasy, shy, and embarrassed in social situations and tends to avoid interactions with other people, except for relatives and close friends. Based on the high score on the D3 (Physical Malfunctioning) subscale, we would infer that he is preoccupied with his health and physical functioning. The high D4 (Mental Dullness) subscale would lead to the additional inferences that he is getting little enjoyment out of life and may have concluded that life is no longer worthwhile.

Scale 3 subscales. Because Jeff's T score on scale 3 is only moderately high (T = 64), we would be cautious about making inferences about the scale 3 subscales. However, the extremely high score (T = 97) on the Hy3 (Lassitude-Malaise) subscale is noteworthy. This high score reinforces earlier inferences that Jeff is somatically preoccupied and is feeling quite depressed.

Scale 4 subscales. As with scale 3, because the scale 4 score (T = 62) is not very high, we would be cautious in making any inferences from these subscales. However, the Pd5 (Self-alienation) subscale is so high (T = 82) that we would note that it suggests considerable emotional turmoil. Jeff is uncomfortable and unhappy. He does not find daily life interesting or rewarding. He has problems in concentrating. He experiences regret and remorse for past deeds but is vague about the nature of his misbehavior. He finds it hard to settle down, and he may use alcohol excessively.

Scale 6 subscales. Although Jeff's scale 6 score is high (T = 68), the scale 6 subscale scores are not very helpful in understanding it because they are all between 55 and 58. Thus, no additional inferences would be made on the basis of these subscales.

Scale 8 subscales. Although Jeff's scale 8 score is within the average range (T = 58), several of the scale 8 subscales are quite high and are noteworthy. The extremely high score (T = 92) on the Sc4 (Lack of Ego Mastery, Conative) subscale suggests that Jeff feels that life is a strain and is experiencing depression and despair. He has difficulty coping with everyday problems and worries excessively. He does not find his daily life very interesting or rewarding and responds to stress by withdrawing into fantasy and daydreaming. He may have given up hope of things getting better and may wish that he were dead. The high score (T = 78) on the Sc2 (Emotional Alienation) subscale also indicates that he is experiencing feelings of depression and despair and may wish that he were dead. He may appear to be apathetic and frightened. He may have sadistic and/or masochistic needs.

Scale 0 Subscales

Examination of the scale 0 subscales sometimes helps clarify the primary reasons for an elevation on scale 0. Jeff's T score on scale 0 is 73, and scores on two of the scale 0 subscales are high. The T score of 74 on Si1 (Shyness/Self-Consciousness) indicates that Jeff feels quite shy, embarrassed, and ill at ease around other people. The T score of 68 on Si3 (Self/Other Alienation) suggests that he has low self-esteem and feels unable to effect changes in his life. The relatively lower T score of 58 on Si2 (Social Avoidance) may mean that, in spite of his social discomfort, Jeff is not totally avoiding social activities.

Content Scales

As stated in Chapter 6, the content scales of the MMPI-2 are much more homogenous than the clinical scales and can be helpful in understanding more clearly Jeff's symptoms, problems, and behaviors. Jeff has eight content scale scores equal to or greater than 65. We would next generate inferences from each of these high content scale scores.

Anxiety (T = 80). The very high score of 80 on the Anxiety scale suggests that Jeff is feeling anxious, nervous, worried, and apprehensive. He may have problems concentrating; his sleep may be disturbed; and he may be uncomfortable making decisions. He also may report feeling that life is a strain and that he is pessimistic about things getting better. He lacks self-confidence and feels overwhelmed by the responsibilities of daily life.

Obsessiveness (T = 77). The high score of 77 on the Obsessiveness scale suggests that Jeff frets, worries, and ruminates about trivial things. He may engage in compulsive behaviors such as counting or hoarding. He has difficulties making decisions. He is rigid and dislikes change. He lacks self-confidence, lacks interest in things, and may feel dysphoric and despondent.

Depression (T = 77). The high score of 77 on the Depression scale indicates that Jeff is experiencing symptoms of depression. He feels sad, blue, and despondent, and he may cry easily. He feels fatigued and lacks interest in things. He feels pessimistic and hopeless. He may recently have been preoccupied with thoughts of death and suicide. He lacks self-confidence and often feels guilty. He feels lonely and empty much of the time. He may be expressing health concerns.

Anger (T = 70). The high score of 70 on the Anger scale suggests that Jeff feels angry and hostile much of the time. He may feel like swearing or smashing things and may at times have temper tantrums during which he is quite verbally aggressive. Others see him as irritable, grouchy, impatient, and stubborn.

Low Self-esteem (T = 72). Based on the high score of 72 on the Low Self-esteem scale, we would expect Jeff to have a very poor self-concept. He anticipates failure and gives up easily. He is overly sensitive to criticism and rejection. He is likely to be passive in relationships, and it may be difficult for him to accept compliments from others. He worries and frets a great deal, and he has difficulty making decisions.

Social Discomfort (T = 68). A high score of 68 on the Social Discomfort scale suggests that Jeff is shy and socially introverted. He would rather be alone than around other people. It is difficult for him to initiate conversations, and he dislikes parties and other group activities.

Work Interference (T = 81). Although the Work Interference scale has not yet been adequately validated, we can make some inferences based on the content of items in the scale. Jeff is reporting some negative attitudes and characteristics that are likely to lead to poor work performance. He seems to have rather negative perceptions of co-workers. He is not very confident about his career or vocational choices and feels that his family does not approve of his choices. He seems to be lacking in ambition and energy. He has a poor self-concept that leads to difficulty making decisions. He tends to be obsessive and to have problems in concentrating.

Negative Treatment Indicators (T = 66). The marginally elevated score of 66 on the Negative Treatment Indicators scale suggests that Jeff has some attitudes that are likely to interfere with psychological treatment. He may have negative attitudes toward doctors and mental health treatment. He seems to have problems that he is reluctant to share with others because he believes that they cannot understand him. He may feel unable to make significant changes in his life. He is a poor problem solver, often shows poor judgment, and gives up easily when problems are encountered.

Supplementary Scales

Jeff's scores on some of the supplementary scales discussed in Chapter 7 can be used to generate additional inferences about him.

Anxiety (T = 81). The very high score of 81 on the Anxiety scale suggests that Jeff is quite maladjusted. He is likely be experiencing considerable emotional turmoil, including anxiety and depression. He has a slow personal tempo and may be apathetic, unemotional, unexcitable, inhibited, and overcontrolled.

He seems to be pessimistic about the future. He is lacking in self-confidence and may be hesitant and vacillating in his behavior. He lacks poise in social situations. He is submissive, compliant, suggestible, cautious, conforming, and accepting of authority.

Repression (T = 65). Because Jeff's score on the Repression scale is only moderately high, not much emphasis should be placed on it in the interpretation of Jeff's MMPI-2. However, a score at this level on this scale indicates that Jeff tends to be passive in relationships. He may be described by others as slow, painstaking, unexcitable, and clear-thinking. He tends to be conventional and formal in his attitudes.

Ego Strength (T = 30). In spite of limited information concerning the meaning of low scores on the MMPI-2 scales, there exist some data suggesting that very low scores on the Es scale are associated with serious psychopathology. In addition, Jeff's very low score of 30 suggests that Jeff is presenting himself as overwhelmed and unable to respond to the demands of his life situation.

MacAndrew Alcoholism Scale—Revised (Raw Score = 24). A raw score of 24 is a marginally high score on the MAC-R screening measure of alcohol and drug problems. The score is not high enough to permit us to have confidence in inferences about Jeff's use or abuse of substances. However, it is at a level where we would want to obtain additional information about his substance use.

Addiction Acknowledgement Scale (T = 70). The high score on the AAS indicates that in responding to the MMPI-2 items, Jeff acknowledged substance abuse problems and behaviors associated with the abuse. Additional information should be obtained about his patterns of substance use and abuse.

Addiction Potential Scale (T = 76). The high score on the APS suggests that Jeff is likely to have problems with the abuse of alcohol and/or other drugs. Additional information about these possible problems should be obtained.

College Maladjustment Scale (T = 81). If Jeff's very high score of 81 on the College Maladjustment scale were encountered in a college setting, one would infer serious psychological problems. However, since Jeff was not evaluated in such a setting, no inferences would be made about him based on this score.

Post-traumatic Stress Disorder Scales (T = 75,79). The PTSD scales are only useful when we know that we are dealing with persons who have been exposed to extraordinary stressors (such as combat experiences). Because we have no information suggesting that Jeff has been exposed to such stressors, we would make no inferences about him based on these scales.

Resolving Inconsistent Inferences

A review of the inferences that were generated concerning Jeff's characteristic traits and behaviors indicates a striking consistency among them. However, several of the inferences appear to be somewhat inconsistent. It may be helpful to discuss how these apparent inconsistencies would be handled.

Jeff's T score of 70 on the Anger content scale led to inferences that he is angry and hostile much of the time, that he may have temper tantrums during which he is verbally aggressive, and that others may describe him as irritable and grouchy. However, based on the 27/72 code type and the high scores on scales 2 and 0, he was described as overcontrolled and not overtly aggressive or belligerent. Because there has been much more research on the 27/72 code type and on scales 2 and 0, we would have greater confidence in inferences based on these sources. It is not really clear at this time just what aspects of anger the Anger content scale is assessing. However, review of Jeff's critical-item endorsement indicates that he admitted to feeling like smashing things and having a strong urge to do something harmful or shocking. He is often said to be hotheaded and easily angered. Based on all of this information, we would probably conclude that Jeff harbors a great deal of anger and resentment, that it does not usually get expressed openly, but that he may occasionally vent the angry feelings in outbursts. Thus, the apparently inconsistent inferences really seem to be describing anger control and expression as important problem areas for Jeff.

Several other apparently inconsistent inferences also were generated. On the basis of the 27/72 code type and the high score on scale 0, it was inferred that Jeff has a strong need for achievement and is likely to enjoy work and get pleasure from productive personal achievement. A high score on the Work Interference content scale led to the inference that Jeff lacks ambition and energy. Because inferences based on code types are likely to be more accurate than those based on individual scales and because there has been much more empirical research concerning the 27/72 code type and scale 0 than concerning the Work Interference content scale, we would have greater confidence in the inferences that Jeff has a strong need to achieve and is likely to enjoy work.

A third area where apparently inconsistent inferences were generated is self-blame versus blaming others. The 27/72 code type and the high score on the Pd5 subscale (Self-alienation) led to inferences that Jeff is likely to blame himself for problems, whereas the high scores on scale 6 and on the Anxiety (A) scale led to inferences that he tends to blame others. Again, because we tend to have more confidence in inferences based on code types than those based on individual scales, we would probably describe Jeff as self-blaming. That the scale 6 score was only moderately high (T = 68) also suggests that we should have more confidence in the inference that he is self-blaming.

Some apparently inconsistent inferences could also be made concerning Jeff's relationships with other people. The 27/72 code type led to an inference that he is passive and dependent in relationships. However, high scores on scales 2 and 0 suggest that he is not likely to be involved with others and is seen by them as hard to get to know. Because numerous inferences were possible that Jeff is introverted, shy, and timid, it is likely that others are misinterpreting shyness and aloofness and that in the few relationships that Jeff forms he is indeed passive and dependent.

A marginally high score on the Repression scale led to an inference that Jeff is described by others as clear-thinking. However, the 27/72 code type and high scores on scale 7 and on the Obsessiveness content scale led to inferences that he has problems in concentration, ruminates excessively, and is quite indecisive. Because the score on the Repression scale was not very high, we would not have much confidence in the descriptors based on it.

Finally, Jeff's high scores on scale 2 and on the Sc4 (Lack of Ego Mastery, Conative) content scale led to inferences that he may wish he were dead and may talk about suicide. Because scores on these scales are extremely high, we would want to include these inferences in a description of Jeff. However, we would also want to note that examination of Jeff's critical-item endorsement indicates that he did not endorse any of the items dealing directly with suicidal ideation or attempts.

Integration of Inferences

Having dealt with apparent inconsistencies in inferences about Jeff, the next step would be to examine all of the inferences concerning characteristic traits and behaviors and to organize them into meaningful categories. Earlier in this chapter one possible list of categories was suggested.

Symptoms. There is clear agreement from many aspects of the MMPI-2 protocol that Jeff is experiencing a great deal of emotional turmoil. He is likely to be clinically depressed; he seems to feel sad and unhappy; and he may be showing signs of psychomotor retardation. He lacks energy and has lost interest in things going on around him. He may cry easily. Life seems to be a strain for him, and he is pessimistic about things ever getting better for him. At times he may wish that he were dead and may talk about committing suicide. However, in responding to the MMPI-2 items, he did not endorse the items dealing directly with suicidal ideation or attempts.

Jeff also appears to feel anxious, tense, and nervous. He is high-strung, jumpy, agitated, and apprehensive. He worries excessively and is vulnerable to real and imagined threat. He may have a sense of dread that something bad is going to happen to him. He is likely to be experiencing difficulties in concentrating and attending. His thinking may be obsessive and his behavior compulsive. He is likely to be quite indecisive, even about very trivial matters. Perceived lack of control over thoughts and emotions may cause him to fear that he is losing his mind. Jeff may be presenting multiple, vague somatic complaints.

Jeff has high scores on all three of the substance abuse scales of the MMPI-2 (MAC-R, AAS, APS), strongly suggesting that he may have problems with alcohol or other drugs. The high score on the Addiction Acknowledgement Scale (AAS) indicates that he is openly admitting to misuse of alcohol or other drugs and the problems associated with the misuse. Clearly, additional information about substance use is needed.

Major needs. Many of the inferences generated from Jeff's MMPI-2 scores suggest that he has very strong unfulfilled dependency needs. He is likely to be passive-dependent in relationships, and he worries about not being popular and socially accepted. Jeff seems to harbor above-average levels of anger and resentment. Most of the time he is overcontrolled and does not express these negative feelings directly. However, occasional temper tantrums and verbal outbursts of anger may occur. Jeff seems to have rather strong abasement needs. He often evaluates himself negatively and compares himself unfavorably with others. He may have some sadistic and/or masochistic tendencies. Jeff seems to have strong needs to achieve and to receive recognition for his accomplishments, but insecurity and fear of failure keep him from placing himself in many directly competitive situations.

Perceptions of the Environment. Jeff sees the world as a demanding place and feels incapable of responding to the demands of his daily life. He has a sense of dread that bad things are going to happen to him. He seems to feel that his needs are not met by others and that he is getting a raw deal from life. He may be cynical, skeptical, and suspicious about the motives of other people.

Reactions to stress. Jeff's MMPI-2 scores suggest that he feels poorly equipped to deal with stress. He is vulnerable to real and imagined threat, tends to anticipate problems before they occur, and often overreacts to minor stress. During times of increased stress he may develop increased somatic symptoms and become increasingly clinging and dependent. Although he prefers to use denial and repression as defenses, these mechanisms do not seem to be working well for him now. As a result, he is overwhelmed with emotional turmoil. At times Jeff may respond to stress by withdrawing into fantasy and daydreaming, and at other times he may seek escape from stress through the use of alcohol and/or other drugs.

Jeff appears to be a responsible, conscientious person. He is likely to be neat, organized, and persistent in his approach to problems. However, he is also likely to be quite cautious, conventional, rigid, and not creative. He is rather indecisive about most things in his life. He is a rather poor problem solver and often may show poor judgment. He tends to give up easily when faced with increasing stress.

Self-concept. Jeff's MMPI-2 scores indicate that he has an extremely negative self-concept. He is plagued by feelings of inadequacy, insecurity, and inferiority. He is quite self-critical and often compares himself unfavorably with other people. He has high expectations for himself and feels guilty when he falls short of his goals. He blames himself for the problems in his life and feels hopeless and unable to effect life changes.

Sexual identification. Not many inferences were generated about Jeff's sexual identification. His average scale 5 score would suggest that his interests and attitudes are likely to be similar to those of most men. His high scale 0 score indicates that he may be especially uncomfortable around women. He did not endorse the critical items that would suggest that he is dissatisfied with his sex life, that he has engaged in unusual sex practices, or that he has been in trouble because of his sexual behavior.

Emotional control. Jeff is likely to be emotionally overcontrolled much of the time. He tends to deny his impulses and is not likely to display feelings openly. He tends to emphasize rational rather than emotional aspects of situations. Although he is not likely to express anger and resentment openly most of the time, during occasional tantrums these feelings may get expressed verbally. He is so emotionally uncomfortable at this time that he may cry easily.

Interpersonal relationships. Jeff is a shy, socially introverted person. Although he has the capacity to form deep emotional ties and very much wants to be involved with others, his poor self-concept causes him to feel quite uncomfortable in social situations. He is likely to avoid large gatherings and may feel most comfortable when with a few close friends. He is troubled by his limited interactions with other people, and he feels lonely much of the time. In relationships Jeff is likely to be passive, submissive, and compliant. He is very unassertive and likely to make concessions to avoid confrontations.

Other people's perceptions of Jeff are quite variable. Sometimes he is seen as sentimental, peaceable, and soft-hearted, and he elicits nurturance and helping behavior from others. At other times he is seen as moody, irritable, dull, and moralistic. His shyness often may be misinterpreted as indicating that he is aloof, cold, and distant.

Jeff has ambivalent feelings about other people. He is drawn to them because they represent sources of gratification for his strong dependency needs. However, he also has negative perceptions of other people. He seems to view them as not being very understanding and supportive. He is quite sensitive to criticism, and his feelings are easily hurt. At times he can be rather blunt and harsh in social interactions. Jeff has some negative attitudes about co-workers that are likely to interfere with productive work.

Psychological resources. Because the MMPI-2 scales tend to emphasize psychopathology and negative characteristics and because Jeff has many high scores on the scales, most of the inferences generated tend to be negative. However, a review of the inferences previously generated reveals a few that could be viewed as psychological resources. Although he often feels like a failure, Jeff has strong needs to achieve and to receive recognition for his accomplishments. He is neat, meticulous, persistent, and reliable. He has the capacity for forming deep emotional ties. Others sometimes see him in positive ways (e.g., sensitive, kind, soft-hearted, and peaceable) and react to him in nurturant and helping ways.

Dynamics and Etiology

As often is the case, the MMPI-2 results do not offer direct inferences concerning dynamics and etiology. We would have to rely on other data sources and/or higher-order inferences to understand how Jeff came to be as he has been described so far. For example, one might hypothesize that Jeff's abuse of alcohol and/or other drugs results from attempts to deal with feelings of inferiority and alienation. His negative self-concept could have resulted from consistently negative reactions from parents or other significant adults in his life. His social introversion could have developed as a way of avoiding the negative reactions that he came to expect from others.

Diagnostic Impressions

As discussed earlier in this chapter, it is not possible to assign psychiatric diagnoses solely on the basis of psychological test data. The criteria for most DSM-III-R diagnoses include information that cannot be obtained from psychological test data. However, we can make statements concerning the diagnoses that are most consistent with Jeff's MMPI-2 results.

The MMPI-2 data consistently suggest that Axis I diagnoses of depressive disorder (major depression or dysthymia) and generalized anxiety disorder should be considered. The 27/72 code type and high scores on scales 2 and 7 and on the Depression (DEP) and Anxiety (ANX) content scales support these diagnoses. The 27/72 code type and a high score on the Obsessiveness (OBS) content scale indicate that an Axis I diagnosis of obsessive-compulsive disorder should be considered. The very high scores on the Addiction Acknowledgment Scale (AAS) and the Addiction Potential Scale (APS) and a marginally high score on the MacAndrew Alcoholism Scale—Revised (MAC-R) are consistent with substance use disorders, and more information about substance use should be obtained. There are no direct empirical data that would lead to inferences about Axis II diagnoses. However, the descriptions of symptoms and personality characteristics that were generated about Jeff are consistent with Axis II diagnoses of dependent and obsessive-compulsive personality disorders.

Implications for Treatment

Many indicators in Jeff's MMPI-2 data suggest that he is in a great deal of emotional turmoil and is not likely to be meeting the responsibilities of daily life in an effective manner. Because of his intense discomfort, he is likely to be motivated for psychotherapy. Although his high score on scale 2 suggests that he might terminate treatment prematurely when the immediate crisis passes, the 27/72 code type indicates that he is likely to remain in treatment longer than most patients. The high score on scale 7 indicates that he is not likely to be very responsive to brief psychotherapy, but the 27/72 code type suggests that he can be expected to show slow but steady progress over time.

There are indications of characteristics that are likely to interfere with effective therapy. He may be so depressed that he will not have the energy to participate effectively in traditional psychotherapy. A referral to assess the appropriateness of antidepressant medication should be considered. His very low score on the Ego Strength (Es) scale suggests that he has limited psychological resources that can be utilized in treatment. Based on the high score on scale 7, we would expect that he will rationalize and intellectualize a great deal in therapy. He probably would be resistant to psychological interpretations and could come to express significant hostility toward a therapist. The high score on the Negative Treatment Indicators (TRT) content scale indicates that he has negative attitudes toward doctors and mental health treatment. He seems to feel that other people cannot really understand his problems, and he seems to believe that he is helpless to change major aspects of his life. His rigidity and tendency to give up easily in stressful situations could be liabilities in treatment.

As stated earlier, Jeff's scores on the substance abuse scales of the MMPI-2 (MAC-R, AAS, APS) indicate that he may have problems with alcohol and/or other drugs. If corroborating information supports the inferences concerning substance abuse, a treatment program should include a substance abuse component.

Summary

The above analysis of this single case is lengthy in its presentation because it is meant as a teaching-learning tool for the beginning MMPI-2 user. The experienced MMPI-2 user would write a much briefer interpretation of the protocol. Specifically, the following is what the author would write about Jeff in a clinic chart, to the referring source, or for his own psychotherapy notes.

The MMPI-2 protocol produced by Jeff appears to be valid. He was not overly defensive in responding to the MMPI-2 items. He admitted to some deviant attitudes and behaviors, but this admission is seen as an accurate reporting of problems.

Jeff appears to be having some significant psychological problems. He is experiencing a great deal of emotional turmoil. He feels overwhelmed and unable to respond to the responsibilities of daily life. He seems to be clinically depressed and anxious. He is pessimistic about the future and probably has been having some suicidal ideation. He may be reporting somatic symptoms and difficulties in concentrating and attending. His thinking may be obsessive and his behavior compulsive.

Jeff sees the world as a demanding place and feels that his needs are not being met by other people. Although he typically engages in denial and repression, these defenses do not seem to be working very well for him at this time.

Jeff has an extremely negative self-concept. He is plagued by feelings of inadequacy, insecurity, and inferiority. He tends to blame himself for the problems in his life.

Although Jeff may be harboring feelings of anger and resentment, he typically does not express feelings openly. However, brief verbal outbursts of anger may occur during temper tantrums.

Jeff is shy, timid, and socially introverted. Although he seems to want to be involved with other people and has the capacity to form deep emotional ties, he tends to withdraw from many social interactions in order to protect himself from the criticism and rejection that he has come to expect. In relationships he is likely to be passive, submissive, and unassertive. He can be expected to make concessions in order to avoid unpleasant confrontations. He seems to have negative perceptions of other people, seeing them as neither understanding nor supportive of him.

Jeff's MMPI-2 data are consistent with DSM-III-R Axis I diagnoses of depressive disorder (major depression or dysthymia), generalized anxiety disorder, and obsessive-compulsive disorder. Substance abuse disorders should be ruled out. His symptoms and personality characteristics are consistent with Axis II diagnoses of dependent and obsessive-compulsive personality disorders.

Jeff is not coping very well with the demands of his life situation and needs psychological treatment. Because of his intense psychological discomfort, he is likely to be receptive to psychotherapy. If his depression is so severe that it interferes with his ability to participate meaningfully in psychotherapy, a medical referral to evaluate the appropriateness of antidepressant and antianxiety medications should be considered. Although he is not likely to respond well to brief psychotherapy, he probably will stay in treatment longer than many patients and may show slow but steady progress. If corroborating information indicates that he is abusing alcohol and/or other drugs, a substance abuse component should be included in his treatment plan.

ADDITIONAL PRACTICE CASES

Brief interpretations of three additional MMPI-2 profiles will now be presented. As a learning exercise, clinicians can write their own interpretations of each profile and then compare their interpretations with the ones presented below.

Jane

Figures 10.5 and 10.6 present the MMPI-2 basic scale and content scale profiles for Jane, a 35-year-old Caucasian female. She was evaluated shortly after her admission to a chronic-pain treatment unit. She was involved in a minor automobile accident several years ago, and periodically since the accident she has experienced pain and discomfort in her neck and back. Comprehensive neurological evaluation has not indicated any physical basis for her continuing pain. Jane is the only child of middle-class parents. Her father is a salesman, and her mother has never worked outside of the home. Jane graduated from college with a degree in elementary education. She has taught third and fourth grades at several different schools since her graduation. She has never been married, but she was engaged while she was a college student. Her fiancé ended the relationship after he became involved with another woman. Jane has been taking pain medication off and on since her accident, but she has denied using the medications except as prescribed by her physician.

Jane's MMPI-2 profile is valid. She omitted no items. Her scores on the validity scales indicate that she was neither overly defensive nor especially self-critical in responding to the MMPI-2 items.

Having three clinical scales with T scores greater than 65 (with two of them greater than 75) indicates that she has significant psychological problems. The most salient feature of the profile is the 1–3 two-point code type. Both scales 1 and 3 are significantly higher than scale 2, forming a conversion valley pattern. This code type indicates that she sees herself as having medical problems and wants to receive medical treatment. The very high score on the Health Concerns (HEA) content scale also indicates somatic preoccupation. She lacks insight into psychological factors underlying her symptoms. She tends to have a Pollyannaish attitude toward life and uses repression and denial excessively. The absence of very high scores on scales 2 and 7 suggests that she is not experiencing disabling anxiety, depression, or other emotional turmoil. However, the moderate elevations on scale 7 and on the Anxiety (ANX) content scale indicate that she may be preoccupied, probably with somatic concerns.

The 1–3 code type indicates that she tends to be immature, egocentric, and selfish. She has strong needs for attention, affection, and sympathy, and she is very demanding in relationships. She is a very dependent person, but she is not

Figure 10.5. MMPI-2 profile for basic scales for practice case (Jane). Reproduced by permission of University of Minnesota Press.

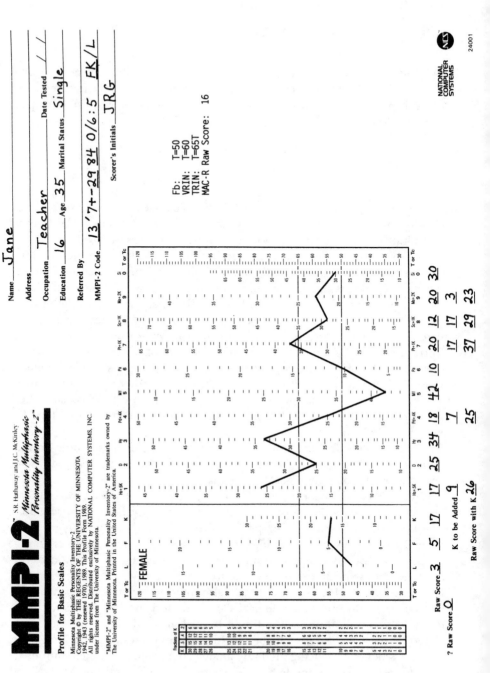

250

Figure 10.6. MMPI-2 profile for content scales for practice case (Jane). Reproduced by permission of University of Minnesota Press.

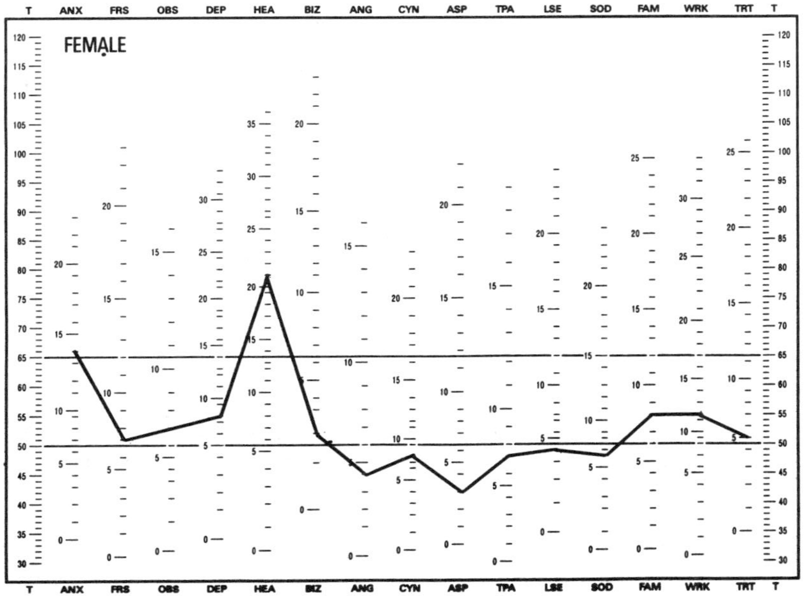

comfortable with the dependency and experiences conflict because of it. Although she may be involved with other people, relationships are likely to be quite superficial. She tends to exploit relationships to satisfy her own needs. She may use her physical symptoms as a way of justifying her demands for attention and support. Although Jane may harbor anger and resentment toward other people who are perceived as not fulfilling her needs for attention, she is not likely to express these feelings openly and directly. Rather, they are likely to get expressed in passive, indirect ways or in periodic angry outbursts.

Her very low scale 5 score indicates that she presents herself as stereotypically feminine. However, the 1–3 code type suggests that she lacks skills in dealing with the opposite sex and may be deficient in heterosexual drive.

Based on the 1–3 code type, the elevated HEA content scale score and her history and presenting complaints, an Axis I diagnosis of somatoform pain disorder seems to be appropriate. Although no direct inference concerning an Axis II diagnosis is indicated, the personality characteristics described above are consistent with a diagnosis of dependent personality disorder.

Because of her unwillingness to acknowledge psychological factors underlying her symptoms, Jane would be difficult to motivate in traditional psychotherapy. If a therapist stresses the link between symptoms and psychological factors, she is likely to terminate therapy. It may be possible to get her to discuss problems in her life situation as long as no direct link to somatic symptoms is suggested. Persons with the 1–3 code type tend to be suggestible, and she may be willing to try activities suggested by a therapist.

Steve

Figures 10.7 and 10.8 present the MMPI-2 basic scale and content scale profiles for Steve, a 30-year-old Caucasian male. The MMPI-2 was administered when he and his wife consulted a community mental health agency for help with marital problems. Steve and his wife had been married for 6 years and had two children. His wife was a high school graduate who was not employed outside of the home. He graduated from high school and had been employed by several different companies as a salesman. Steve's wife complained that he was selfish, insensitive, and inconsiderate. She said that he used alcohol excessively and that he often stayed out late at night without telling her where he was. Although Steve felt that his wife's concerns were exaggerated, he agreed to seek professional help when she threatened to divorce him if he did not do so.

It would appear that Steve completed the MMPI-2 in a valid manner. He omitted no items, and his scores on the three validity scales are not suggestive of defensiveness, exaggeration, or other response sets.

The MMPI-2 profile suggests that Steve is not presenting himself as experiencing psychological distress or as being psychologically maladjusted. Only two T scores are higher than 65, and neither is higher than 70. The F-scale T score of approximately 65 suggests that he is admitting to some deviant attitudes and behaviors, but it is not high enough to be suggestive of serious psychopathology. The scores on scales 2 and 7 are about average, suggesting that he is not overwhelmed by depression, anxiety, or other emotional turmoil. However, a moderate elevation on the Depression (DEP) content scale indicates that he may be dissatisfied with his current life situation.

When the 4–9 code type is significantly higher than it is in Steve's profile, it is suggestive of marked disregard for social standards and values and of asocial or antisocial behavior. At this less elevated level, the code type suggests rebellion, nonconformity, and resentment of authority. However, the very elevated Antisocial Practices (ASP) content scale score raises concern that he may act out in asocial or antisocial ways. Steve's history would be helpful in deciding to what extent we might expect him to act out in the future. The possibility of substance abuse is raised by the 4–9 code type and is consistent with his wife's allegation that he uses alcohol excessively. His MAC-R raw score of 29 is high enough to raise concerns about substance abuse.

The 4–9 code type and the very low scale 0 score (T = 37) lead us to infer that Steve is an ambitious and energetic person and that he may be restless and overactive. He is likely to seek out excitement and emotional stimulation. He can be expected to be extroverted and talkative and to create good first impressions. This configuration of scores also suggests that Steve is narcissistic and self-indulgent. He may show poor judgment, often acting without considering the consequences of his actions. He is unwilling to accept responsibility for his own behavior and is quick to blame failures and difficulties on other people. He has a low tolerance for frustration and harbors intense feelings of anger and hostility that may get expressed in occasional emotional outbursts. Because of his self-centeredness, relationships with other people are likely to be superficial. He seems incapable of forming deep emotional ties, and he keeps others at an emotional distance. The high score on the Cynicism (CYN) content scale suggests that he sees other people as dishonest, selfish, and untrustworthy. The 4–9 code type suggests that beneath an outward facade of self-confidence and security, he may be a rather immature, insecure, and dependent person. The moderately elevated score on the Low Self-esteem (LSE) content scale reinforces the inference concerning self-doubts. The very low scale 5 score (T = 37) indicates that Steve is presenting himself in very stereotypically masculine ways.

Based on the high MAC-R raw score and his wife's allegations of excessive alcohol abuse, the appropriateness of an Axis I substance abuse diagnosis should be considered. A very elevated 4–9 profile would indicate an Axis II diagnosis of antisocial personality disorder. Steve's moderately elevated 4–9 profile does not justify such a diagnosis, but many of the personality characteristics associated with this personality disorder may be present.

Several inferences about treatment implications are appropriate. Because he is not acknowledging psychological problems, is not experiencing much emotional turmoil, and does not accept responsibility for his own behavior,

Figure 10.7. MMPI-2 profile for basic scales for practice case (Steve). Reproduced by permission of University of Minnesota Press.

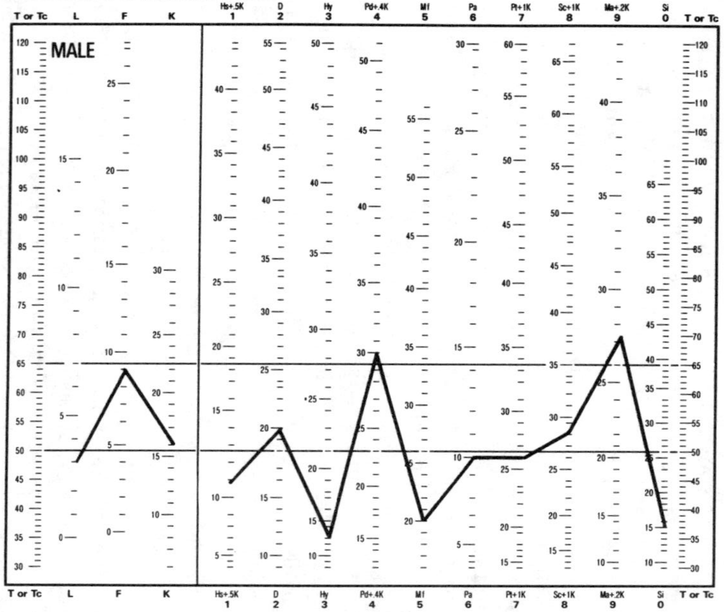

MMPI-2
S.R. Hathaway and J.C. McKinley
Minnesota Multiphasic Personality Inventory-2™

Profile for Basic Scales

Minnesota Multiphasic Personality Inventory-2
Copyright © by THE REGENTS OF THE UNIVERSITY OF MINNESOTA
1942, 1943 (renewed 1970), 1989. This Profile Form 1989.
All rights reserved. Distributed exclusively by NATIONAL COMPUTER SYSTEMS, INC.
under license from The University of Minnesota.

"MMPI-2" and "Minnesota Multiphasic Personality Inventory-2" are trademarks owned by
The University of Minnesota. Printed in the United States of America.

Name _Steve_
Address _____
Occupation _Salesman_ Date Tested _/ /_
Education _12_ Age _30_ Marital Status _Married_
Referred By _____
MMPI-2 Code _94+- 28/671: 503 F-K/L_
Scorer's Initials _JRG_

Fb: T=70
VRIN: T=61
TRIN: T=65T
MAC-R Raw Score: 29

MALE

	L	F	K	Hs+.5K 1	D 2	Hy 3	Pd+.4K 4	Mf 5	Pa 6	Pt+1K 7	Sc+1K 8	Ma+.2K 9	Si 0
Raw Score	3	9	16	3	20	13	24	20	10	16	12	24	15
? Raw Score _0_													
K to be Added _8_				8			6			16	16	3	
Raw Score with K _11_				11			30		26	28	27		

NATIONAL
COMPUTER
SYSTEMS

24001

Figure 10.8. MMPI-2 profile for content scales for practice case (Steve). Reproduced by permission of University of Minnesota Press.

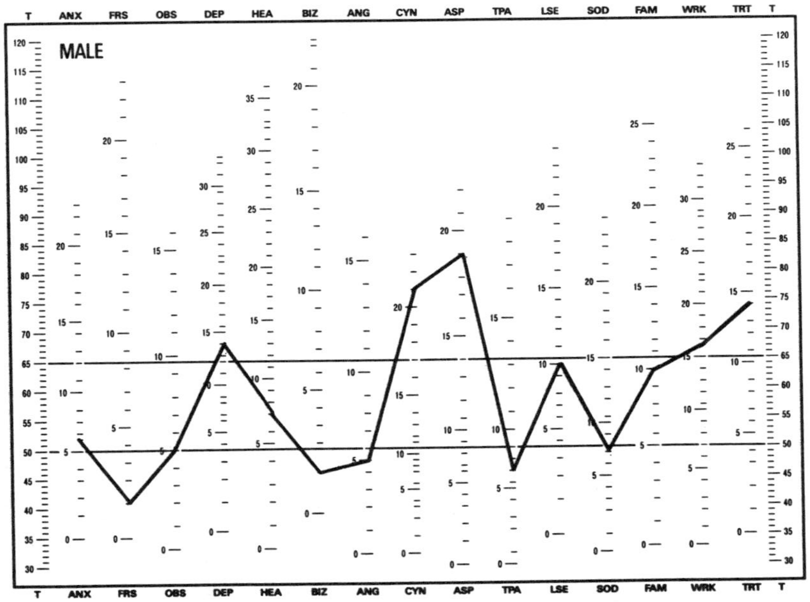

255

Steve is not likely to benefit much from traditional psychotherapy. The elevated score on the Negative Treatment Indicators (TRT) content scale suggests that he is likely to have negative attitudes toward mental health treatment, to feel that others cannot understand his problems, and to believe that he is unable to make significant changes in his life. He may agree to treatment to avoid something more negative (e.g., divorce), but he is likely to terminate prematurely after the immediate crisis passes. If corroborating information supports the inferences about substance abuse problems, a treatment program should include a substance abuse component.

Greg

Figures 10.9 and 10.10 present the MMPI-2 basic scale and content scale profiles for Greg, a 22-year-old Caucasian male. He had never been married and was living with his parents. He graduated from a vocational high school, where his grades were about average. He did not participate in sports or other school activities, and he did not have any close friends. He had never had a serious relationship with a woman. After completing high school, he held a number of temporary and part-time jobs, including janitorial and restaurant kitchen work. At the time of the current hospital admission, he was unemployed.

The MMPI-2 was administered to Greg approximately one week following his admission to a state-supported psychiatric hospital. He was cooperative, and he completed the MMPI-2 in about two hours. Greg had a history of several prior psychiatric hospitalizations during which he was diagnosed as paranoid schizophrenic. At the time of the current admission, Greg was quite anxious, fearful, and agitated. He reported that people were following him and trying to kill him, but he did not state any reasons why they might want to do so. He admitted that he had been hearing voices telling him that he was a bad person. During his hospitalization, Greg was treated with antipsychotic medications and supportive psychotherapy, and his symptoms improved. After several weeks he was discharged to his family as improved, and outpatient treatment at a local mental health center was recommended.

The validity scales indicate that Greg approached the MMPI-2 in a valid manner. His L- and K-scale scores are not indicative of defensiveness. His T score of 89 on the F scale suggests that he was admitting to a large number of deviant attitudes and behaviors. However, the F-scale score is not high enough to suggest random or fake-bad responding. The VRIN and TRIN scale scores rule out random responding and a true response bias.

The high score on the F scale, two clinical scale T scores higher than 75, and the 8–6 two-point code type are indicative of serious psychopathology. This impression is reinforced by significantly elevated scores on six of the content scales and by the positive slope (left side low, right side high) of the clinical scale profile. The absence of high scores on scales 2 and 7 suggests that Greg is not feeling overwhelmed by emotional turmoil.

The 8–6 code type, with both scales quite high and both higher than the scale 7 score, indicates that Greg is likely to be experiencing frankly psychotic symptoms. This inference is reinforced by the very high score (T = 84) on the Bizarre Mentation (BIZ) content scale. His thinking is likely to be autistic, fragmented, tangential, and circumstantial, and thought content is likely to be bizarre. Difficulties in concentrating and attending, deficits in memory, and poor judgment are likely. Delusions of persecution or grandeur and hallucinations may be present, and feelings of unreality may be reported. Greg is likely to have difficulty handling the responsibilities of daily life, and he may withdraw into fantasy and daydreaming during times of increased stress. The high score on the Fears (FRS) content scale suggests that he feels fearful and uneasy much of the time and that he may report multiple specific fears or phobias.

The 8–6 code type and the high score (T = 77) on the Low Self-esteem (LSE) content scale indicate that Greg harbors intense feelings of inferiority and insecurity. He lacks self-confidence and feels guilty about perceived failures. He is likely to withdraw from everyday activities and may have suicidal ideation. He is not likely to be emotionally involved with other people. He is suspicious and distrustful of others and avoids deep emotional ties. In general, his lifestyle can be characterized as schizoid.

The MMPI-2 data are consistent with an Axis I diagnosis of schizophrenic disorder, paranoid type. No direct inferences can be made concerning Axis II diagnoses for Greg. However, the symptoms and personality characteristics suggested by the MMPI-2 scores are consistent with a diagnosis of schizoid personality disorder.

Both direct and second-order inferences can be made concerning implications for treatment. Based on other inferences made about Greg, it is clear that he has very serious psychological problems, is likely to have considerable difficulty meeting the demands of his daily life, and is in need of psychological or psychiatric treatment. If he were not already in an inpatient psychiatric facility, one would infer that he should be. The 8–6 code type suggests that a medical consultation is indicated to determine the appropriateness of antipsychotic medication. The prognosis for psychological treatment is not good. Greg's problems are serious and long-standing. Although he may agree to enter treatment because he is feeling overwhelmed, he is likely to have difficulty establishing a trusting relationship with a therapist and accepting responsibility for his own behavior. However, he is likely to remain in treatment longer than many patients. Although he may want to deal with vague and abstract issues in therapy, he is likely to benefit more from treatment that focuses on specific, concrete problems in his immediate life situation.

GIVING FEEDBACK TO CLIENTS

Although most persons who take the MMPI-2 are interested in their test performance and many expect to receive feedback about their test results, often such feedback is not given at all or is not given in a comprehensive and sys-

Figure 10.9. MMPI-2 profile for basic scales for practice case (Greg). Reproduced by permission of University of Minnesota Press.

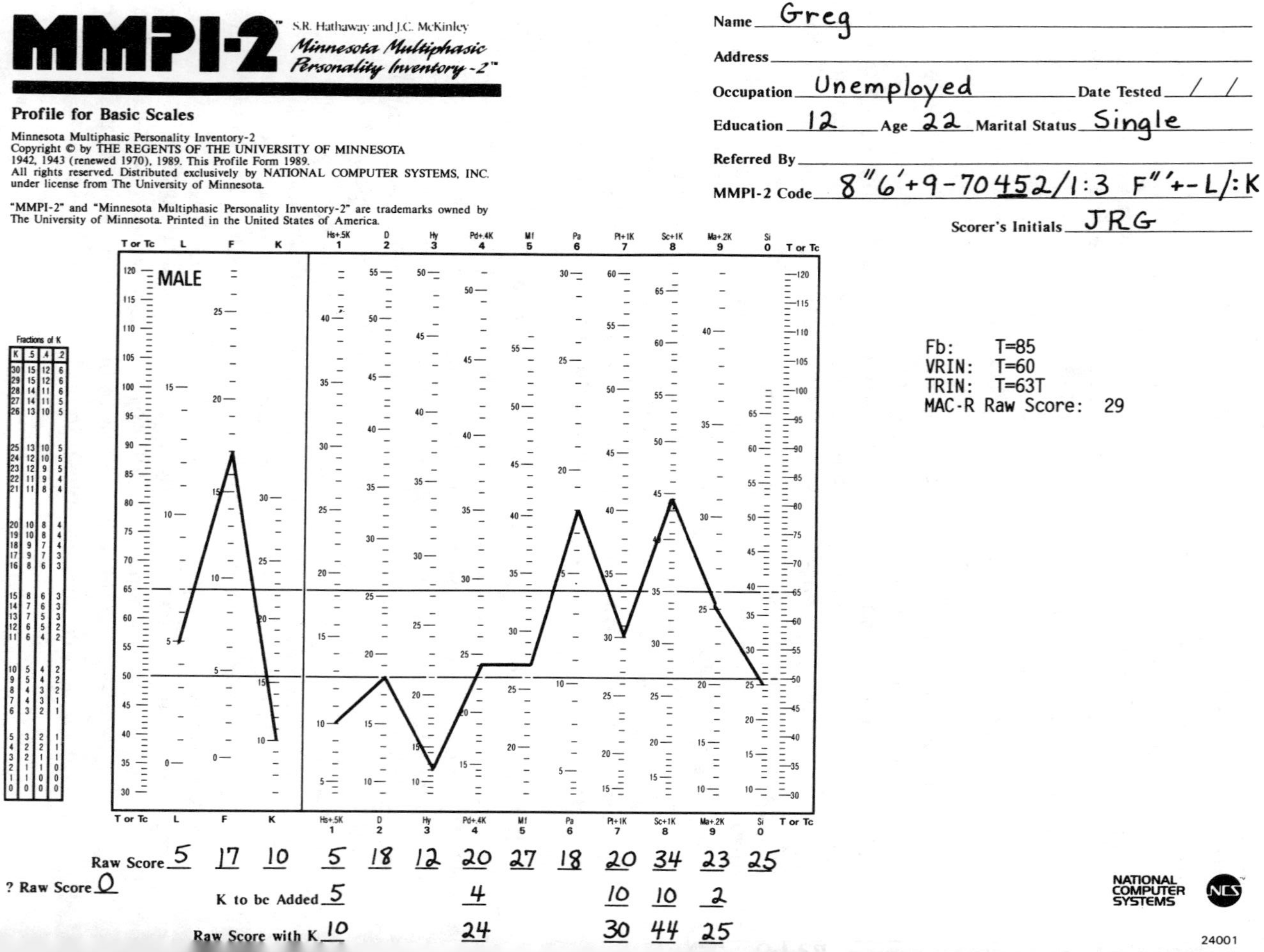

MMPI-2

S.R. Hathaway and J.C. McKinley

Minnesota Multiphasic Personality Inventory-2™

Profile for Basic Scales

Minnesota Multiphasic Personality Inventory-2
Copyright © by THE REGENTS OF THE UNIVERSITY OF MINNESOTA
1942, 1943 (renewed 1970), 1989. This Profile Form 1989.
All rights reserved. Distributed exclusively by NATIONAL COMPUTER SYSTEMS, INC.
under license from The University of Minnesota.

"MMPI-2" and "Minnesota Multiphasic Personality Inventory-2" are trademarks owned by
The University of Minnesota. Printed in the United States of America.

Name Greg

Address

Occupation Unemployed Date Tested / /

Education 12 Age 22 Marital Status Single

Referred By

MMPI-2 Code 8"6'+9-70452/1:3 F"+- L/:K

Scorer's Initials JRG

Fb: T=85
VRIN: T=60
TRIN: T=63T
MAC-R Raw Score: 29

MALE

	L	F	K	Hs+.5K 1	D 2	Hy 3	Pd+.4K 4	Mf 5	Pa 6	Pt+1K 7	Sc+1K 8	Ma+.2K 9	Si 0
Raw Score	5	17	10	5	18	12	20	27	18	20	34	23	25
K to be Added				5			4			10	10	2	
Raw Score with K				10			24			30	44	25	

? Raw Score 0

24001

NATIONAL COMPUTER SYSTEMS

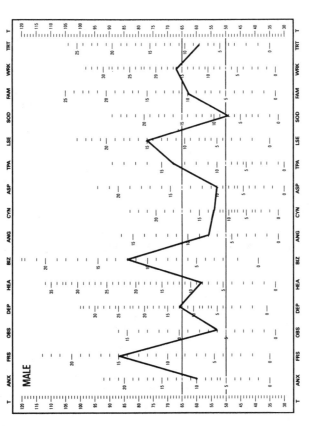

Figure 10.10. MMPI-2 profile for content scales for practice case (Greg). Reproduced by permission of University of Minnesota Press.

259

tematic way. In addition to client expectations, there are other important reasons for routinely giving feedback about MMPI-2 results (Butcher, 1990b; Finn, 1990; Pope, 1992). In many circumstances clients have legal rights to information about their test results. Also, according to the *Ethical Principles of Psychologists* (American Psychological Association, 1990), psychologists have the professional responsibility to provide clients with information about test results in a manner that can be easily understood.

In addition, giving feedback about MMPI-2 results to clients can be clinically very beneficial. If handled well by clinicians, feedback can be a vehicle for establishing good rapport with clients. Feedback also can be helpful to clients in understanding why treatment is being recommended, suggesting possible problem areas to be explored in treatment, and identifying resources that can be utilized in treatment. In fact, some preliminary data suggest that receiving MMPI-2 feedback is associated with a significant decline in symptomatic distress and a significant increase in self-esteem (Finn & Tonsager, 1992). When the MMPI-2 is repeated during treatment, discussion of changes in results can help the client and therapist assess treatment progress and define additional treatment goals.

As discussed in Chapter 2, one is more likely to obtain valid and interpretable test results if, prior to test administration, the examiner explains why the MMPI-2 is being administered, who will have access to the test results, and why it is in the client's best interest to cooperate with the testing. At this time the examiner can also explain that the client will be given feedback about test results and will have an opportunity to ask general questions about the test and specific questions about test results. Clients often ask for a written report of test results and interpretations. In most cases, it is not a good idea to provide such a report, because the clinician cannot be sure that the client will understand everything in the report. Instead, indicate that you will be willing to meet with the client after the testing has been completed, discuss the results, and give the client an opportunity to ask questions and make comments.

General Guidelines

Graham (1979) suggested some important general guidelines to keep in mind when giving feedback to clients:

1. You should communicate in a manner that is easily understood by the client. Some clients will be able to understand rather complex and technical explanations, whereas others will require greatly simplified explanations.
2. Use vocabulary that the client can understand. Avoid psychological jargon. If you use psychological terms, take the time to explain exactly what they mean.
3. Present both positive and negative aspects of the client's personality and functioning. Finding positive things to say may be rather difficult in the case of certain clients, but one can usually do so. Clients are much more likely to accept

interpretations if you maintain a balance between positive and negative characteristics.

4. Avoid terms such as "abnormal," "deviant," or "pathological." It is helpful to explain that most symptoms and negative characteristics are shared by most people but perhaps are not possessed by individuals to the same extent as by a particular client.

5. Do not overwhelm the client with a long list of adjectives. Instead, limit your interpretations to a few of the most important things you want the client to hear and understand, and explain each as fully as possible to ensure that the client understands what you are trying to communicate.

6. Encourage clients to make comments and ask questions about what you have said. This often provides additional information about clients and makes them feel that they are a part of the process.

7. Do not argue with clients or otherwise try to convince them that your interpretations are correct. This will increase their defensiveness and can jeopardize your future role as therapist or counselor.

8. When you have completed your discussion of the test results, ask the client to summarize the major points that have been covered. This will increase the likelihood that the client will remember what has been discussed and will give you an opportunity to clarify any misunderstandings that clients might have about what was discussed.

Explaining the MMPI-2

To increase the likelihood that the client will have confidence in your interpretations and take the feedback seriously, you should indicate that the MMPI-2 has been used by psychologists for more than fifty years, that there have been thousands of research studies concerning what MMPI-2 scores mean, and that the test was revised and updated in 1989. You could also point out that the test is the most widely used psychological test in the world.

Before giving specific feedback about a client's MMPI-2 results, you should spend some time explaining in general how the test was developed and how it is interpreted. You may want to illustrate your explanation by referring to an MMPI-2 profile. It is a good idea not to use the client's own profile at this time. Clients may be so anxious about their own scores that it will be difficult for them to attend to what you are saying.

Scale Development

Typically, it is not very difficult to explain to clients how the MMPI-2 scales were developed. For most people the empirical construction of the basic scales is easy to understand and makes sense. Having an understanding of these procedures increases the client's confidence in the test and in your interpretations.

You should explain that the MMPI-2 is made up of a large number of statements to which true or false responses are given. The basic MMPI-2 scales were developed by comparing the item responses of a group of patients having specific kinds of problems (e.g., anxiety and depression) with the responses of patients who did not have these specific problems and with the responses of persons who were not having any serious psychological problems. You should avoid mentioning that the groups were defined according to psychiatric diagnoses. If a client persists in asking what the scale abbreviations (e.g., Sc) stand for, you should give honest answers but emphasize that these labels are not important in the way that the MMPI-2 is used today. You should then point out that the client's responses were scored for each of these scales. It is helpful to show the client a profile sheet (not his or her own yet) and point out that each number at the top and bottom of the sheet corresponds to one of these original scales.

As was mentioned in Chapter 6, the content scales of the MMPI-2 represent direct communication between the client and the clinician. High scores on any of these scales indicate that the client wants you to know about certain symptoms, problems, and characteristics. Because you may want to present scores on some of the content scales to support inferences about a client, you should mention how these scales were developed. It probably will be sufficient to indicate that each of these scales is made up of items in the test with similar content. For example, the items in the Anxiety content scale all have to do with various aspects of anxiety.

Norms

You should then indicate that the client's score on each of these scales is compared with scores of a large group of persons living in communities throughout the United States. Avoid referring to "normals" or "norms," because clients who have scores different from this group may label themselves as "abnormal." You can again refer to a profile sheet, indicating that most people obtain scores near the lower heavy black line on the profile sheet. If clients ask about the meaning of the numbers along the left or right sides of the profile (T scores) you should explain in very simple terms what they mean. Otherwise, it probably is best not to deal directly with the T scores. You may want to indicate along the side of a profile sheet the percentage of community subjects scoring above various T-score levels (as illustrated in Figure 10.11). You can also point out that, because scores above the upper heavy black line (T > 65) are rare, we tend to emphasize them in our interpretations. These high scores indicate the likelihood of problems similar to those of persons involved in the original development of the scales (e.g., anxiety and depression). Because the meaning of low scores is not very clear at this time, you can simply indicate that scores below the lower heavy black line (T = 50) indicate that the client is not likely to have problems similar to those of people involved in the development of these scales. Finally, you can point

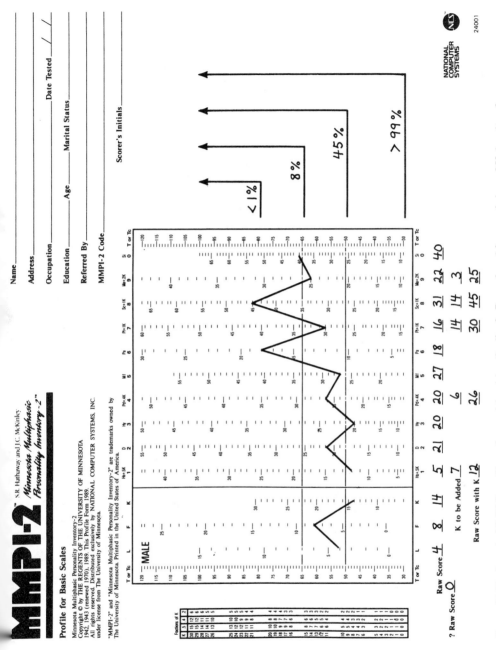

Figure 10.11. Sample MMPI-2 profile to be used in giving feedback. Reproduced by permission of University of Minnesota Press.

out that, because high scores on the scales are rare, they are emphasized in interpreting test results.

Validity Scales

You should emphasize that we can gain useful information about clients from their MMPI-2 results only if they have followed the test instructions (i.e., read each item and responded honestly as the item applies to them). Indicate that there are special MMPI-2 scales to help us determine if the instructions have been followed. Mention that these scales tell us if a person left too many items unanswered (?), responded to the items without really reading them (VRIN), was defensive and denying (L and K), or exaggerated problems and symptoms (F). You should spend more time explaining the validity scales if the client's validity scale scores suggest that the results may be invalid or of questionable validity.

Sources of Interpretive Statements

After you are confident that the client understands how the scales were developed and what is meant by high scores, you should explain how interpretive inferences are made on the basis of these scores. Emphasize that the inferences are based on extensive study of persons who have obtained high scores on the various scales. For example, when we have studied persons who have had high scores on scale 2, we have found that they have reported feeling more depressed than persons scoring lower on that scale. You can also mention that such studies have been done concerning persons who obtain high scores on several of the scales at the same time (e.g., 2 and 7). In interpreting a specific client's scores, we infer that the client will have problems and characteristics similar to persons we have studied who have had similar scores. Clients often ask about the meaning of specific MMPI-2 items. You should acknowledge that their responses to such items are important because they represent something that the clients want us to know but that we tend to emphasize scale scores rather than individual item responses in our interpretations.

Organizing Feedback

Once you are confident that the client has a basic understanding of how the MMPI-2 was developed and how it is interpreted, you are ready to present feedback about his or her specific results. There is no correct or incorrect way of organizing your feedback. However, you may find it helpful to organize it by using some of the categories described earlier in this chapter: (1) test-taking attitudes; (2) overall adjustment level; (3) characteristic traits and behaviors (e.g., symptoms, needs, self-concept, interpersonal relationships, psychological resources); and (4) treatment implications. In most instances you should not talk specifically about diagnosis, and discussion of dynamics and etiology can be deferred until later interactions with the client.

Illustration

Assuming that a general explanation of the MMPI-2 has been given, we can illustrate the specific feedback that we might give to Jeff, the person whose MMPI-2 was interpreted previously. You will remember that Jeff is a 23-year-old Caucasian who took the MMPI-2 at a mental health center where he had been referred after emergency treatment following ingestion of a large number of aspirin.

You should begin by pointing out to Jeff that he was very cooperative in completing the MMPI-2. He omitted no items, and his average scores on the L, F, and K scales (and on several additional validity scales not included on the standard profile sheet) suggest that he carefully read each item and responded to it honestly and thoughtfully as it applied to him. Therefore, you have confidence that his scores on the other scales will give an accurate picture of what he is like.

Next, you should indicate that he seems to be in a great deal of emotional turmoil. His high score on scale 2 indicates that he is likely to be depressed and dissatisfied with his current life situation, and his high score on scale 7 indicates that he is likely to feel anxious, tense, and nervous much of the time. The impression of turmoil is reinforced by high scores on several of the content scales (Anxiety, Depression, and Obsessiveness). You could add that he seems to be lacking in self-confidence (high scores on scales 2 and 7 and on the Low Self-esteem content scale). High scores on scale 0 and on the Social Discomfort content scale suggest that he is shy, introverted, and uncomfortable around people that he does not know well. The high score on the Work Interference content scale suggests that his psychological problems are likely to interfere with work performance.

You could point out that several scores suggest that he feels that he cannot cope with the demands of his everyday life situation. The most important source of this inference is the very low score on the Ego Strength (Es) scale. You could explain that this means that he is feeling that he just does not know how to cope with everything that is happening in his life at this time.

You probably then would want to point out the high scores on scales assessing substance abuse problems (MAC-R, AAS, APS), emphasizing that in his responses to items on the AAS scale he clearly admitted to such problems. You could speculate that he might be using alcohol and/or other drugs as a way of trying to handle some of the discomfort that he is feeling.

You could indicate that the high score on scale 6 could mean that he sees the world as a rather demanding place and feels that he gets a raw deal from life. He may be somewhat suspicious and skeptical of the motives of other people. His high score on the Anger content scale suggests that he is resentful of this perceived mistreatment.

Based on the inferences that have been made about psychological turmoil and feeling overwhelmed by the demands of everyday life, you might indicate that he probably is feeling the need for some professional psychological help. You could emphasize that when people are in so much turmoil, they generally are willing to get involved in treatment and often show positive changes as a result of treatment. However, you would probably want to add that the moderately high score on the Negative Treatment Indicators content scale suggests that he may have reservations about sharing his feelings with other people because he does not think that they will understand them and that he may feel unable to bring about changes in his own life.

The concerns about substance abuse suggest that additional assessment of this possibility is indicated. If he readily admits to such problems, you would want to encourage him to consider involvement with Alcoholics Anonymous or some other program designed to help individuals deal with substance abuse problems.

At this point, having pointed out many of Jeff's problem areas and negative characteristics, you should balance the feedback by mentioning some positive aspects of his MMPI-2 results. You could point out that his high scores on scales 2 and 7 indicate that he is likely to be a persistent, conscientious, and reliable person. These high scores, coupled with his high score on scale 0, indicate that he has a strong need to achieve, probably enjoys work, and gets pleasure from productive personal achievement.

The reader will note that many of the inferences made about Jeff earlier in this chapter have not been addressed in the feedback session. In keeping with the recommendation made earlier in this section, feedback should be limited to a few of the most important things that you want to communicate to Jeff. Although you should have encouraged Jeff's questions and reactions throughout the feedback session, near the end of the session you should ask very directly if he has any questions or comments about the feedback or any aspect of his MMPI-2 results. You would end the feedback session by having Jeff summarize what had been discussed, which would give you the opportunity to determine if he had misunderstood or misinterpreted some aspects of the feedback. If he had, you should carefully restate your interpretations and again ask Jeff to repeat what he heard.

If you were going to be Jeff's therapist or counselor, you might want to mention that you and he probably will discuss the MMPI-2 results again at various times during treatment. If you plan to readminister the MMPI-2 during treatment, you should mention this possibility to him and indicate that it will give you and him an opportunity to examine possible changes as treatment progresses.

11

Computerized Administration, Scoring, and Interpretation

In an era of almost unbelievable advances in computer technology, it is not at all surprising that there have been increasing efforts to automate the assessment process. Automation in psychological assessment occurs whenever a computer performs functions previously carried out by clinicians. Computers can be used to administer, score, and interpret tests. Although automation has been applied to a variety of psychological tests, objective personality inventories, such as the MMPI-2, lend themselves most readily to automation. Butcher (1987) has pointed out that increasing applications of computers in psychology are altering the ways psychologists function.

Computers have some definite advantages over human clinicians in the assessment enterprise. First, they are very efficient. Operations that would take a person minutes, or even hours, to complete can be performed by computers in seconds. Second, computers are accurate and reliable. Assuming that accurate information has been programmed into a computer, there is almost perfect reliability in the functions that it performs. Third, the computer has far greater storage capacity than the human clinician. A virtually infinite number of bits of information can be stored by the computer, to be called upon as needed in the assessment process. Fourth, the flexibility provided by computer technology offers the possibility of developing tests or sets of test items tailored to the individual examinee. Programs can be written to interact with the examinee so that the answer to any specific item determines the next item to be presented, skipping irrelevant data and/or exploring some areas in more depth. Although such computer-tailored assessment procedures have been employed in ability testing and diagnostic interviewing, this potential has not been widely exploited in personality assessment (Butcher, Keller, & Bacon, 1985). However, there have been several demonstrations that adaptive testing procedures can be applied to the MMPI and the MMPI-2 (Ben-Porath, Slutske, & Butcher, 1989; Butcher et al., 1985; Roper, Ben-Porath, & Butcher, 1991).

AUTOMATED ADMINISTRATION

Instead of using the traditional test booklets and answer sheets, subjects can complete the MMPI-2 on a computer terminal or personal computer. The typical procedure is for subjects to sit at the computer while the test items are individually displayed on a monitor. A response is made to each item by pressing designated keys on the keyboard. Each response is recorded automatically and saved in the computer's memory for later processing. Programs can permit subjects to change responses after they are made and to omit items and consider them again at the end of the test.

Automated administration has several advantages. First, many subjects find it more interesting than the traditional procedure and are more motivated to complete the task (Carr, Ghosh, & Ancill, 1983; Evan & Miller, 1969; Greist & Klein, 1980; Honaker, Harrell, & Buffaloe, 1988; Lucas, Mullin, Luna, & McInroy, 1977). Second, because subjects enter response data into the computer, no professional or clerical time is used for this purpose. However, a major disadvantage is that each administration consumes an hour or more of computer time. Also, some subjects, particularly upset and confused clients or patients, may be overwhelmed by the task.

As discussed previously, the MMPI-2 is a very robust instrument. Various forms of the original MMPI (e.g., booklet, audio tape recording) yielded essentially equivalent results. However, we must be concerned about the equivalence of the computer-administered and more traditional versions of the test. Guidelines for computer-based testing developed by the American Psychological Association (1986) clearly specify that the equivalence of traditional and computer-administered procedures must be demonstrated.

Moreland (1985a) reviewed data concerning the equivalence of MMPIs administered using computers and in traditional ways and concluded that differences due to type of administration typically are small and probably of little practical consequence. Honaker (1988) reviewed nine studies and concluded that available research did not justify the conclusion that computer administration yields equivalent scores to those obtained in the traditional manner. Honaker indicated that most studies of computerized and standard administrations of the MMPI have found differences on one or more scales. However, computer administration formats that provide response options that closely approximate those available for the traditional version are more likely to yield results that are comparable to those obtained from standard administration. Honaker also concluded that the reliability of scores derived from computer-administered MMPIs is comparable to or better than that of scores derived from standard administration. There has not yet been a study addressing the relative validity of scores derived from computer and standard administrations of the MMPI, MMPI-2, or MMPI-A.

Roper et al. (1991) administered the MMPI-2 twice to a group of college students. Each student completed the test once using a booklet and standard instructions and once using computerized administration. On the standard validity and clinical scales, men had significantly different scores for scales F, K, and Si, and women had no significantly different scores. Roper et al. concluded that the differences that were found were very small and would not warrant a different clinical interpretation. Because Roper et al. were studying an adaptive version of the MMPI-2, the computerized administration presented items in a different order from that used in the standard administration. The different order of items could have contributed to the small differences that were found in the study.

Based on the MMPI studies that have been published and on the MMPI-2 study by Roper et al., it would appear that differences between computerized and standard administrations are small and not clinically important. This is especially true when the procedures used for computerized administrations offer test subjects response options similar to those available in the standard administration. What is needed at this time are studies comparing MMPI-2 scores from standard test administrations and from computerized administrations using the software that is available from the test publisher. These studies should be conducted with both clinical and nonclinical subjects. The issue of validity of scores derived from the two forms of administration could be addressed by comparing relationships between scores derived from the two forms of administration and external information about subjects. If these studies demonstrate the equivalence of scores from standard and computer administrations, clinicians opting to use computer administration should use only the software provided by the test publisher. The use of nonstandardized software for test administration increases the likelihood that resulting scores will not be comparable to those that would have been obtained if the test had been administered using standard procedures.

AUTOMATED SCORING

Programming a computer to score the MMPI-2 is not a very difficult task. The computer's memory stores information that determines which items and which responses are scored for each scale. Dozens, and even hundreds, of scales can be scored in a matter of seconds. Because test norms also can be stored in the computer's memory, raw scores on the various scales can be converted easily to T scores, and profiles based on these T scores can be printed. This entire process can be accomplished in a fraction of the time that it would take to score and plot even the basic validity and clinical scales by hand. The use of MMPI-2 scoring information and norms is covered by copyright standards, so permission to develop scoring programs must be obtained from the test publisher.

Several different procedures for computer scoring of the MMPI-2 are offered by National Computer Systems (NCS). Software is available for scoring using personal computers. If the test is administered on-line, the individual responses are saved in the computer's memory and are ready for immediate scoring. If the subject uses a test booklet and answer sheet, an examiner or clerk can enter the subject's responses from an answer sheet via a computer keyboard. With a little practice, the 567 items can be entered in a very short time (5 to 10 minutes). For high-volume users special answer sheets can be processed by relatively inexpensive scanners that can be attached to personal computers. Another available alternative is for response data to be entered on a keyboard and transmitted via teleprocessing equipment to Minneapolis, where the scoring is done on a mainframe computer operated by the test distributor. The resulting scores are printed on the test user's computer printer, so they are available almost immediately. A final option is for special answer sheets to be used and mailed to Minneapolis for scoring. Ordinarily the resulting scores are sent by return mail within 24 hours of receipt. Persons who antici-pate using computerized scoring should consult with the test distributor to determine the most convenient and economical method for their particular circumstances and to make sure that appropriate materials are used.

AUTOMATED INTERPRETATION

Automated interpretation of the MMPI-2 is not as simple and straightforward as are administration and scoring. Interpretive statements are written for vari-ous scores and patterns of scores, and these statements are stored in the com-puter's memory. When a test is administered to a specific subject and scored, the computer searches its memory to find interpretive statements that previ-ously were judged to be appropriate for these particular scores and patterns of scores. These statements are then selected and printed.

It is important to distinguish between automated and actuarial interpreta-tion (Graham & Lilly, 1984). Automation refers to the use of computers to store interpretive statements and to select and assign particular statements to particular scores and patterns of scores. The decisions about which state-ments are to be assigned to which scores may be based on research, actuarial tables, or clinical experience. Regardless of how the decisions are made ini-tially, they are made automatically by the computer after the test is adminis-tered and scored.

Actuarial interpretations are ones in which the assignment of interpretive statements to scores and patterns of scores is based entirely on previously established empirical relationships between test scores and the behaviors included in the interpretive statements. Experience and intuition play no part in actuarial interpretation. This is procedure that Meehl seemed to have in mind when he made his pleas for a good cookbook for test interpre-tation (Meehl, 1956).

The MMPI-2 interpretive services currently available are not actuarial in nature. They are what Wiggins (1973) called automated clinical prediction.

On the basis of published research, clinical hypotheses, and clinical experience, clinicians generate interpretive statements judged to be appropriate for particular sets of test scores. The statements are stored in the computer and called upon as needed. It should be made perfectly clear that the accuracy (validity) of these kinds of interpretations depends on the knowledge and skill of the clinicians who generated the interpretive statements. The validity of these interpretations should not be assumed and needs to be demonstrated every bit as much as the validity of a test (Moreland, 1985c).

Unlike scoring programs, interpretive programs are not controlled by copyright procedures. As long as scoring and generation of T scores are not included, individuals are free to develop programs for interpreting MMPI-2 scores. Some programs are written by persons with considerable knowledge of the research literature and adequate experience with the original and revised instruments. Others are written by persons knowing very little about the tests. Because most interpretive services do not make available the algorithms underlying their programs, it is difficult to evaluate their accuracy. Ben-Porath and Butcher (1986) have suggested that potential users of computerized interpretation services should ask the following questions before deciding which service, if any, will be used:

1. To what extent has the validity of the reports been studied?
2. To what extent do the reports rely on empirical findings in generating interpretations?
3. To what extent do the reports incorporate all of the currently available validated descriptive information?
4. Do the reports take demographic variables into account?
5. Are different versions of the reports available for various referral questions (e.g., employment screening versus clinical diagnosis)?
6. Do the reports include practical suggestions?
7. Are the reports periodically revised to reflect newly acquired information?

These seem to be reasonable questions for potential users to ask of the services they are considering. Additionally, they should find out who actually wrote the interpretive programs and determine the professional qualifications of the writers to do so. Services that cannot or will not provide answers to these questions should not be considered seriously.

SAMPLE OF COMPUTERIZED INTERPRETIVE REPORT

To illustrate computerized interpretation of the MMPI-2, the answer sheet of the person discussed in detail in Chapter 10 (Jeff) was sent to the test distributor, NCS, for processing, and the resulting computer report is presented in its entirety in an appendix to this chapter. The Minnesota Report was written almost entirely by one interpretive expert, Dr. James N. Butcher at the University of Minnesota, incorporating both his knowledge of research data and his personal clinical experience. The program is built hierarchically around code type interpretations (Butcher, 1989b; Butcher et al., 1985). If a profile fits an established code type, a standard report for that configural pattern is printed. Additions and modifications are based on scores on other scales. If the profile does not fit an established code type, the report is based on a scale-by-scale analysis.

The Minnesota Report relies heavily on research findings on the original MMPI and the MMPI-2 and the report author's clinical experience with the original and revised versions of the test (Butcher, 1989b). According to the user's guide for the Minnesota Report (Butcher, 1989b), several demographic and situational variables are taken into consideration in generating interpretive statements about scores. Gender, age, educational level, marital status, and type of setting in which the test was completed are all considered.

Several different options are available for persons who want to use the computerized services provided by NCS. It is possible to obtain scoring only for the basic validity and clinical scales or for these scales plus a large number of supplementary scales and indices. In addition to the Minnesota Adult Clinical Report, which is the one illustrated in this chapter, there are reports available for use with adolescents, for personnel selection, and for use in alcohol and drug treatment settings. Readers wanting more information about these services should consult a recent NCS catalog or contact NCS.

As the reader can see in the sample report, the Minnesota Adult Clinical Report includes scores for the validity and clinical scales and for a large number of supplementary scales. Some of these scores are profiled, and others are listed along with corresponding T-score values. Two profile codes are reported for each case. The old Welsh code is based on K-corrected T scores resulting from a comparison of the subject's raw scores with data from the original normative sample. The new Welsh code is based on K-corrected T scores resulting from a comparison of the subject's raw scores with data from the MMPI-2 normative sample. Other measures, such as mean profile elevation, percentage of true and false responses, and the F-minus-K Index, are calculated and reported.

The narrative portion of the report contains several sections. The first section deals with the validity of the profile. If the profile is judged to be invalid, scores are reported, but no further interpretive statements are made. The second section provides a description of the most salient symptoms suggested by the test results. The third section describes the most likely manner in which the subject interacts with other people. The fourth section gives statements concerning the likelihood that the symptoms and other characteristics described in other sections are likely to remain stable or change over time. This section states whether or not the scores suggest code type definition. The fifth section provides the most likely DSM-III-R diagnoses for the subject. The sixth and final section of the narrative addresses treatment considerations. Following the narrative sections of the report are listings of the Koss-Butcher and Lachar-Wrobel critical items and items that the examinee has omitted. The last information presented in the report is a listing of individual item responses to the 567 items in the MMPI-2.

It is important to note that the Minnesota Report includes cautionary statements indicating that the descriptions, inferences, and recommendations contained in the report need to be verified by other sources of clinical information and that the information in the report should be most appropriately used by a trained, qualified test interpreter.

COMPARISON OF COMPUTERIZED AND CLINICIAN-GENERATED INTERPRETATIONS

A comparison of the Minnesota Report and the clinician-generated interpretation for the same case that was presented in Chapter 10 (Jeff) reveals considerable agreement between the interpretations. Both interpretations conclude that the MMPI-2 results are valid and interpretable. The Minnesota Report notes a tendency to exaggerate symptoms. This tendency is not mentioned in the clinician-generated interpretation because the F-scale score is not as high as one would expect from persons who are exaggerating symptoms and problems. Both interpretations indicate that the configuration of clinical scale scores is well defined and is likely to be stable over time. The computerized interpretation adds that the social introversion that characterizes Jeff is stable over long periods of time.

The two interpretations are remarkably similar in describing Jeff as experiencing emotional turmoil and feeling overwhelmed by the responsibilities of daily life. Both interpretations describe depression, anxiety, and somatic symptoms. The possibility of suicidal ideation is raised in both interpretations, but the clinician-generated interpretation adds that Jeff did not endorse the critical items dealing directly with suicidal ideation. Both interpretations mention lack of energy, problems with concentration, obsessiveness, guilt, and indecision, and they state that at times Jeff may feel out of control of his thought processes.

The two interpretations are quite similar concerning diagnoses that should be considered. On Axis I these include depressive and anxiety disorders. Both reports indicate that alcohol and drug use disorders also should be considered. On Axis II both interpretations indicate that Jeff's symptoms and behaviors are consistent with dependent and obsessive-compulsive personality disorder diagnoses.

Descriptions of Jeff's personality characteristics and behaviors are remarkably similar in the two interpretations. Jeff is described as a shy, insecure, passive-dependent person who is not appropriately assertive in relationships. Although he is capable of forming intimate relationships, his insecurity often interferes with them. Both interpretations note that Jeff is likely to be experiencing anger and resentment. The clinician-generated interpretation infers that the anger and resentment may result from perceptions of other people as not very understanding, supportive, or trustworthy. The clinician-generated interpretation also indicates that Jeff is not likely to express negative feelings openly most of the time but may do so in occasional emotional outbursts. Both interpretations describe Jeff as shy and socially uncomfortable. The clinician-generated interpretation raises the possibility that other people may misinterpret Jeff's shyness as aloofness and emotional distancing.

Both interpretations indicate that Jeff is in need of mental health services. Because of his intense discomfort, he is likely to be receptive to treatment, but he has a low capacity for self-change. Both interpretations suggest the possibility of medication to alleviate symptoms of depression. The computerized interpretation suggests several specific therapeutic approaches, cognitive behavioral therapy and assertiveness training, and suggests that negative work attitudes should be addressed in treatment. Whereas the clinician-generated interpretation indicates that a substance abuse component to treatment be considered, this is not mentioned in the computerized interpretation.

The clinician-generated interpretation indicates that Jeff is likely to stay in treatment longer than many patients, will not respond well to brief therapy, and will show slow but steady progress over time. The clinician-generated interpretation also mentions several characteristics that may interfere with effective treatment. Jeff is described as having limited psychological resources that could be utilized in therapy. He also can be expected to rationalize and intellectualize excessively and to resist psychological interpretations. He tends to be rigid and to give up easily when things are not going well. He may come to express considerable hostility toward the therapist.

There is at least one apparent inconsistency in the computerized interpretation. On the one hand Jeff is described as passive, unassertive, and unlikely to confront others. On the other hand, he also is described as competitive, uncooperative, and critical of others. The clinician-generated interpretation hypothesizes that Jeff harbors negative attitudes but does not express them for fear of losing the support of others.

The clinician-generated interpretation includes inferences about Jeff's personality and behavior not directly addressed by the computerized interpretation. Jeff is described as trying to handle stress through the use of denial, repression, and withdrawal into fantasy and daydreaming. He is described as cynical and suspicious of the motives of other people. He has strong needs for achievement and recognition for his accomplishments. Although the WRK scale suggests characteristics that could interfere with work performance, Jeff seems to derive satisfaction from productive work. The clinician-generated interpretation suggests some psychological resources not included in the computerized interpretation. Jeff is described as neat, meticulous, conscientious, persistent, and reliable.

In summary, the overall agreement between the computerized and clinician-generated interpretations is remarkable. Although there are some specific differences in emphasis, the descriptions that emerge from the two interpretations are very similar. The agreement is not really unexpected. The clinician who developed the interpretive program for the Minnesota Report and the clinician who did the interpretation in Chapter 10 were basing their interpretations on basically the same research data. For this sample case, the clinician-generated interpretation is more detailed than the computerized interpretation.

The comparison of the two interpretations for this sample case should not be considered as general support for the validity of computerized interpretations of the MMPI-2. However, it should be noted that the clinician-generated interpretation was completed before the author had access to the Minnesota Report. One must consider that the case used for this comparison is not a very difficult or complicated one. A clearly defined two-point code type emerges that has been researched extensively. Few internal inconsistencies mark the test scores. Of course, there very well might be less agreement between computerized and clinician-generated interpretations of less clearcut MMPI-2 profiles as these.

EVALUATION OF AUTOMATED SERVICES

As there were many different automated programs and services available for the original MMPI, it was difficult to evaluate the accuracy and usefulness of all of them. Some scoring programs used incorrect item numbers for certain scales and inappropriate procedures for converting raw scores to T scores. Because of recent clarifications of copyright standards, however, the number of computerized scoring programs available for the MMPI-2 has been significantly reduced, allowing the test publisher to monitor their accuracy more adequately. Nevertheless, the ultimate responsibility for assessing the accuracy of computerized services rests with the clinicians who use them (American Psychological Association, 1986).

Copyright standards do not apply to the interpretation of MMPI-2 scores. Therefore, several different companies offer interpretive programs and services for the MMPI-2. Although evaluating the adequacy of these services is an extremely difficult task, clinicians using the services have the responsibility for doing so.

There are important issues involved in the use of automated MMPI-2 interpretations. One has to do with the extent to which the automated reports are integrated by adequately trained clinicians with other data available about test subjects. Although most of the services advise users that inferences in the test reports need to be verified by other data sources, in practice some clinicians use the automated reports instead of a comprehensive assessment. This is not responsible clinical practice.

Another concern is that, because they are computer-generated, the automated reports are seen as valid and questions rarely are asked about research demonstrating their validity. In fact, Ziskin (1981) has recommended the use of automated reports in forensic cases because the computer-generated profiles and reports are viewed by judges and jurors as more scientific. Despite these perceptions, the validity of automated interpretations must be demonstrated empirically.

The qualifications of users of automated services are very important. Although services purport to assess the qualifications of potential users to make sure that they use the interpretive reports appropriately, many users of the services are not qualified professionally to use them. Some psychologists who are trained and licensed to practice psychology do not know enough about the MMPI-2 to evaluate the appropriateness of the automated interpretations. The services also are available to physicians, social workers, and others who are licensed to offer mental health services. Typically, the members of these other professions are not adequately trained to evaluate the appropriateness of the automated interpretations.

Most computer services list scores for large numbers of supplementary scales, many of which clearly are experimental in nature. Although information about the development, reliability, and validity of these scales sometimes is provided in materials supplied to users, the typical reader of computerized reports does not have readily accessible information permitting informed decisions about the relative importance to attach to the various scores provided. It would be very helpful if reports would indicate which scales are considered to be experimental so users could exercise appropriate caution in interpreting them.

Most services also provide numerous indexes, such as the Goldberg index and the Henrichs modification of the Meehl-Dahlstrom rules. The guidelines of the American Psychological Association indicate that when such indexes are presented in computerized interpretations, information concerning hit rates and other validity data should be available to users. Such information typically has not been provided by computer services.

Automated services for the MMPI-2 typically list critical items that test subjects have endorsed in the scored direction. Although critical-item endorsements represent an important additional source of inferences concerning test subjects, many users of the services overinterpret such endorsements. Each critical item is, in fact, a single item scale whose reliability is very questionable. Thus, extreme caution must be exercised in interpreting the critical items.

As Butcher (1978) noted, automated interpretive systems often become fixed at a rather naive level. Although the potential exists for modifying the systems as new interpretive data become available, there has been a tendency not to change systems that have been operating smoothly and producing a profit for companies. Butcher discussed several instances where, in response to critics, only minor cosmetic changes or no changes at all were made in existing interpretive programs.

PROFESSIONAL GUIDELINES FOR COMPUTERIZED ASSESSMENT

In an effort to try to address some of the potential problems involved in the use of automated assessment services, in 1966 the American Psychological Association developed some interim standards for such services (Fowler, 1969a). The standards made it clear that organizations offering the services have primary responsibility for ensuring that the services are used by qualified persons and for demonstrating the reliability and validity of the interpretations included in the reports. In 1986 the Committee on Professional Standards and the Committee on Psychological Tests and Assessment of the American Psychological Association published updated guidelines for computer-based tests and interpretations (American Psychological Association, 1986). These updated guidelines state that it is the responsibility of test developers of computer-based test services to demonstrate the equivalence of computerized and conventional versions of a test. Developers offering interpretations of test scores should describe how the interpretive statements are derived from the original scores and should make clear the extent to which interpretive statements are based on quantitative research versus clinical opinion. When statements in an interpretive report are based on expert clinical opinion, users should be given information that will allow them to weigh the credibility of the opinion. Developers are expected to provide whatever information is needed to permit review by qualified professionals engaged in scholarly review of their interpretive services.

The updated guidelines make it very clear that professionals are responsible for any use they make of computer-administered tests or computer-generated interpretations. Users should be aware of the method employed in generating the scores and interpretations and be sufficiently familiar with the test to be able to evaluate its applicability to the purpose for which it will be used. The user should judge, for each test taker, the validity of the computerized test report based on the user's professional knowledge of the total context of testing and the test taker's performance and characteristics.

Clearly, the developers of computerized testing services and the clinicians who use them share responsibility for ensuring that the results are valid and are used appropriately. In the past, it has seemed that each of these parties has assumed that the other has major responsibility.

EVALUATING THE VALIDITY OF COMPUTERIZED INTERPRETATIONS

It would be desirable to demonstrate empirically the validity of computer-generated interpretations of MMPI-2 results. To date no study has been published concerning the validity of computerized MMPI-2 interpretations. Several different approaches were used to try to determine the validity of automated interpretations of the original MMPI (Moreland, 1985c). Early

efforts involved asking users of the automated services to provide overall ratings of the accuracy of each automated report (e.g., Fowler, 1969b; Klett,
1971; Webb, 1970; Webb, Miller, & Fowler, 1969, 1970). Other studies were a
bit more sophisticated. Ratings were made of the accuracy of individual para-
graphs or individual statements within each report (e.g., Lachar, 1974a;
Lushene & Gilberstadt, 1972). Not surprisingly, the users rated most reports
as being accurate. A major problem with these studies is that judgments
about what patients really were like and impressions from the MMPI-based
reports were confounded. In addition, it is not possible to judge to what
extent reports were rated as accurate because of the inclusion of glittering
generalities (Baillargeon & Danis, 1984) in the reports.

The most acceptable validity studies have been those in which external criterion information has been collected by persons who had no knowledge of
MMPI results or of the automated reports. The interpretive statements in the
reports were compared with the external criterion information about patients
(e.g., Anderson, 1969; Hedlund, Morgan, & Master, 1972). The results of
studies of this kind suggested, at best, moderate validity for the MMPI automated reports. However, the level of accuracy of the automated reports was
not much different from that previously reported when clinicians examined
MMPI profiles of subjects and generated descriptions of the subjects (e.g.,
Graham, 1967; Little & Shneidman, 1959; Sines, 1959).

In a well-designed study, Moreland and Onstad (1985) attempted to overcome some of the problems of earlier studies. They asked clinicians to rate
the accuracy of sections of automated reports for sixty-six patients. To assess
discriminant validity, clinicians also rated "bogus" reports that were not actually those of the patients indicated but were for other profiles of the same
general code types and elevations as those of the patients. These investigators
found that the genuine reports were rated as more accurate than the bogus
reports overall and for five of the six sections of the reports.

Eyde, Kowal, and Fishburne (1991) studied the accuracy of statements in
computerized interpretations of MMPIs provided by seven different computer services. Experienced clinical psychologists rated the accuracy of individual statements in computerized reports for six military test subjects, some
of whom were patients and some of whom were subjects in a normative study.
Case history materials were used as the criterion. The results indicated that
the accuracy of statements varied considerably among the services. When the
seven services were rank ordered according to overall accuracy, the Minnesota Report was ranked as most accurate. However, it should be noted that
the accuracy for even the highest-ranked report was only modest. The percentages of Minnesota Report statements rated as accurate ranged from 25
percent to 58 percent for the various subjects studied. An interesting feature
of the study was that data for African-American and Caucasian test subjects
were analyzed separately. The authors concluded that differences in the accuracy of computerized interpretations for African-American and Caucasian
subjects were quite minimal. Although the methodology used in the Eyde et

al. study was an improvement over most earlier studies, it had several important limitations. The most important limitations were the small number of subjects studied and the limited number of MMPI code types represented.

CONCLUSIONS AND RECOMMENDATIONS

The use of computers for the administration of the MMPI-2 is not likely to become a widespread practice. The computer time required for subjects to take the test on-line is so great that in most settings this form of administration is not appropriate. MMPI-2 users who intend to utilize computerized administration procedures are cautioned that only limited data exist concerning the equivalence of MMPI-2 scores for standard and computerized administration procedures. Based on data from the original MMPI, it seems likely that equivalence can be demonstrated.

The use of computerized scoring of the MMPI-2 permits the clinician to obtain scores on numerous MMPI-2 scales in a very efficient manner. As long as the accuracy of the scoring programs can be determined, their use is recommended. Because copyright standards apply to computerized scoring of the MMPI-2, it seems likely that the test publisher and test distributor will be able to have adequate quality control of scoring programs.

Computerized interpretation of MMPI-2 scores is not covered by copyright standards. Persons are free to develop computer programs for interpreting scores. As a result, many different interpretive services are available to clinicians. The accuracy (validity) of these interpretations needs to be demonstrated empirically and cannot be assumed, and it is the responsibility of the developers of the computerized interpretations to do so. It is the responsibility of the users of the services to obtain information needed to make intelligent decisions concerning the use of the interpretations.

With all of the problems and concerns about computerized interpretations, what conclusions can be reached concerning their use with the MMPI-2? This author is in agreement with Matarazzo (1986) and Fowler and Butcher (1986) that computerized interpretations should not serve as the equivalent of or substitute for a comprehensive psychological assessment conducted by a properly trained clinician. The automated reports are intended as professional-to-professional consultations, and the hypotheses generated are to be considered in the context of other information available about the test subjects. It is the responsibility of the users of these reports to determine to what extent the interpretations apply to particular test subjects. When used in this manner by qualified professionals, the automated interpretive reports have considerable potential. When used instead of a comprehensive psychological assessment conducted by a qualified clinician, their use is irresponsible and not recommended.

Based on data for the original MMPI, it would appear that the accuracy of interpretive statements in automated reports varies considerably from service to service and from code type to code type within each service. The level of accuracy of the interpretive statements is at best modest, but in many cases is no less than the accuracy of clinician-generated descriptions of clients based on the same data. Moreland (1984) suggested that clinicians be especially skeptical of automated reports when the profiles involved are rare ones. In these instances, authors of the automated systems have had to rely less on empirical data and more on clinical lore and individual experience in generating interpretive statements.

178 000002

THE MINNESOTA REPORT:™ ADULT CLINICAL SYSTEM INTERPRETIVE REPORT

By James N. Butcher

Minnesota Multiphasic Personality Inventory -2™

S.R. Hathaway and J.C. McKinley

Client ID: 000000001
Report Date: 13-MAY-92
Age: 23
Sex: Male
Setting: Outpatient Mental Health
Education: 12
Marital Status: Never Married

PROFILE VALIDITY

This is a valid MMPI-2 profile. The client was quite cooperative with the
evaluation and appears to be willing to disclose personal information.
There may be some tendency on the part of the client to be overly frank and
to exaggerate his symptoms in an effort to obtain help. He may be open to
the idea of psychological counseling since his clinical scale pattern
reflects psychological symptoms in need of attention.

SYMPTOMATIC PATTERNS

The client's MMPI-2 profile reflects much psychological distress at this
time. He has major problems with anxiety and depression. He tends to be
high-strung and insecure, and may also be having somatic problems. He is
probably experiencing loss of sleep and appetite, and a slowness in personal
tempo.

Individuals with this profile often have high standards and a strong need to
achieve, but they feel that they fall short of their expectations and then
blame themselves harshly. This client feels quite insecure and pessimistic
about the future. He also feels quite inferior, has little self-confidence,
and does not feel capable of solving his problems.

In addition, the following description is suggested by the content of this
client's responses. He has endorsed a number of items suggesting that he is
experiencing low morale and a depressed mood. He is preoccupied with
feeling guilty and unworthy. He feels that he deserves to be punished for
wrongs he has committed. He feels regretful and unhappy about life, and
seems plagued by anxiety and worry about the future. He feels hopeless at
times and feels that he is a condemned person. He has difficulty managing
routine affairs, and the item content he endorsed suggests a poor memory,
concentration problems, and an inability to make decisions. He appears to
be immobilized and withdrawn and has no energy for life. He views his
physical health as failing and reports numerous somatic concerns. He feels
that life is no longer worthwhile and that he is losing control of his
thought processes. According to his self-report, there is a strong
possibility that he has seriously contemplated suicide. The client's recent
thinking is likely to be characterized by obsessiveness and indecision.

INTERPERSONAL RELATIONS

He appears to be quite passive and dependent in interpersonal relationships
and does not speak up for himself even when others take advantage. He
avoids confrontation and seeks nurturance from others often at the price of
his own independence. He forms deep emotional attachments and tends to be

quite vulnerable to being hurt. He also tends to blame himself for interpersonal problems.

He appears to be rather shy and inhibited in social situations, and may avoid others for fear of being hurt. He has very few friends, and is considered by others as "hard to get to know." He is quiet, submissive, conventional, and lacks self-confidence in dealing with other people. Individuals with this passive and withdrawing lifestyle are often unable to assert themselves appropriately, and find that they are frequently taken advantage of by others.

The content of this client's MMPI-2 responses suggests the following additional information concerning his interpersonal relations. He feels intensely angry, hostile, and resentful of others, and would like to get back at them. He is competitive and uncooperative, tending to be very critical of others.

BEHAVIORAL STABILITY

Individuals with this profile are often experiencing psychological distress in response to stressful events. The intense affect may diminish over time or with treatment.

The relative scale elevation of the highest scales in his clinical profile reflects high profile definition. If he is retested at a later date, the peak scores on this test are likely to retain their relative salience in his retest profile pattern. Social introversion tends to be very stable over time. His generally reclusive interpersonal behavior, introverted life style, and tendency toward interpersonal avoidance is likely to be prominent in any future test results.

DIAGNOSTIC CONSIDERATIONS

Individuals with this profile tend to be considered neurotic, and receive diagnoses such as Dysthymic Disorder or Anxiety Disorder. They may also receive an Axis II diagnosis of Dependent or Compulsive Personality Disorder. The alcohol or drug problems he has acknowledged in his responses should be taken into consideration in any diagnostic evaluation.

His extremely high score on addiction proneness indicators suggests great proneness to the development of an addictive disorder. Further evaluation of substance use or abuse problems is strongly recommended.

In his responses to the MMPI-2, he has acknowledged some problems with excessively using or abusing addictive substances.

TREATMENT CONSIDERATIONS

```
            TM                                                        page 3
MMPI-2                      TM
THE MINNESOTA REPORT:                    ID: 000000001   REPORT DATE: 13-MAY-92
ADULT CLINICAL SYSTEM
INTERPRETIVE REPORT
```

Individuals with this MMPI-2 pattern are usually feeling a great deal of discomfort and tend to want help for their psychological problems. His self-esteem is low and he tends to blame himself too much for his difficulties. Although he worries a great deal about his problems, he seems to have little energy left over for action to resolve them.

Symptomatic relief for his depression may be provided by antidepressant medication. Psychotherapy, particularly cognitive-behavioral treatment, may also be beneficial.

The passive, unassertive personality style that seems to underlie this disorder might be a focus of behavior change. Individuals with these problems may learn to deal with others more effectively through assertiveness training.

The item content he endorsed indicates attitudes and feelings that suggest a low capacity for self-change. His potentially high resistance to change might need to be discussed with him early in treatment in order to promote a more treatment-expectant attitude. In any intervention or psychological evaluation program involving occupational adjustment, his negative work attitudes could become an important problem to overcome. He holds a number of attitudes and feelings that could interfere with work adjustment.

NOTE: This MMPI-2 interpretation can serve as a useful source of hypotheses about clients. This report is based on objectively derived scale indexes and scale interpretations that have been developed in diverse groups of patients. The personality descriptions, inferences and recommendations contained herein need to be verified by other sources of clinical information since individual clients may not fully match the prototype. The information in this report should most appropriately be used by a trained, qualified test interpreter. The information contained in this report should be considered confidential.

ID: 000000001   RPT DATE: 13-MAY-92
SEX: Male   EDUC: 12
AGE: 23   MARS: Never Married
SETTING: Outpatient Mental Health

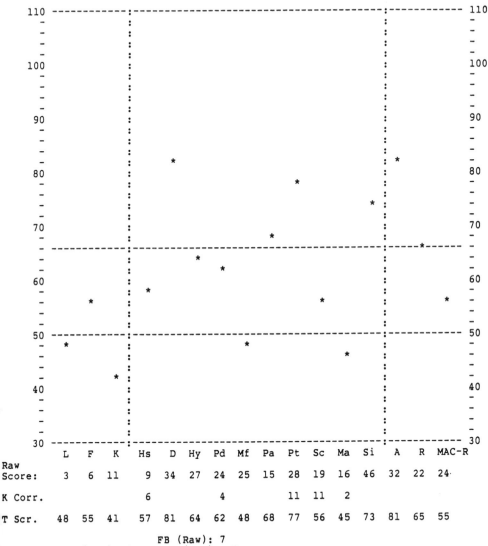

	L	F	K	Hs	D	Hy	Pd	Mf	Pa	Pt	Sc	Ma	Si	A	R	MAC-R
Raw Score:	3	6	11	9	34	27	24	25	15	28	19	16	46	32	22	24
K Corr.				6			4			11	11	2				
T Scr.	48	55	41	57	81	64	62	48	68	77	56	45	73	81	65	55

FB (Raw): 7
? Cannot Say (Raw): 0   F-K (Raw): -5

Welsh Code (new): 2"70'6+34-18/59:   F/LK:
Welsh Code (old): 2*7"046'385-19/   F-/KL?:

Percent True : 44
Percent False: 56
Profile Elev.: 63.8

285

MMPI-2

ID: 000000001   RPT DATE: 13-MAY-92
SEX: Male   EDUC: 12
AGE: 23   MARS: Never Married
SETTING: Outpatient Mental Health

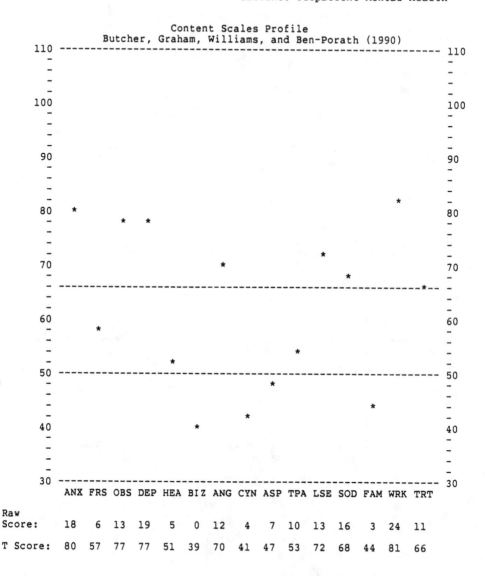

Content Scales Profile
Butcher, Graham, Williams, and Ben-Porath (1990)

	ANX	FRS	OBS	DEP	HEA	BIZ	ANG	CYN	ASP	TPA	LSE	SOD	FAM	WRK	TRT
Raw Score:	18	6	13	19	5	0	12	4	7	10	13	16	3	24	11
T Score:	80	57	77	77	51	39	70	41	47	53	72	68	44	81	66

286

SUPPLEMENTARY SCORE REPORT

	Raw Score	T Score
*Marital Distress (MDS)	--	--
Addiction Potential (APS)	33	76
Addiction Admission (AAS)	7	70
Ego Strength (Es)	25	30
Dominance (Do)	8	30
Social Responsibility (Re)	16	39
Overcontrolled Hostility (O-H)	8	35
PTSD - Keane (PK)	23	75
PTSD - Schlenger (PS)	34	79
True Response Inconsistency (TRIN)	9	50
Variable Response Inconsistency (VRIN)	4	46

Depression Subscales (Harris-Lingoes):

Subjective Depression (D1)	20	85
Psychomotor Retardation (D2)	8	65
Physical Malfunctioning (D3)	6	75
Mental Dullness (D4)	8	77
Brooding (D5)	5	68

Hysteria Subscales (Harris-Lingoes):

Denial of Social Anxiety (Hy1)	0	30
Need for Affection (Hy2)	8	55
Lassitude-Malaise (Hy3)	13	97
Somatic Complaints (Hy4)	2	48
Inhibition of Aggression (Hy5)	2	40

Psychopathic Deviate Subscales (Harris-Lingoes):

Familial Discord (Pd1)	2	51
Authority Problems (Pd2)	5	61
Social Imperturbability (Pd3)	0	30
Social Alienation (Pd4)	5	57
Self-Alienation (Pd5)	10	82

Paranoia Subscales (Harris-Lingoes):

Persecutory Ideas (Pa1)	3	58
Poignancy (Pa2)	3	55
Naivete (Pa3)	6	56

*No score is calculated for MDS (Marital Distress Scale) because the
individual either omitted their marital status or indicated they were
never married or were widowed.

	Raw Score	T Score

Schizophrenia Subscales (Harris-Lingoes):

	Raw Score	T Score
Social Alienation (Sc1)	4	55
Emotional Alienation (Sc2)	4	78
Lack of Ego Mastery, Cognitive (Sc3)	3	60
Lack of Ego Mastery, Conative (Sc4)	10	92
Lack of Ego Mastery, Def. Inhib. (Sc5)	3	61
Bizarre Sensory Experiences (Sc6)	2	51

Hypomania Subscales (Harris-Lingoes):

	Raw Score	T Score
Amorality (Ma1)	1	42
Psychomotor Acceleration (Ma2)	8	63
Imperturbability (Ma3)	0	30
Ego Inflation (Ma4)	3	50

Social Introversion Subscales (Ben-Porath, Hostetler, Butcher, & Graham):

	Raw Score	T Score
Shyness / Self-Consciousness (Si1)	13	74
Social Avoidance (Si2)	5	58
Alienation--Self and Others (Si3)	11	68

Uniform T scores are used for Hs, D, Hy, Pd, Pa, Pt, Sc, Ma, and the Content
Scales; all other MMPI-2 scales use linear T scores.

CRITICAL ITEMS

The following critical items have been found to have possible significance in analyzing a client's problem situation. Although these items may serve as a source of hypotheses for further investigation, caution should be taken in interpreting individual items because they may have been inadvertently checked.

Acute Anxiety State (Koss-Butcher Critical Items)

3. I wake up fresh and rested most mornings. (F)
10. I am about as able to work as I ever was. (F)
15. I work under a great deal of tension. (T)
39. My sleep is fitful and disturbed. (T)
140. Most nights I go to sleep without thoughts or ideas bothering me. (F)
172. I frequently notice my hand shakes when I try to do something. (T)
218. I have periods of such great restlessness that I cannot sit long in a chair. (T)
223. I believe I am no more nervous than most others. (F)
469. I sometimes feel that I am about to go to pieces. (T)

Depressed Suicidal Ideation (Koss-Butcher Critical Items)

9. My daily life is full of things that keep me interested. (F)
38. I have had periods of days, weeks, or months when I couldn't take care of things because I couldn't "get going." (T)
65. Most of the time I feel blue. (T)
71. These days I find it hard not to give up hope of amounting to something. (T)
95. I am happy most of the time. (F)
130. I certainly feel useless at times. (T)
233. I have difficulty in starting to do things. (T)
273. Life is a strain for me much of the time. (T)
388. I very seldom have spells of the blues. (F)
411. At times I think I am no good at all. (T)
454. The future seems hopeless to me. (T)
485. I often feel that I'm not as good as other people. (T)
518. I have made lots of bad mistakes in my life. (T)

Threatened Assault (Koss-Butcher Critical Items)

37. At times I feel like smashing things. (T)
85. At times I have a strong urge to do something harmful or shocking. (T)
213. I get mad easily and then get over it soon. (T)

389. I am often said to be hotheaded. (T)

Situational Stress Due to Alcoholism (Koss-Butcher Critical Items)

264. I have used alcohol excessively. (T)
487. I have enjoyed using marijuana. (T)
489. I have a drug or alcohol problem. (T)
518. I have made lots of bad mistakes in my life. (T)

Mental Confusion (Koss-Butcher Critical Items)

 31. I find it hard to keep my mind on a task or job. (T)
299. I cannot keep my mind on one thing. (T)
325. I have more trouble concentrating than others seem to have. (T)

Persecutory Ideas (Koss-Butcher Critical Items)

124. I often wonder what hidden reason another person may have for doing
     something nice for me. (T)
251. I have often felt that strangers were looking at me critically. (T)

Antisocial Attitude (Lachar-Wrobel Critical Items)

 27. When people do me a wrong, I feel I should pay them back if I can, just
     for the principle of the thing. (T)
 35. Sometimes when I was young I stole things, (T)
105. In school I was sometimes sent to the principal for bad behavior. (T)
266. I have never been in trouble with the law. (F)

Family Conflict (Lachar-Wrobel Critical Items)

 21. At times I have very much wanted to leave home. (T)

Somatic Symptoms (Lachar-Wrobel Critical Items)

175. I feel weak all over much of the time. (T)
229. I have had blank spells in which my activities were interrupted and I
     did not know what was going on around me. (T)
464. I feel tired a good deal of the time. (T)

12

Assessing Adolescents: The MMPI-A

The MMPI-2 should not be used with subjects younger than 18 years of age. The adolescent version of the instrument, the MMPI-A, should be used with subjects between the ages of 14 and 18. Because both the MMPI-2 and the MMPI-A can be used with 18-year-olds, clinicians must decide which version is most appropriate for each 18-year-old. Generally, it is advisable to use the MMPI-2 for mature 18-year-olds who are in college or otherwise living independently of their parents and the MMPI-A for less mature persons who have not adopted independent lifestyles.

DEVELOPMENT OF THE MMPI-A[1]

Rationale

Although the original MMPI was designed for use with adult subjects, almost from the time of its publication it also was used with adolescents (e.g., Hathaway & Monachesi, 1953, 1957, 1961, 1963). A recent survey (Archer et al., 1991) revealed that the MMPI was the most widely used objective instrument for assessing adolescents. However, significant concerns have been expressed about assessing adolescents using the MMPI, which had been developed with and normed on adults. Some of the items in the MMPI were awkward and inappropriate for many adolescents (e.g., "My sex life is satisfactory"; "I worry over money and business"). Additionally, the MMPI item pool contained very few items dealing with experiences unique to adolescence (e.g., school, peer-group issues).

There were very few MMPI scales developed specifically for use with adolescents. Rather, scales developed using adult samples were adapted for use with adolescents. For many of the supplementary scales that had been developed for use with adults, no adolescent normative data were available.

---

[1]Unless otherwise indicated, information concerning the development of the MMPI-A was abstracted from Williams, Butcher, Ben-Porath, and Graham (1992) and Butcher et al. (1992).

It was not at all clear whether clinicians using the MMPI to assess adolescents should use the adult norms or adolescent norms to convert raw scores to T scores. Use of the adult norms tended to overpathologize adolescent subjects (e.g., Archer, 1984, 1987; Klinge, Lachar, Grisell, & Berman, 1978). Use of adolescent norms (Dahlstrom et al., 1972; Marks et al., 1974) tended to underpathologize disturbed adolescents assessed in clinical settings (e.g., Archer, Stolberg, Gordon, & Goldman, 1986; Klinge & Strauss, 1976).

Another concern in using the MMPI with adolescents had to do with the source of interpretive statements about scores and configurations of scores of adolescent subjects. Should one use descriptors developed specifically for adolescents or those developed on the basis of adult data? Several studies suggested that the use of adult descriptors resulted in more accurate descriptions of adolescent subjects (Ehrenworth & Archer, 1985; Lachar, Klinge, & Grisell, 1976). Other studies (e.g., Wimbish, 1984) concluded that the most accurate inferences about adolescent subjects came from the use of adolescent-based descriptors.

Because of these kinds of issues, concurrently with the development of the MMPI-2, data were collected that later would be used to develop the MMPI-A, a version of the test designed specifically for use with adolescents (Butcher et al., 1992). Although there are marked similarities between the MMPI-A and MMPI-2, there also are some important differences.

Development of the MMPI-A Booklet

An experimental form (Form TX) was developed initially. It included the 550 items from the original MMPI with some items rewritten to eliminate awkward and archaic language. New items were added to assess treatment compliance, attitudes toward changing one's own behavior, treatment-related characteristics, and problems with alcohol and drugs. In addition, items were added dealing with adolescent-specific areas (e.g., school behavior, attitudes toward teachers, peer-group influences, eating problems, and relationships with parents and other adults).

Based on several criteria, items from the 704-item TX booklet were selected for inclusion in the MMPI-A. Most of the items scored on the standard MMPI scales were included (see Table 12.1). Several items from the standard scales were eliminated because they were judged to be inappropriate for this age group or had objectionable item content. In order to keep the length of the MMPI-A as short as possible, additional items that were scored only on scales 5 and 0 were eliminated. Items were also included that were needed to score some of the more promising supplementary scales and newly developed adolescent content scales.

Because of data indicating that the standard F scale did not work very well with adolescent subjects, the scale underwent some significant modifications. The problem had been that adolescents, even normal subjects taking the test

Table 12.1. Number of Items in the Basic Scales of the MMPI
          and MMPI-A

	Number of Items	
Scale	MMPI	MMPI-A
L	15	14
F	64	66
K	30	30
Hs	33	32
D	60	57
Hy	60	60
Pd	50	49
Mf	60	44
Pa	40	40
Pt	48	48
Sc	78	72
Ma	46	46
Si	70	62

in a valid manner, tended to produce high F-scale scores. Careful analysis indicated that many of the standard F-scale items were not endorsed as infrequently by adolescents as by adults. The F scale of the MMPI-A includes 66 items that were endorsed in the deviant direction by no more than 20 percent of the boys and girls in the normative sample. Only 37 of the 64 original F-scale items also are in the MMPI-A F scale. The F scale was divided into two parts: F1 consists of items in the first part of the test booklet; F2 consists of items in the latter half of the test booklet.

After the 478 items were selected for the MMPI-A booklet, several items were rewritten to take into account that the wording of some of the items carried over from the MMPI was not appropriate for adolescent subjects. Items were changed from past to present or present-perfect tense. For example, "As a youngster, I was often sent to the principal for misbehaving in school" was changed to "In school I have sometimes been sent to the principal for bad behavior." Several studies have demonstrated that the rewritten items are essentially equivalent to the original ones (Archer & Gordon, 1992; Williams, Ben-Porath, & Hevern, 1991).

MMPI-A Norms

There are 805 boys and 815 girls in the MMPI-A normative sample. They were solicited randomly from schools in seven states throughout the United States. They ranged in age from 14 to 18 years. Some preliminary data were collected for 13-year-old subjects, but because later analyses indicated that these younger subjects had considerable difficulty completing the experimental booklet, they were eliminated from the sample (Butcher et al., 1992). Data presented in the MMPI-A manual (Butcher et al., 1992) indicate that the normative sample is approximately representative of the U.S. population in terms of ethnicity and socioeconomic status. Most of the adolescent normative subjects were living in intact homes.

Unlike earlier MMPI adolescent norms that differed by age levels, only one set of norms is presented for boys and another for girls. This is because preliminary data analyses indicated that there were only minor differences between age groups within the normative sample. To maintain consistency with the MMPI-2, raw scores on the eight clinical scales (1, 2, 3, 4, 6, 7, 8, 9) and the content scales are transformed to uniform T scores using the same composite distribution developed for the MMPI-2. The rationale for the use of uniform T scores and the procedures used to develop them were described in Chapter 8 of this book. Scores on all other scales are transformed to linear T scores.

MMPI-A VALIDITY SCALES

The L, F, and K scales from the original MMPI were maintained in the MMPI-A, although, as described above, there were some significant changes made in the F scale. Two additional validity scales, which have corresponding scales in the MMPI-2, were also developed. The Variable Response Inconsistency Scale (VRIN) assesses inconsistent responding and is useful in identifying random protocols. The True Response Inconsistency Scale (TRIN) assesses true and false response biases.

MMPI-A CLINICAL SCALES

As stated earlier, the 10 standard clinical scales of the MMPI were maintained in the MMPI-A. Although some of the scales have several fewer items than the original ones (see Table 12.1), and include some slightly rewritten items, the clinical scales of MMPI-A are essentially the same as the corresponding scales of the MMPI and the MMPI-2. Consistent with traditional use of the MMPI with adolescents, none of the MMPI-A clinical scales is K-corrected.

MMPI-A SUPPLEMENTARY SCALES

Several supplementary scales that had been developed from the MMPI item pool were carried over to the MMPI-A and several new scales were developed using the new MMPI-A item pool.

Anxiety (A) Scale

The Anxiety (A) scale was developed to assess the most important dimension that emerged when the validity and clinical scales of the MMPI were factor analyzed (Welsh, 1956). The MMPI-A version of this scale has thirty-five of its original thirty-nine items. Archer, Gordon, Anderson, and Giannetti (1989) found correlate patterns that were consistent with those derived from studies of adults. Adolescents scoring high on the A scale were anxious, fearful, guilt-prone, and self-critical.

Repression (R) Scale

The Repression (R) scale was developed to assess a second dimension that emerged when the validity and clinical scales of the MMPI were factor analyzed (Welsh, 1956). The MMPI-A version of this scale has thirty-three of the original forty items. In adults, higher scorers on the R scale were conventional, submissive people who try to avoid unpleasant confrontations and disagreeable situations. Archer et al. (1989) reported that adolescents who scored high on the R scale were described as overcontrolled, inhibited, and less spontaneous than other adolescents.

MacAndrew Alcoholism Scale—Revised (MAC-R)

The MacAndrew Alcoholism scale (MAC; MacAndrew, 1965) was developed using items from the original MMPI and became widely used as a way to screen for substance abuse problems. A revised version of the scale (MAC-R), with four items added to replace those deleted because of objectionable content, was included in the MMPI-2. The MMPI-A includes the revised version of the scale from the MMPI-2 minus one item, which has been replaced on the MMPI-A by another. Thus, the MAC-R scale of the MMPI-A has forty-nine items.

Previous research demonstrated that the MAC scale of the original MMPI was effective in identifying alcohol or other drug problems for adolescents as well as for adults (Graham, 1990). It seems likely that the MMPI-A version of the scale will be equally effective. However, there are some preliminary research data indicating that two other substance abuse scales developed from the MMPI-A item pool may be even more effective than the MAC-R in identifying adolescent substance abusers (Weed, Butcher, & Williams, 1992).

Alcohol/Drug Problem Proneness (PRO) Scale

The Alcohol/Drug Problem Proneness (PRO) scale is an empirically derived scale developed to assess the likelihood of alcohol or other drug problems in adolescents (Weed, Butcher, & Williams, 1992). Items in the scale differenti-ated adolescent boys and girls in alcohol/drug treatment from adolescent boys and girls who were in treatment for psychological problems other than alcohol/drug problems. The items in the scale deal with peer-group influ-ence, stimulus seeking, rule violation, negative attitudes toward achievement, problems with parents, and bad judgment (Butcher et al., 1992). The MMPI-A manual (Butcher et al., 1992) suggests that T scores equal to or greater than 60 indicate a potential for developing alcohol or drug problems. Obvi-ously, one should not conclude from test data alone that an adolescent has problems with alcohol or other drugs. Elevated scores on this scale of the MMPI-A should be viewed as an indicator that more information should be obtained. Although preliminary data suggest that this will be a useful scale, more research is needed before we can determine if it works as well as, or perhaps better than, the more familiar MAC-R in identifying alcohol or other drug problems in adolescents.

Alcohol/Drug Problem Acknowledgment (ACK) Scale

The Alcohol/Drug Problem Acknowledgment (ACK) scale is a face-valid scale consisting of items in which adolescents admit to using alcohol and/or other drugs and having some of the symptoms and problems associated with such use (Weed, Butcher, & Williams, 1992). Because of the obvious nature of the items in the ACK scale, adolescents who are denying alcohol or drug problems can easily avoid detection. However, elevated scores on the ACK scale (T ≥ 60) indicate that adolescent subjects are readily admitting to signif-icant alcohol or other drug problems (Butcher et al., 1992).

Immaturity (IMM) Scale

The Immaturity (IMM) scale was developed using items in the MMPI-A item pool (Archer, Pancoast, & Gordon, 1992). The scale is thought to assess the degree to which adolescents report behaviors, attitudes, and perceptions of self and others that reflect immaturity in terms of interpersonal style, cogni-tive complexity, self-awareness, judgment, and impulse control (Butcher et al., 1992). Tentative items for the scale were identified by correlating MMPI-A items with scores from a sentence completion test of ego maturation. The items were then rated by judges concerning their relevance to the construct of ego maturation. The scale was refined by deleting items that did not add to its internal consistency and by adding items from the MMPI-A item pool that demonstrated both a conceptual and statistical relationship to the immaturity construct.

Preliminary data have indicated that higher IMM scores were associated with a higher incidence of academic difficulties and with disobedient, defiant, and antisocial behaviors (Butcher et al., 1992). As one would expect for a scale of this kind, scores decrease with increasing chronological age. Although these preliminary data are somewhat promising, further research is needed before this scale can be recommended for routine clinical use. Of special interest will be data indicating to what extent the IMM scale is not simply a measure of general maladjustment.

MMPI-A CONTENT SCALES

Using procedures similar to those applied in developing content scales for the MMPI-2, Williams et al. (1992) developed a set of fifteen scales to assess the basic content dimensions represented in the MMPI-A item pool. The authors identified tentative items, and statistical procedures were used to refine the scales. Table 12.2 lists the MMPI-A content scales and their internal-consistency and test-retest reliability coefficients. Some of the scales are slightly modified versions of scales in the MMPI-2 (e.g., Anxiety, Depression,

Table 12.2. Internal Consistency and Temporal Stability of the MMPI-A Content Scales

| | Internal Consistency | | | | |
| | Clinical Sample | | Normative Sample | | |
Scales	Boys ($n = 420$)	Girls ($n = 293$)	Boys ($n = 805$)	Girls ($n = 815$)	Temporal Stability[a] ($n = 154$)
Anxiety (A-anx)	.80	.86	.76	.80	.81
Obsessiveness (A-obs)	.76	.80	.72	.72	.70
Depression (A-dep)	.83	.89	.80	.83	.82
Health Concerns (A-hea)	.78	.85	.81	.82	.76
Bizarre Mentation (A-biz)	.73	.76	.75	.75	.68
Anger (A-ang)	.75	.79	.69	.66	.72
Cynicism (A-cyn)	.78	.83	.79	.81	.73
Alienation (A-aln)	.72	.75	.69	.75	.62
Conduct Problems (A-con)	.74	.79	.72	.72	.62
Low Self-esteem (A-lse)	.73	.80	.71	.75	.78
Low Aspirations (A-las)	.63	.63	.55	.59	.66
Social Discomfort (A-sod)	.78	.85	.77	.78	.76
Family Problems (A-fam)	.82	.82	.81	.82	.82
School Problems (A-sch)	.70	.74	.69	.69	.64
Negative Treatment Indicators (A-trt)	.77	.80	.72	.75	.68

[a]1-week test-retest of subset of sample, boys and girls combined.

Source: Williams, C.L., Butcher, J.N., Ben-Porath, Y.S., & Graham, J.R. (1992). *MMPI-A content scales: Assessing psychopathology in adolescents.* Minneapolis: University of Minnesota Press. Copyright © 1992 by University of Minnesota. Reproduced by permission.

and Health Concern), whereas others were specifically developed to deal with adolescent problems and concerns (e.g., Conduct Problems, School Problems, and Low Aspirations).

Examination of the data in Table 12.2 reveals that the internal consistencies of the MMPI-A content scales are adequate but somewhat lower than for the corresponding MMPI-2 content scales. The temporal stability of the MMPI-A content scales also is a bit lower than for corresponding MMPI-2 scales but comparable to that usually reported for adolescent personality scales.

In an initial effort to demonstrate the validity of the MMPI-A content scales, Williams et al. (1992) reported extratest correlates for the MMPI-A normative subjects and for subjects from several different clinical settings. Sources of extratest data included adolescent self-report, parent ratings, clinician ratings, and information obtained from records of clinical subjects. Available data seem to indicate validity for the MMPI-A content scales for which construct-relevant external measures were available. However, the validity of some of the scales remains unknown because relevant external measures were not available in the Williams et al. research. Although preliminary validity data concerning the MMPI-A content scales are encouraging, more research is needed to determine to what extent these content-based scales add to the basic validity and clinical scales in understanding adolescents who complete the MMPI-A.

The content scales should be used in conjunction with the basic validity and clinical scales. High content scale scores often clarify the reasons for elevated scores on these other scales and help clinicians determine which of the many descriptors typically associated with these other scales should be emphasized. Scores on the content scales should be viewed as direct communications from adolescent test subjects to examiners. Because the items of the content scales are obvious and face-valid, adolescents who are trying to manage impressions of adjustment or maladjustment can do so on these scales. Therefore, the scales are most useful when the validity scales do not indicate defensiveness or exaggeration.

MMPI-A HARRIS-LINGOES SUBSCALES

The Harris-Lingoes subscales were discussed in some detail in Chapter 6 of this book. These scales, in slightly modified form, are available in the MMPI-A. Two of the subscales have one less item in the MMPI-A than in the MMPI-2, and some of the items have been slightly rewritten. However, these changes are so minor that the scales are comparable in the two tests.

As discussed in Chapter 6, the Harris-Lingoes subscales can be useful in understanding why adolescent subjects obtain high scores on the clinical scales. Analysis of subscale scores can help to determine which of a variety of descriptors for elevated clinical scales should be emphasized for a particular test subject. High scores (T > 65) on the Harris-Lingoes subscales should only be interpreted when the T score on the parent scale is greater than 60. Little is known about external correlates of the Harris-Lingoes subscales in adolescent subjects. Therefore, the meaning of high scores must be inferred from the content of the items in the scales. Interpretive inferences for the Harris-Lingoes subscales were presented in Chapter 6 of this book. As with the MMPI-2, the Harris-Lingoes subscales of the MMPI-A probably are too short and unreliable to be used as independent scales on which to base interpretation. Likewise, because little is known about the meaning of below-average scores on the Harris-Lingoes subscales, low scores should not be interpreted.

MMPI-A SCALE 0 SUBSCALES

The subscales for scale 0 that were described in Chapter 6 also are available for the MMPI-A. These subscales have the same number of items as for the MMPI-2. As with the Harris-Lingoes subscales, analysis of elevated scores (T > 65) on these subscales can help clinicians understand why adolescent test subjects obtained elevated scores on scale 0. Like the Harris-Lingoes subscales, these subscales should not be used as independent scales for clinical interpretation, and low scores on the subscales should not be interpreted. High scores (T > 65) on these subscales should be interpreted only when the T score on scale 0 is greater than 60.

RELIABILITY AND VALIDITY OF THE MMPI-A

Table 12.3 reports internal-consistency and temporal-stability coefficients for the basic validity and clinical scales of the MMPI-A for boys and girls in the normative sample (Butcher et al., 1992). The internal-consistency coefficients are slightly lower than the corresponding values for the MMPI-2, but most of the scales have relatively good internal consistency, especially considering that no efforts were made to ensure internal consistency when the scales were developed originally. The temporal stability of most scales also is slightly lower than for the MMPI-2. This is in keeping with generally lower stability of personality test scores for adolescents than for adults. As is the case with the MMPI-2, scale 6 has the lowest test-retest reliability coefficient of any of the clinical scales.

There are several ways in which the validity of the MMPI-A can be assessed. To the extent that scores and configurations of scores on the MMPI-A are congruent with corresponding scores on the MMPI, interpretive information that has accumulated for the MMPI can be generalized to the MMPI-A. In addition, it is important to demonstrate that the MMPI-A scales have relevant external correlates.

Table 12.3. Internal Consistency and Temporal Stability of the MMPI-A Validity and Clinical Scales in the Normative Sample

| Scale | Internal Consistency | | Temporal Stability |
	Boys (n = 805)	Girls (n = 815)	Boys & Girls[a] (n = 154)
L	.64	.58	.61
F	.90	.82	.55
F1	.80	.73	.49
F2	.85	.84	.55
K	.72	.70	.75
Hs	.78	.79	.79
D	.65	.66	.78
Hy	.63	.55	.70
Pd	.63	.68	.80
Mf	.43	.40	.82
Pa	.57	.59	.65
Pt	.84	.86	.83
Sc	.88	.89	.83
Ma	.61	.61	.70
Si	.79	.80	.84

[a]1-week test-retest interval.

Source: Butcher, J.N., Williams, C.L., Graham, J.R., Archer, R.P., Tellegen, A., Ben-Porath, Y.S., & Kaemmer, B. (1992). *Minnesota Multiphasic Personality Inventory-Adolescent (MMPI-A): Manual for administration, scoring, and interpretation.* Minneapolis: University of Minnesota Press. Copyright © 1992 by University of Minnesota. Reproduced by permission.

The MMPI-A manual (Butcher et al., 1992) reported data concerning congruence between scores and code types based on the MMPI-A and code types based on the MMPI using the Marks et al. (1974) adolescent norms. When code types are well defined (at least five T-score points between the lowest scale in the code type and the next-highest scale), code types from the two versions of the test are quite congruent. In the MMPI-A normative sample 95.1 percent of the boys and 87.8 percent of the girls had the same one-point code type. Corresponding rates for the clinical sample were 95.4 percent and 94.4 percent. In the normative sample, 95.2 percent of the boys and 81.8 percent of the girls had the same two-point code type when the definition criterion of five T-score points was used. The corresponding rates for the clinical boys and girls were 95.0 and 91.0, respectively. Clearly, profile configurations are quite similar for the MMPI and the MMPI-A.

The data concerning congruence do not address the important question of which code type more accurately reflects what a test subject is really like when different code types are found for the two versions of the test. To answer this question we will need to compare the inferences based on the two code types with external information about each subject and determine which inferences are more accurate. Graham, Timbrook, Ben-Porath, and Butcher (1991) demonstrated a methodology for assessing relative accuracy for adults and reported that MMPI-2 code types yielded inferences that were at least as accurate, and perhaps more so, than inferences based on the MMPI code types.

Although configurations are quite similar for the two versions of the test with adolescents, average T scores are somewhat different. For most scales, T scores on the MMPI-A are about five points lower than on the MMPI with adolescent norms. For boys, scale 4 is about ten T-score points lower on the MMPI-A, and for girls scale 9 is about ten T-score points lower on the MMPI-A. For both boys and girls, scores on scale 0 are about the same for the two versions of the test. There are several possible reasons for these normative differences. It may be that contemporary normative subjects endorse more items in the scored direction on most scales because they are more deviant. However, it may also be that they are endorsing more items in the scored direction because current test instructions encourage subjects to answer all the items if they possibly can. When the earlier MMPI normative data were collected, subjects were not given as much encouragement to answer all the items, and they typically omitted more items than contemporary subjects. When subjects answer more items, they are likely to endorse some of them in the scored direction.

Because of the considerable congruence between the MMPI and MMPI-A, when MMPI-A results are being interpreted, it is appropriate to use much of the information that has been reported for the MMPI with adolescents (e.g., Hathaway & Monachesi, 1963; Marks et al., 1974; Williams & Butcher, 1989a, 1989b; Wrobel & Lachar, 1992). This information has been summarized by Archer (1992a) and by Butcher and Williams (1992). Although the MMPI-A only recently has been published, some research has identified reliable external correlates for MMPI-A scores for normal adolescents and adolescents in a variety of clinical settings (Butcher et al., 1992).

INTERPRETING THE MMPI-A VALIDITY SCALES

Cannot Say (?) Scale

As with the MMPI-2, excessive item omissions lead to artificially lowered scores on the MMPI-A scales. Every effort should be made to encourage adolescents to answer as many of the items as possible. If more than thirty items are omitted, the protocol should be considered invalid and no scales should be interpreted (Butcher et al., 1992). If more than ten items are omitted, the protocol should be interpreted with caution, and the omitted items should be examined to make sure than they are not concentrated on specific scales.

L Scale

Higher scores on the L scale indicate that adolescents have approached the test in a defensive manner. They have tried to present themselves in an unrealistically favorable light, denying minor flaws and weaknesses and claiming excessive virtue (Butcher et al., 1992). T scores equal to or greater than 65 on the L scale suggest that adolescents have approached the test in such a defensive manner that the resulting scores may not reflect accurately what they really are like. If the high L-scale score is accompanied by high scores on other scales, these scales can be interpreted, but it should be acknowledged that the scores may underrepresent symptoms and problems. If the L-scale score is high and there are no high scores on the problem-oriented scales, it is not possible to determine if the person has serious problems and is covering them up or if the person has average adjustment and is simply overstating virtues. T scores between 60 and 64 on the L scale indicate some defensiveness, which should be taken into account in interpreting the other scales.

F, F1, and F2 Scales

As discussed in Chapter 3, the MMPI-A F scale is divided into two parts (F1 and F2). The F1 scale gives information about the first half of the test, and the F2 scale gives information about the second half of the test. As long as T scores are used, scores on the F, F1, and F2 scales can be interpreted similarly. The F-scale score is the sum of the F1 and F2 scales.

T scores between 70 and 74 suggest a possible problematic response pattern (Butcher et al., 1992). T scores equal to or greater than 75 indicate even greater likelihood of problematic responding. When a subject obtains a T score of 70 or greater on one of the F scales, the first thing that should be done is to rule out random or careless responding. The Variable Response Inconsistency Scale (see below) is very useful in this regard. Next, the possibility of true response bias should be ruled out. The True Response Inconsistency Scale (see below) is helpful in evaluating this possibility. If the F scale is elevated and the VRIN or TRIN scale suggests invalidity, none of the MMPI-A scales should be interpreted. However, if the F1 scale is within acceptable limits and the F2 scale suggests invalid responding, the basic validity and clinical scales can be interpreted but none of the supplementary or content scales can. If response bias is ruled out based on the VRIN- and/or TRIN-scale scores, then an elevated score on one or more of the F scales is likely to be suggestive of either serious psychopathology or exaggeration of symptoms and problems.

K Scale

Although scores on the MMPI-A clinical scales are not K-corrected, the K scale can be considered a measure of defensiveness in adolescent subjects. T scores

equal to or greater than 65 suggest the possibility that the adolescent test subject has approached the test in an invalid manner. The TRIN scale (see below)
should be consulted to rule out the possibility of a false response bias. The
MMPI-A manual (Butcher et al., 1992) recommends that an adolescent MMPI-
A profile should not be invalidated solely on the basis of the K scale. However,
it should be acknowledged that when the K-scale score is high (T ≥ 65), scores
on the other scales may underrepresent problems and symptoms.

Variable Response Inconsistency (VRIN) Scale

The Variable Response Inconsistency (VRIN) scale assesses the extent to
which test subjects have responded consistently to the items in the inventory.
Although the VRIN scale of the MMPI-2 has proved to be very useful in
detecting invalid protocols of adult subjects, the scale's usefulness with adolescent subjects remains to be demonstrated (Butcher et al., 1992). The
MMPI-A manual recommends that the possibility of random responding be
considered when VRIN-scale T scores for adolescents are equal to or greater
than 75.

True Response Inconsistency (TRIN) Scale

The True Response Inconsistency (TRIN) scale assesses the extent to which
adolescent test subjects have responded to MMPI-A items with indiscriminant
true or false responding. As with the VRIN scale, this scale has not been validated with adolescent subjects and should be used cautiously until more data
are available. The MMPI-A manual (Butcher et al., 1992) recommends that T
scores equal to or greater than 75 be considered as indicating the possibility
of a true or false response bias. As with the TRIN scale of the MMPI-2, T
scores on the TRIN scale are always 50 or higher, with T or F added to the T
score to indicate either a true or false response bias.

INTERPRETING THE MMPI-A CLINICAL SCALES

The MMPI-A manual (Butcher et al., 1992) suggests that T scores of 65 or
greater on the clinical scales should be considered clinically significant. T
scores between 60 and 64 on the clinical scales also can be interpreted as
high scores, although one should have less confidence in inferences based
on scores at this lower level. Because very little is known about the meaning
of low scores on the clinical scales for adolescents, low scores should not be
interpreted at this time. The interpretations of high scores on the clinical
scales are based on research with the original MMPI and with the MMPI-A.
As with the MMPI-2, it should be acknowledged that not every interpretive

statement will be characteristic of every adolescent who has a high score on a particular scale. The inferences should be treated as hypotheses to be validated on the basis of other information available about each adolescent test subject.

Scale 1 (Hs)

High scores on scale 1 generally indicate preoccupation with health, illness, and bodily functioning. Although adolescents with bona fide physical disorders may obtain moderately elevated scores (T = 60–64), higher scores suggest somatic complaints that have a psychological component. High-scoring boys and girls may report fatigue and lack of energy, and high-scoring girls may report eating problems. Adolescents who score high on scale 1 tend to have histories of poor academic performance in school. They are insecure, fear failure, and blame themselves for problems. They are self-centered, demanding, and attention-seeking. They tend to be somewhat isolated from other people.

Scale 2 (D)

In clinical settings, high scores on scale 2 tend to be associated with depression, guilt, pessimism, and suicidal ideation. High scorers frequently report somatic symptoms, anxiety, fear, worry, and eating problems (girls). They lack self-confidence and perform poorly in school. They tend to be shy, timid, and socially withdrawn.

Scale 3 (Hy)

High scores on scale 3 are associated with somatic symptoms and sleep disturbance. In clinical samples, high scorers may report depression and suicidal ideation. High scorers are immature, self-centered persons who are socially involved and demand attention and affection. They tend to deny difficulties and lack insight into their own motives.

Scale 4 (Pd)

High scores on scale 4 are associated with a variety of acting-out and delinquent behaviors. High scorers resent authority and have academic and behavior problems in school and conflicts with their families. They often admit to alcohol or other drug problems. High scorers often are impulsive and aggressive. They are self-centered persons who do not accept responsibility for their own behavior. Adolescents in clinical settings may have histories of being physically abused (boys) or sexually abused (girls).

Scale 5 (Mf)

For boys high T scores on scale 5 suggest stereotypically feminine interest patterns and for girls stereotypically masculine interest patterns. Boys with high T scores on scale 5 are less likely than other boys to get into trouble with the law. Girls with high T scores on scale 5 are more likely than other girls to have academic and behavior problems in school, to have histories of learning disability, and to act out in a variety of ways. It should be noted that Hathaway and Monachesi (1963) reported that both boys and girls with higher T scores on scale 5 were less likely to act out in delinquent ways.

Scale 6 (Pa)

Adolescents who score high on scale 6 often have academic and behavior problems in school. They are moody, changeable, and unpredictable. They are suspicious, guarded, evasive, and withdrawn interpersonally and are very sensitive to criticism and rejection. They tend to blame others for their own problems and difficulties.

Scale 7 (Pt)

High scores on scale 7 tend to be associated with anxiety, depression, and other emotional turmoil. High scorers often have difficulty concentrating and attending, and they are indecisive. They have poor self-concepts, are self-critical and perfectionistic, and feel guilty about perceived failures. They tend to be shy and socially introverted. Clinical records of high-scoring boys may indicate histories of being sexually abused.

Scale 8 (Sc)

In clinical settings high scorers on scale 8 may manifest psychotic symptoms (e.g., delusions, hallucinations, or ideas of reference). They may appear to be fearful, confused, and disorganized. Somatic symptoms and drug use also may be characteristic of high scorers on this scale. High scorers tend to have academic and behavior problems in school. They resort to excessive fantasy during times of increased stress. They have low self-esteem and feel inferior to others. They are described as isolated, aloof, and uninvolved with others. High scorers may have histories of being sexually abused.

Scale 9 (Ma)

High scorers on scale 9 often have histories of acting-out behaviors that include school problems or drug use. They are resentful of authority and often behave quite impulsively. They are rather socially extroverted persons. They are self-centered and may have unrealistic self-appraisal.

Scale 0 (Si)

High scorers on scale 0 are introverted, shy, and timid, and they have difficulty making friends. They lack self-confidence, are emotionally overcontrolled, and blame themselves for problems. They have strong needs for attention, affection, and support. In clinical settings they may report feeling depressed and suicidal. They tend not to use alcohol or drugs or to be involved in delinquent behaviors.

INTERPRETING THE MMPI-A CONTENT SCALES

The MMPI-A content scales can be very helpful in understanding patterns of scores on the basic clinical scales. A rather wide variety of descriptors can be applied to an adolescent subject who has a high score on a clinical scale. Examination of the content scales can offer direction concerning which of these descriptors should be emphasized in interpreting the high clinical scale score. For example, descriptors for high scores on scale 4 for adolescents include family problems, school problems, and delinquency. If the content scales indicate elevation on the Family Problems (A-fam) content scale but not on the School Problems (A-sch) or Conduct Problems (A-con) scales, interpretation of the scale 4 elevation would emphasize the adolescent's problems and conflicts with family, rather than school difficulties or problems with the law.

Scores on the MMPI-A content scales give important information about what the adolescent wants the examiner to know about him or her. The content of the scales is very obvious, and adolescents may be aware of what information they convey as they respond to the items. Adolescents who are interested in creating particular impressions of themselves can do so by manipulating their responses to the items in these scales. Therefore, scores on the content scales will be most informative when test subjects approach the test in an honest, candid manner. The content scales should not be interpreted when the validity scales suggest that the adolescent test subject has been quite defensive or has exaggerated symptoms and problems in responding to test items.

Scores on the MMPI-A content scales are expressed as uniform T scores, as are scores on the eight basic clinical scales. Inferences can be made about T scores equal to or greater than 60, but greatest emphasis should be places on T scores that are equal to or greater than 65. The descriptions of high scorers on the MMPI-A content scales that follow are based on consideration of the items in each scale and on empirically derived descriptors that have been reported for the scales (Williams et al., 1992). As with the clinical scales, not every descriptor associated with a content scale will apply to every adolescent who has a high score on the scale. Inferences should be treated as hypotheses that are to be validated based on other information available about the adolescent.

Anxiety (A-anx)

Adolescents who score high on the MMPI-A Anxiety (A-anx) scale are reporting many symptoms of anxiety, including tension, frequent worrying, and sleep disturbances. They report problems with concentration, confusion, and inability to stay on a task. They may believe that life is a strain and that their difficulties are insurmountable. They may worry about losing their minds or feel that something dreadful is about to happen to them. They seem to be aware of their problems and how they differ from others. Correlate data indicate that scores on the A-anx scale are related to general maladjustment as well as to specific symptoms such as depression and somatic complaints.

Obsessiveness (A-obs)

Adolescents who score high on the MMPI-A Obsessiveness (A-obs) scale report worrying beyond reason, often over trivial matters. They may have ruminative thoughts about bad words or counting unimportant things. They have times when they are unable to sleep because of their worries. They report difficulty making decisions and dread making changes in their lives. They report that others sometimes lose patience with them, and they often are regretful about things they have said or done. Correlate data suggest that the A-obs scale is related to general maladjustment, and also dependent, anxious behaviors in clinical boys and suicidal ideation and/or gestures in clinical girls. Further research with subjects known to be exhibiting obsessive symptoms is needed to clarify the interpretation of the scale.

Depression (A-dep)

Adolescents who score high on the MMPI-A Depression (A-dep) scale report many symptoms of depression, including fatigue and crying spells. They are dissatisfied with their lives and feel that others are happier than they are. They may be expressing many self-depreciative thoughts. They may feel that life is neither interesting nor worthwhile. They may report feeling blue and wishing they were dead, and suicidal ideation is possible. They report feeling lonely even when with other people. The future seems too uncertain for them to make serious plans, and they may have periods when they are unable to "get going." They may be characterized by a sense of hopelessness and not caring. Correlate data indicate that for normal girls and for clinical boys and girls, high scores on the A-dep scale are indicative of depression and dysphoria. For clinical subjects suicidal ideation and/or gestures are possible.

Health Concerns (A-hea)

Adolescents who score high on the MMPI-A Health Concerns (A-hea) scale report numerous physical complaints and indicate that physical problems interfere with enjoyment of school activities and contribute to school absence. They worry about their health and may feel that their problems would disappear if only their health would improve. Correlate data offer considerable support for the A-hea scale as a measure of somatic complaints in clinical subjects. For normal boys and girls higher scores on this scale seem to be related to misbehavior, poor academic performance, and other school problems.

Bizarre Mentation (A-biz)

Adolescents scoring high on the MMPI-A Bizarre Mentation (A-biz) scale report very strange thoughts and experiences, which may include auditory, visual, or olfactory hallucinations. They may feel that something is wrong with their minds. They may feel that they are being plotted against or that someone is trying to poison them. They may believe that others are trying to steal their thoughts, possibly through hypnosis. They may believe that evil spirits or ghosts possess or influence them. Correlate data indicate that the A-biz scale is a measure of general maladjustment for nonclinical subjects, with higher-scoring subjects having problems in school and receiving low marks. For boys and girls in clinical settings, high scores on the A-biz scale are suggestive of bizarre sensory experiences and other symptoms and behaviors that may be indicative of psychosis.

Anger (A-ang)

Adolescents who score high on the MMPI-A Anger (A-ang) scale report anger control problems. They often feel like swearing or smashing things. They may get into fights, especially when drinking. They may get into trouble for breaking or destroying things. They report feeling irritable and impatient with others, and they may throw temper tantrums to get their way. They do not like being hurried or having people get ahead of them in line. Correlate data for clinical subjects indicate that higher scorers on the A-ang scale may have histories of assault and other acting-out behaviors. In clinical settings boys may have histories of having been sexually abused.

Cynicism (A-cyn)

Adolescents scoring high on the MMPI-A Cynicism (A-cyn) scale are reporting misanthropic attitudes. They believe that others are out to get them and will use unfair means to gain an advantage. They look for hidden motives whenever someone does something nice for them. They believe that it is safer to trust nobody because people make friends in order to use them. Others are seen as inwardly disliking to help another person. They feel misunderstood by others and see others as jealous of them. Correlate data indicate that the A-cyn scale is not related to acting-out behavior. It may be more of an attitudinal than a behavioral measure.

Alienation (A-aln)

High scorers on the MMPI-A Alienation (A-aln) scale are reporting considerable emotional distance from others. They believe that they are getting a raw deal from life and that no one, including parents, cares about or understands them. They do not believe that they are liked by others nor do they get along well with peers. They have difficulty self-disclosing and report feeling awkward when having to talk in a group. They do not appreciate hearing others give their opinions. They do not believe that others are sympathetic and feel that other people often block their attempts at success. Correlate data indicate that, for both normative and clinical subjects, the A-aln scale is related to feelings of emotional distance from others. The relationship was stronger for clinical subjects, perhaps because there were fewer relevant potential correlates for normative subjects.

Conduct Problems (A-con)

Adolescents scoring high on the MMPI-A Conduct Problems (A-con) scale are reporting a number of different behavioral problems, including stealing, shoplifting, lying, breaking or destroying things, being disrespectful, swearing, and being oppositional. They say that their peers are often in trouble and frequently talk them into doing things they know they should not do. At times they try to make others afraid of them, just for the fun of it. They are entertained by criminal behavior and do not blame people for taking advantage of others. They admit doing bad things in the past that they cannot tell anybody about. Correlate data were similar across normative and clinical samples, across genders, and across measures. High scorers on the A-con scale tended to have various kinds of behavior problems. Not surprisingly, the problems were more numerous for clinical than for normal subjects.

Low Self-esteem (A-lse)

Adolescents scoring high on the MMPI-A Low Self-esteem (A-lse) scale are reporting very negative opinions of themselves, including being unattractive, lacking self-confidence, feeling useless, having little ability, having several faults, and not being able to do anything well. They say that they are likely to yield to others' pressure, changing their minds or giving up in arguments. They let other people take charge when problems have to be solved, and they do not feel that they are capable of planning their own future. They become uncomfortable when others say nice things about them, and at times they may become confused and forgetful. Correlate data indicate that higher scores on the A-lse scale are related to negative views of self and poor school performance. For clinical girls higher scores were related to depression.

Low Aspirations (A-las)

High scorers on the MMPI-A Low Aspirations (A-las) scale are reporting disinterest in being successful and do not expect to be. They do not like to study and read about things, dislike science and lectures on serious topics, and prefer work that allows them to be careless. They avoid newspaper editorials and believe that the comic strips are the only interesting part of the newspaper. They report difficulty starting things and give up quickly when things go wrong. They let other people solve their problems, and they avoid facing difficulties. They believe that others block their success. They report that others tell them they are lazy. Correlate data support the A-las scale as a measure of poor achievement and limited participation in school activities. In addition, the scale is related to antisocial tendencies such as running away, truancy, and sexual acting out (girls).

Social Discomfort (A-sod)

Adolescents with high scores on the MMPI-A Social Discomfort (A-sod) scale report that they find it very difficult to be around others. They report being shy and prefer to be alone. They dislike having people around them and frequently avoid others. They do not like parties, crowds, dances, or other social gatherings. They tend not to speak unless spoken to, and others have told them they are hard to get to know. They have difficulty making friends and do not like to meet strangers. Correlate data suggest that the A-sod scale is a measure of social discomfort and social withdrawal. In addition, for girls higher scores are related to depression and eating problems and contraindicative of aggressive and irresponsible acting-out behaviors.

Family Problems (A-fam)

Adolescents with high scores on the MMPI-A Family Problems (A-fam) scale are reporting serious problems with their parents and other family members. Family discord, jealousy, fault-finding, anger, beatings, serious disagreements, lack of love and understanding, and limited communication characterize the families. These adolescents do not seem to feel that they can count on their families in times of trouble. They wish for the day when they are able to leave home. They feel their parents frequently punish them without cause and treat them more like children than adults. They report that their parents dislike their peer group. Correlate data indicate that high scores on the A-fam scale are related to disagreements with and between parents and with a variety of delinquent and neurotic symptoms and behaviors.

School Problems (A-sch)

High scorers on the MMPI-A School Problems (A-sch) scale are reporting numerous difficulties in school. High scorers report poor grades, suspension, truancy, negative attitudes toward teachers, and dislike of school. The only pleasant aspect of school is being with friends. High scorers say that they do not participate in school activities, and they feel that school is a waste of time. They report frequent boredom and sleepiness at school, and they have been told that they are lazy. Some high scorers may report being afraid to go to school. Correlate data indicate that high scores on the A-sch scale are indicative of both academic and behavioral problems at school. The scale may also be a measure of general maladjustment.

Negative Treatment Indicators (A-trt)

High scorers on the MMPI-A Negative Treatment Indicators (A-trt) scale report negative attitudes toward doctors and health professionals. They do not believe that others are capable of understanding their problems and difficulties and do not care what happens to them. They are unwilling to take charge and face their problems and difficulties and do not assume responsibility for the negative events in their lives. They do not feel that they can plan their own futures. They report having some faults and bad habits that they feel are insurmountable. They report great unwillingness to discuss their problems with others and indicate that there are some issues that they would never be able to share with anyone. They report feeling nervous when others ask them personal questions and have secrets best kept to themselves. Treatment outcome studies with adolescents are needed to establish the external validity of the A-trt scale. Correlate data do not indicate that the scale is simply a measure of general maladjustment.

MMPI-A INTERPRETIVE STRATEGY

The MMPI-2 interpretive strategy can be applied to the MMPI-A, although modifications must be made to take into account the special life circumstances of adolescent subjects. Butcher and Williams (1992) have suggested that the following questions should be addressed in interpreting MMPI-A results:

1. Are there extratest factors that can explain the MMPI-A results?
2. What are the individual's response attitudes?
3. What are the individual's reported symptoms and behaviors? What is the likelihood of acting-out behaviors? If present, are the acting-out problems likely to be seen across settings or in specific settings? How severe is the acting out likely to be?
4. Do problems in school play a significant role in the adolescent's clinical picture? What, if any, are they likely to be?
5. Does the adolescent admit to having a problem with alcohol or other drugs? Does she or he have the potential for developing such a problem?
6. What are the individual's interpersonal relationships like? Are there negative peer-group influences? Are family problems significant? How does she or he respond to authority? Are alienation, cynicism, or isolation a factor?
7. Does the MMPI-A suggest a need for evaluation of possible physical or sexual abuse?
8. What strengths or assets are apparent in the individual?
9. What are the diagnostic implications of the MMPI-A profile?
10. What treatment implications or recommendations are suggested on the basis of the MMPI-A?

Information concerning the above questions comes from examination of the individual validity, clinical, supplementary, and content scales of the MMPI-A. As suggested earlier in this chapter, T scores of 60 or greater on these scales should be considered interpretable. Greatest confidence should be placed in inferences based on scales with T scores equal to or greater than 65. Because of limited information concerning the meaning of low scores on the MMPI-A, below-average scores should not be interpreted except as indi-

cating that test subjects did not admit to the problems and symptoms associated with higher scores on those scales. We cannot know if the low scorers do
not have the problems and symptoms, are consciously denying them, or simply are not aware of them.

The use of code types, which is given prominence in MMPI-2 interpretation, is not recommended at this time for the MMPI-A. Although the early
work of Marks et al. (1974) suggested that MMPI code types of adolescents
could be used to make inferences about them, later work by Williams and
Butcher (1989a, 1989b) did not support the use of code types with adolescents. Ehrenworth and Archer (1985) found that descriptions of adolescents
based on code types were not very accurate. It may be that future research
will demonstrate that MMPI-A code types can be used to generate accurate
inferences about adolescent subjects, but until such research is available, it is
recommended that inferences be limited to high scores on individual scales.

Sometimes the inferences based on high scores on individual scales will
appear to be inconsistent, and it will be necessary to reconcile these apparent
inconsistencies. The same approach recommended in Chapter 10 of this
book for the MMPI-2 is appropriate for the MMPI-A. Consider the possibility
that the apparently inconsistent inferences are accurately reflecting different
facets of the adolescent's personality and behavior. If it seems that some of
the inferences are indeed inconsistent, a decision must be made concerning
which inferences should be emphasized in the interpretation. In general,
greater emphasis should be placed on inferences based on higher scores.
Also, because the standard validity and clinical scales have been more thoroughly researched, greater emphasis should be placed on inferences based on
these scales than on inferences based on the supplementary or content scales.

As with the MMPI-2, sometimes inferences are not based directly on scores
from the MMPI-A scales. Rather, it is necessary to make higher-order inferences about adolescent test subjects based on a basic understanding of that
person. Such higher-order inferences are quite acceptable, but the test interpreter should acknowledge that they do not come directly from the MMPI-A
scales.

Extratest Factors

Several factors external to the MMPI-A may affect test performance and need
to be taken into account in interpreting test results. A test subject's gender is
an important variable, because research has indicated that somewhat different descriptors are appropriate for boys and girls who have high scores on
the MMPI-A scales (Butcher et al., 1992; Williams & Butcher, 1989a; Wrobel
& Lachar, 1992). Reading ability of test subjects is also an important factor to
be considered. The MMPI-A requires about a sixth-grade reading ability. Limited education, English as a second language, or other factors that affect
reading ability should be taken into account. When such factors are present,

special attention should be given to the validity indicators. Circumstances of the testing may also be important. An adolescent who is forced into the testing situation by parents or teachers may respond differently from an adolescent who takes the test because she or he is acknowledging problems and seeking help. Finally, extraordinary life circumstances that could affect test results should be considered. For example, having recently experienced a catastrophic event, such as sexual assault or the death of a parent, could lead to extreme responding on the MMPI-A that would not reflect typical functioning of the adolescent.

Response Attitudes

Before generating inferences concerning the adolescent's symptoms, personality, and behavior, it is necessary to consider response attitudes. Are there indications that the adolescent approached the test in a manner that invalidates the results? Are there response attitudes that are not extreme enough to invalidate the results, but that should be taken into account in interpreting the results? The standard validity indicators of the MMPI-A should be consulted in trying to answer these questions.

MMPI-A results should be considered invalid and uninterpretable if more than thirty items are omitted, if the VRIN T score is 75 or higher, or if the TRIN T score is 75 or higher (in either the true or false direction). T scores equal to or greater than 75 on the L or K scales suggest that the test subject approached the test in a defensive, denying manner and that the scores on the other scales probably underrepresent problems and symptoms. Scores equal to or greater than 65 on the F, F1, or F2 scales suggest the possibility of problematic responding. In such cases the VRIN and TRIN scores should be consulted to rule out the possibility of random responding or a true or false response bias. If these possibilities are ruled out, the elevated F scores may be indicating either serious maladjustment or exaggeration of symptoms and problems. Other information, especially the circumstances of the testing, should be considered in determining the most likely reasons for the high F-scale scores. If it seems likely that the adolescent is exaggerating, it should be understood that the scores on the other scales probably overestimate problems and symptoms.

Symptoms and Behaviors

Assuming that it has been concluded that the MMPI-A results are valid and interpretable, the individual clinical, supplementary, and content scale scores can be consulted to generate inferences concerning symptoms and behaviors. Although inferences can be generated from all high scores, several scales and subscales are especially important in relation to particular symptoms and problem behaviors.

Anxiety. Anxiety and excessive worry are suggested by high scores on scale 7, on the Anxiety (A-anx) and Obsessiveness (A-obs) content scales, and on the Anxiety (A) supplementary scale. Sometimes high scale 2 scorers also report symptoms of anxiety.

Depression. High scores on Scale 2 and the Depression (A-dep) content scale are the major indicators of depression and possibly suicidal ideation, particularly in clinical settings. However, high scores on scales 3, 7, and 0, on the Low Self-esteem (A-lse), Obsessiveness (A-obs), Social Discomfort (A-sod), and Anxiety (A-anx) content scales, and on the Sc2 (Emotional Alienation) and Sc4 (Lack of Ego Mastery, Conative) Harris-Lingoes subscales also can be suggestive of depression and/or suicidal ideation.

Somatic Symptoms. High scores on scale 1 and on the Health Concerns (A-hea) content scale are the clearest indicators of somatic symptoms. However, high scores on scales 2, 3, and 8 and on the Anxiety (A-anx) content scale also can be indicative of somatic symptoms. Eating problems in girls have been associated with high scores on scales 1 and 2 and on the Social Discomfort (A-sod) content scale. Sleep difficulties are suggested by high scores on scale 3, on the Obsessiveness (A-obs) and Anxiety (A-anx) content scales, and on the D1 (Subjective Depression) Harris-Lingoes subscale.

Psychotic Symptoms. High scores on scale 8 and on the Bizarre Mentation (A-biz) content scale indicate the possibility of psychotic behaviors, including delusions, hallucinations, ideas of reference, or disorganized thinking. High scores on the Pa1 (Persecutory Ideas), Sc3 (Lack of Ego Mastery, Cognitive), and Sc6 (Bizarre Sensory Experiences) Harris-Lingoes subscales are additional indicators of psychosis.

Poor Self-esteem. High scores on the Low Self-esteem (A-lse) content scale suggest problems with self-confidence. High scores on several clinical scales (1, 2, 7, 8, 0) also tend to be associated with lack of self-confidence. High scores on scale 9 are indicative of positive self-perceptions that may be somewhat grandiose.

Anger. High scores on scale 4 are suggestive of anger and resentment. High scores on the Anger (A-ang) and Conduct Problems (A-con) content scales indicate problems with anger control.

Acting-Out Behaviors. Delinquent and acting-out behaviors are suggested by high scores on several scales and subscales. Probably the best predictors of such behavior are high scores on scales 4 and 9 and on the Conduct Problems (A-con) content scale. High scores on scale 2 tend to contraindicate acting-out behavior. Preliminary data indicate that girls who have higher T scores on scale 5 tend to act out in various ways, whereas boys with higher T scores on scale 5 tend not to act out. Girls with high scores on the Social Discomfort (A-sod) content scale tend not to act out. Acting out also is characteristic of adolescents who have high scores on the Anger (A-ang) and Low Aspirations (A-las) content scales and on the Pd2 (Authority Problems) and Ma1 (Amorality) Harris-Lingoes subscales.

School Problems. The School Problems (A-sch) content scale was developed specifically to assess academic and behavioral problems in school. In addition, academic problems are suggested by high scores on scales 1, 2, 4, 5 (girls), 6, and 8 and on the Health Concerns (A-hea), Bizarre Mentation (A-biz), Low Self-esteem (A-lse), and Low Aspirations (A-las) content scales. Behavioral problems in school are suggested by high scores on scales 4, 5 (girls), 6, 8, and 9 and on the Health Concerns (A-hea) and Bizarre Mentation (A-biz) content scales. The relationship between school problems and so many MMPI-A scales seems to indicate that adolescents develop school problems for a variety of reasons.

Alcohol/Drug Problems. Adolescents who have moderate (T = 60–64) or high (T ≥ 65) scores on the Alcohol/Drug Problem Acknowledgment (ACK) scale are openly admitting to the use of alcohol and/or drugs and the problems associated with such use. By contrast, moderate (T = 60–64) or high (T ≥ 65) scores on the MacAndrew Alcoholism Scale—Adolescent (MAC-R) and the Alcohol/Drug Problem Proneness (PRO) scale suggest problems with alcohol and/or drugs whether or not the adolescents are acknowledging such problems. Clearly, the likelihood of abuse is greatest if an adolescent has high scores on all three of these scales. Other indicators of possible alcohol or other drug problems include high scores on scales 4, 8, and 9. High scores on scale 2 tend not to be associated with substance abuse. As with the MMPI-2, one should not conclude from MMPI-A data that an adolescent is a substance abuser. However, high scores on the scales mentioned above should alert the clinician to the possibility of substance abuse problems that should be evaluated more completely.

Interpersonal Relationships. Adolescents who have high scores on scale 0 and/or the Social Discomfort (A-sod) content scale are likely to be shy, introverted, and uncomfortable in social situations. High scores on scales 2 and 7 and the Si1 (Shyness/Self-Consciousness) subscale further support the impression of shyness and introversion. On the other hand, high scores on the Hy1 (Denial of Social Anxiety) and Pd3 (Social Imperturbability) Harris-Lingoes subscales and on scale 9 (girls) suggest extroversion and gregariousness. Social withdrawal and isolation are suggested by high scores on scales 1, 6, and 8, on the Alienation (A-aln) and Social Discomfort (A-sod) content scales, on the Pd4 (Social Alienation) and Sc1 (Social Alienation) Harris-Lingoes subscales, and on the Si2 (Social Avoidance) subscale. Adolescents who have high scores on scale 6, on the Cynicism (A-cyn) content scale, or on the Pa1 (Persecutory Ideas) Harris-Lingoes subscale are likely to be cynical, guarded, and untrusting in relationships. High scores on the Anger (A-ang) content scale are associated with irritability and lack of anger control.

High scores on scale 4, the Family Problems (A-fam) content scale, and the Pd1 (Familial Discord) Harris-Lingoes subscale indicate that adolescents are describing their family circumstances very negatively. They do not see their families as loving or supportive, and they tend to be angry, resentful, and rebellious toward family members. Adolescents who have high scores on the Conduct Problems (A-con) or Low Self-esteem (A-lse) content scales or on the Alcohol/Drug Problem Proneness (PRO) scale tend to be influenced easily by peers to become involved in antisocial or delinquent acts.

Physical or Sexual Abuse. It is not possible to determine from test data whether or not an adolescent has been physically or sexually abused. However, some MMPI-A data suggest that adolescents with high scores on certain scales are more likely than other adolescents to have histories of abuse. High scores on scale 4 or on the Family Problems (A-fam) content scale for boys suggest that the possibility of physical abuse be considered carefully. High scores on scales 8 (boys and girls), 4 (girls), and 7 (boys) suggest that the possibility of sexual abuse should be considered carefully. Histories of sexual abuse may be associated with high scores on the Family Problems (A-fam) content scale for girls and with high scores on the Depression (A-dep), Anger (A-ang), Low Self-esteem (A-lse), or School Problems (A-sch) content scales for boys.

Strengths and Assets. Although the scales of the MMPI-A were designed to measure problems and symptoms, it is important to try to address adolescents' strengths and assets when giving feedback to the adolescents, their parents, or school staff. As stated earlier in this chapter, we cannot know if low scores on the MMPI-A scales are indicative of a lack of problems and symptoms or simply denial of them. However, it is appropriate to emphasize in an interpretation that low scores indicate that the adolescents are not reporting the problems and symptoms relevant to the scales, for example: low scorers on scale 2 are not reporting depression and sadness; low scorers on scale 4 are not espousing antisocial attitudes; and low scorers on scale 0 are not describing themselves as shy and introverted. Similarly, low scorers on the Low Aspirations (A-las) content scale are indicating that they have interest in achieving; low scorers on the Conduct Problems (A-con) content scale are claiming that they do not engage in acting-out behaviors, and low scorers on the Family Problems (A-fam) content scale are denying negative attitudes about their families.

Although the limited empirical data available for MMPI-A scales do not suggest positive characteristics associated with high scores, research conducted on the original MMPI scales suggest some characteristics of high scorers that could be viewed as positive (Archer, 1987). High scorers on scale 3 are described as achievement-oriented, socially involved, and friendly. High scorers on scale 4 tend to be sociable, gregarious persons who create favorable first impressions on other people. High-scoring boys on scale 5 are seen by others as intelligent and attaining higher levels of academic achievement. High scorers on scale 7 can be described as conscientious. High scale 8 scorers tend to approach problems in creative ways. High scale 9 scorers are likely to be energetic, extroverted, and self-confident. High scorers on scale 0 are less likely to be involved in delinquent activities.

High scores on some of the Harris-Lingoes subscales also suggest some positive characteristics. For example, high scores on Hy1 (Denial of Social Anxiety) or Pd3 (Social Imperturbability) indicate adolescents who say they are comfortable and confident in social situations. High Ma4 (Ego Inflation) scorers tend to have very favorable self-concepts (perhaps unrealistically so at times).

Diagnostic Implications

Although there is not much research available that directly addresses the most likely diagnoses for adolescents who produce particular scores or patterns of scores on the MMPI or MMPI-A, some inferences can be made on the basis of high scores on certain scales. High scores on scale 1 and/or the Health Concerns (A-hea) content scale suggest somatoform disorder diagnoses; high scores on scale 2 or the Depression (A-dep) content scale suggest depression or anxiety disorder diagnoses; high scores on scales 4 and/or 9 or the Conduct Disorders (A-con) content scale often are associated with conduct disorder diagnoses; high scores on scale 7 and/or the Anxiety (A-anx) content scale suggest anxiety disorder diagnoses; high scores on scale 8 and/or the Bizarre Mentation (A-biz) content scale indicate the possibility of psychotic disorders. High scores on one or more of the alcohol/drug problem scales (MAC-R, PRO, ACK) indicate that the possibility of a substance use disorder should be explored carefully.

It should be emphasized that diagnoses cannot and should not be assigned to adolescents solely on the basis of MMPI-A data. The criteria for most diagnostic categories include information that must be obtained from sources other than test data (e.g., observation, interview, history). Often the best approach is to generate a comprehensive description of the adolescent test subject's symptoms, personality, and behavior and compare this description with the various categories described in the most current version of the diagnostic manual.

Treatment Implications

There are very limited empirical data concerning relationships between MMPI or MMPI-A scores and treatment-related characteristics. The MMPI-A

Negative Treatment Indicators (A-trt) scale was designed to assess characteristics of adolescents that would interfere with effective psychological treatment. High scorers on this scale typically express negative attitudes toward doctors and health professionals, believe that others do not care what happens to them, accept little responsibility for their own behavior, and feel unable and/or unwilling to change. Obviously, characteristics such as these have very negative implications for treatment. However, it must be emphasized that there have not yet been any outcome studies conducted to evaluate the validity of the A-trt scale.

It is often necessary and appropriate to make higher-order inferences concerning treatment-related issues. For example, it is reasonable to infer that an adolescent who is admitting to a great deal of psychological turmoil (as indicated by high scores on scales 2 and 7 and various content scales) will be more willing to cooperate in treatment than someone who is not admitting to such turmoil. Likewise, one might safely infer that an adolescent who produces a very defensive pattern on the validity scales of the MMPI-A would not be very open and cooperative in a therapeutic relationship. Adolescents whose MMPI-A scores (on scales such as 4, 9, A-ang, or A-cyn) suggest considerable anger and resentment will be likely to be rather uncooperative in therapy and will often test limits in relationships with therapists.

ILLUSTRATIVE CASE

To illustrate the interpretive strategy described above, the MMPI-A scores of an adolescent girl will be interpreted.[2] Her MMPI-A profiles are presented in Figures 12.1, 12.2, and 12.3. It should be understood that the interpretive inferences derived from her MMPI-A scores should be treated as hypotheses to be evaluated on the basis of other information available about her.

Background and Referral Concerns

Kara was a 15-year-old Caucasian girl who was in the ninth grade. She was referred for an inpatient psychiatric evaluation because of acting-out problems, truancy, family discord, angry outbursts, irresponsibility, and eating problems. She reported that she recently had been sexually assaulted (inappropriately touched) by an employee of her school.

At the time of hospital admission, Kara received a diagnosis of post-traumatic stress disorder with depression and conduct problems. Hospital staff described Kara as defiant and resentful. She was bossy with peers, became easily

---

[2]The author would like to thank Dr. Carolyn L. Williams for providing the case materials and offering suggestions concerning the interpretation.

Figure 12.1. MMPI-A profile for basic scales for Kara. Reproduced by permission of University of Minnesota Press.

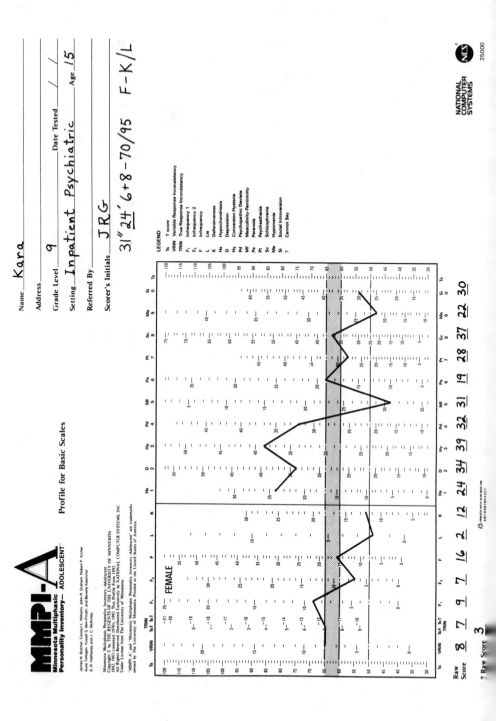

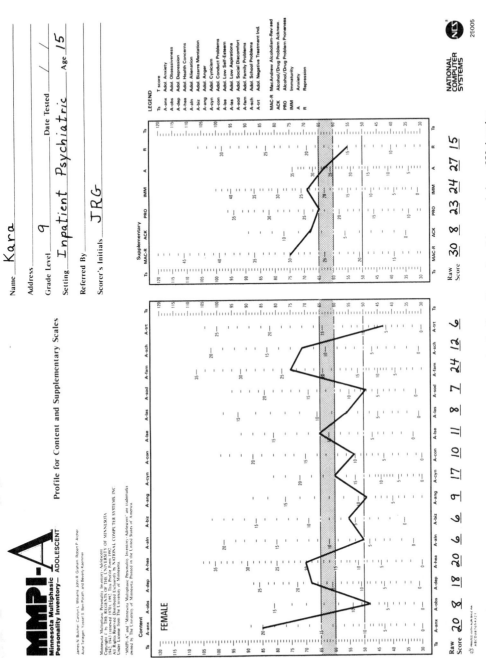

Figure 12.2. MMPI-A profile for content and supplementary scales for Kara. Reproduced by permission of University of Minnesota Press.

Figure 12.3. MMPI-A profile for Harris-Lingoes subscales for Kara. Reproduced by permission of University of Minnesota Press.

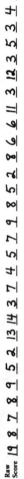

upset, and had problems with anger control. She was described as having a heightened interest in sex. Kara reported six previous suicide attempts, eating binges, and vomiting and laxative use to lose weight. After three and a half months of hospitalization, Kara was discharged with a final diagnosis of conduct disorder, socialized, nonaggressive.

Kara's mother described numerous behavior problems, including vandalism, assaultiveness, running away, lying, and stealing. She indicated that Kara's moods were quite variable, that she displayed nervous mannerisms such as biting her nails, and that she frequently had somatic complaints. According to the mother, there were financial problems and other discord in the family. Kara's father had a history of alcoholism and manic-depressive disorder.

Extratest Factors

The psychologist who administered the MMPI-A to Kara felt that she clearly had the reading ability required to complete the test. Although Kara had been referred for evaluation by her parents, she readily admitted to having problems and symptoms. Thus, there was no reason to believe that she would approach the test in an especially defensive manner. If Kara had been sexually assaulted by a school employee shortly before the evaluation, she could have been experiencing a great deal of acute turmoil that would be reflected in elevations on many of the MMPI-A scales.

Response Attitudes

Kara omitted only three items, suggesting that she was cooperative with the testing. Her VRIN and TRIN T scores of 65 indicate that she did not respond to the items inconsistently and did not approach the test with a true or false response set. Her L- and K-scale T scores are near 50, suggesting that she did not respond in a denying or defensive manner. Her moderately elevated T scores on the F and F1 scales indicate that she endorsed more than an average number of deviant items. Because her scores on the F scales are not very high and we are able to rule out random and true response sets, we interpret these scores as indicative of significant psychopathology.

Symptoms and Behaviors

Kara has T scores at or above 65 on five of the clinical scales (1, 2, 3, 4, 6) and between 60 and 64 on another scale (8). These scores indicate significant psychopathology and a variety of symptoms and problem behaviors.

The high scale 1 score indicates preoccupation with health, illness, and bodily functioning. Kara is likely to report somatic symptoms, particularly during times of increased stress. This impression is confirmed by high scores on the Health Concerns (A-hea) content scale, on scale 2, and on the Hy4 (Somatic Complaints) Harris-Lingoes subscale. The high scale 2 score also raises the possibility of eating problems.

The high score on scale 2 leads us to infer that Kara has symptoms of depression. She is likely to feel sad, pessimistic, and guilty, and she may have suicidal ideation. The high score on the Depression (A-dep) content scale is consistent with these inferences. High scores on the Hy3 (Lassitude-Malaise), Pd5 (Self-Alienation), Sc2 (Emotional Alienation), and Sc4 (Lack of Ego Mastery, Conative) Harris-Lingoes subscales lend further support to inferences concerning depression and suicidal ideation.

Several of Kara's scores indicate that she is likely to be experiencing considerable emotional turmoil. Her high scale 2 score suggests anxiety, fear, and worry. A very high score on the Anxiety (A-anx) content scale and a moderately high score on the Anxiety (A) supplementary scale support inferences concerning anxiety.

Kara's high scale 2 score and her moderately high scores on scale 8 and on the Low Self-esteem (A-lse) content scale suggest that she has many self-doubts and insecurities. She is likely to feel inferior to other people.

Kara's high score on scale 4 leads us to expect a variety of acting-out and delinquent behaviors, academic and behavior problems in school, and conflicts with her family. The absence of high scores on the Conduct Problems (A-con) content scale and the Harris-Lingoes Pd2 subscale (Authority Problems) suggests that she is not likely to act out in ways that get her into trouble with the law. However, her very high scores on the Family Problems (A-fam) and School Problems (A-sch) content scales and the Harris-Lingoes Pd1 (Familial Discord) subscales indicate that she is likely to be acting out at home and at school.

Kara's score on scale 8 is not high enough to suggest psychotic symptoms and behaviors. Likewise, the average-range score on the Bizarre Mentation (A-biz) content scale contraindicates psychosis. High scores on the Harris-Lingoes Sc6 (Bizarre Sensory Experiences) subscale sometimes are indicative of psychotic behavior. However, high scores on this subscale also can reflect unusual symptoms and experiences associated with drug use. Based on her high score on scale 8, we also might infer that Kara resorts to excessive fantasy during times of increased stress.

School Problems

Many of Kara's high scores are indicative of academic and behavioral problems in school. The combination of high scores on scale 4 and the School Problems (A-sch) content scale is strongly indicative of such problems. In addition, high scores on scales 1, 2, 6, and 8 and on the Health Concerns (A-hea) and Low Self-esteem (A-lse) content scales reinforce concerns about school problems.

Alcohol/Drug Problems

Several aspects of Kara's MMPI-A performance lead to concerns about possible problems with alcohol or other drugs. The high score on the ACK scale indicates that in responding to the MMPI-A items she admitted to use of alcohol and/or drugs and the problems associated with such use. A very high score on the MAC-R and a relatively high score on the PRO scale reinforce concerns about alcohol or other drug problems. Although one should not diagnose these problems from MMPI-A data alone, Kara's scores suggest that her use of alcohol or other drugs is a definite possibility that should be assessed more completely.

Interpersonal Relationships

Kara's MMPI-A scores give somewhat conflicting information concerning social adjustment. Although her high score on scale 2 suggests that she might be shy, introverted, and uncomfortable in social situations, her average scores on scale 0 and on the Social Discomfort (A-sod) content scale indicate that these descriptors probably do not apply to her. Her high scores on scales 1 and 6 and her moderate score on scale 8 raise the possibility that she is suspicious and guarded in relationships and that others may see her as distant and difficult to get to know. She is likely to be self-centered (high scores on scales 2 and 4) and demanding of attention and affection (high scale 3 score). She probably is quite sensitive to perceived criticism and rejection and may feel that others really do not care about her (high scale 6 score and moderately high Cynicism (A-cyn) content scale score). Although a moderately high score on the Low Self-esteem (A-lse) scale could indicate that she is easily influenced by peers, the absence of a high score on the Conduct Problems (A-con) content scale suggests that this is not particularly characteristic of her.

Clearly, Kara has some very negative perceptions of her family. High scores on scale 4, the Pd1 (Familial Discord) subscale, and the Family Problems (A-fam) content scale indicate that she sees her family as unloving, not supportive, and critical. She does not feel that she can count on them when she really needs them. Because she does not have an elevated score on the Anger (A-ang) content scale, it is not likely that she is openly expressing anger toward her family.

Physical or Sexual Abuse

Several of Kara's scores raise concerns about possible sexual abuse. Her high score on scale 4 and her moderately high score on scale 8 are similar to those of clinical girls whose records indicate histories of being sexually abused. Although one should not reach conclusions about sexual abuse from MMPI-A scores, this possibility should be assessed more fully, particularly given her reported sexual assault by a school employee.

Strengths

Although Kara's scores suggest many symptoms and problem behaviors, there are positive features to be considered. Her high scores on scales 3 and 4 and the absence of a high score on scale 0 indicate that she is likely to be somewhat active and outgoing. Because she does not have a high score on the Conduct Problems (A-con) content scale, she probably does not act out in ways that will lead to trouble with the law. The average score on the Low Aspirations (A-las) content scale suggests that, despite the problems that she is having at school, she is still interested in learning and achieving.

Diagnostic Implications

There are not adequate data concerning diagnoses associated with MMPI-A scores to permit direct inferences concerning the most likely diagnoses for an adolescent subject who has completed the test. However, having generated a detailed symptomatic and behavioral description of the test subject, one can consult descriptions for various diagnostic categories and determine which seems to apply to an adolescent test subject. Often the criteria include information that cannot be obtained from test data. Therefore, the most appropriate procedure is to state inferences concerning diagnoses that are consistent with the MMPI-A scores. These inferences need to be validated using other information available about adolescent subjects (e.g., history and observation).

As often is the case with adolescents in clinical settings, the diagnostic picture for Kara is rather mixed. Her MMPI-A scores suggest several different kinds of emotional turmoil. Her high scores on scale 2, on the Anxiety (A-anx) content scale, and on the Anxiety (A) supplementary scale indicate that an anxiety disorder diagnosis (e.g., overanxious disorder) should be considered. The high scores on scales 2 and 3, coupled with high scores on the Depression (A-dep) and Low Self-esteem (A-lse) content scales and on the Sc2 (Emotional Alienation) and Sc4 (Lack of Ego Mastery, Conative) Harris-Lingoes subscales, support an inference that a mood disorder diagnosis (e.g., major depression or dysthymia) might be appropriate. Kara's high scores on scales 1, 2, and 3 and on the Health Concerns (A-hea) and Anxiety (A-anx) content scales are consistent with a diagnosis of somatoform disorder. There are several indications (i.e., high scores on scales 1 and 2) that an eating disorder diagnosis (e.g., bulimia) should be ruled out.

Although Kara's high scores on scale 4 and on the Family Problems (A-fam) and School Problems (A-sch) content scales indicate that she may act out in some circumstances, more direct indicators of acting out that would be consistent with a conduct disorder diagnosis are absent. For example, her scores on the Conduct Problems (A-con), Anger (A-ang), and Low Aspirations (A-las) content scales are all in an average range. Likewise, she does not have high scores on the Harris-Lingoes Pd2 (Authority Problems) or Ma1 (Amorality) subscales. The high scale 2 score is contraindicative of acting out. The impression is that Kara may act out in certain circumstances (i.e., at school and at home) but probably will not get into trouble with the law.

Because of elevated scores on the three alcohol/drug problem scales (MAC-R, ACK, and PRO), alcohol- or drug-related diagnoses should be considered. This would require additional information from others (e.g., parents) who might be aware of such problems.

Treatment Implications

Kara's high scores on many of the MMPI-A scales indicate that she is having serious psychological problems and is in need of professional help. Because she is experiencing a great deal of emotional turmoil (high scores on scales 1 and 2 and on the Anxiety (A-anx) and Depression (A-dep) content scales), she is likely to feel the need for help. Her low score on the Negative Treatment Indicators (A-trt) scale suggests that she does not have very negative attitudes toward mental health professionals. However, some of her scores suggest that she will not respond very well to psychological intervention. Her high scale 3 score indicates that she may lack insight into her own motives and behaviors. She probably will not accept responsibility for her own behavior (high scale 4), blaming other people for her problems and difficulties (high scale 6). If inferences concerning alcohol or other drug problems are confirmed from other data, a treatment program should include a component to address these problems.

Summary

Interpretation of Kara's MMPI-A scores adds significantly to an understanding of her problems. She is in a great deal of emotional turmoil (e.g., depression, anxiety, suicidal ideation). She seems to have very negative attitudes toward her family and school, and these attitudes are likely to lead to some acting out in these environments. However, there are indications that she will not act out in ways that will get her into trouble with the law. The possibilities of an eating disorder and of alcohol or other drug problems need to be evaluated further. Although her symptoms and problems could be related to recent stressors (e.g., the alleged sexual assault), the picture that emerges from the MMPI-A interpretation is that of a young girl with long-standing problems. Family circumstances are likely to be very important, and any treatment program should address her family difficulties. Kara seems to be uncomfortable enough that she may recognize the need for professional help, but in therapy she is likely to be denying and have difficulty accepting responsibility for her own behavior. In spite of her many problem behaviors, Kara has assets that can be utilized in therapy. She seems to have strong needs to be involved with other people. Although she has had academic and behavior problems in school, she still has a desire to achieve. It is not possible from the MMPI-A scores to determine if Kara actually was sexually assaulted by a school employee as she claimed. However, several of her scores were similar to those of clinical girls who have histories of having been sexually abused. Therefore, her allegation should be taken seriously and investigated thoroughly.

References

Adler, T. (1990, April). Does the "new" MMPI beat the "classic"? *APA Monitor*, pp. 18–19.

Alker, H.A. (1978). Minnesota Multiphasic Personality Inventory. In O.K. Buros (Ed.), *Eighth mental measurements yearbook* (pp. 931–935). Highland Park, NJ: Gryphon.

American Psychological Association (1986). *Guidelines for computer-based tests and interpretations.* Washington, DC: APA.

American Psychological Association (1990). Ethical principles of psychologists. *American Psychologist, 45,* 390–395.

Anderson, B.N. (1969). *The utility of the Minnesota Multiphasic Personality Inventory in a private psychiatric hospital setting.* Unpublished master's thesis, Ohio State University, Columbus, OH.

Anderson, W., & Bauer, B. (1985). Clients with MMPI high D-Pd: Therapy implications. *Journal of Clinical Psychology, 41,* 181–189.

Anthony, N. (1971). Comparison of clients' standard, exaggerated, and matching MMPI profiles. *Journal of Consulting and Clinical Psychology, 36,* 100–103.

Apfeldorf, M., & Hunley, P.J. (1975). Application of MMPI alcoholism scales to older alcoholics and problem drinkers. *Journal of Studies on Alcohol, 37,* 645–653.

Archer, R.P. (1984). Use of the MMPI with adolescents: A review of salient issues. *Clinical Psychology Review, 41,* 241–251.

Archer, R.P. (1987). *Using the MMPI with adolescents.* Hillsdale, NJ: Lawrence Erlbaum.

Archer, R.P. (1992a). *MMPI-A: Assessing adolescent psychopathology.* Hillsdale, NJ: Lawrence Erlbaum.

Archer, R.P. (1992b). Minnesota Multiphasic Personality Inventory-2. In J.J. Kramer & J.C. Conoley (Eds.), *Eleventh mental measurements yearbook* (pp. 558–562). Lincoln, NE: Buros Institute of Mental Measurements.

Archer, R.P., & Gordon, R.A. (1992). *Psychometric stability of MMPI-A item modifications.* Unpublished manuscript.

Archer, R.P., Gordon, R.A., Giannetti, R., & Singles, J.M. (1988). MMPI scale clinical correlates for adolescent inpatients. *Journal of Personality Assessment, 52,* 707–721.

Archer, R.P., Gordon, R.A., Anderson, G.L., & Giannetti, R. (1989). MMPI special scale clinical correlates for adolescent inpatients. *Journal of Personality Assessment, 53,* 654–664.

Archer, R.P., Maruish, M., Imhof, E.A., & Piotrowski, C. (1991). Psychological test usage with adolescent clients: 1990 survey findings. *Professional Psychology: Research and Practice, 22,* 247–252.

Archer, R.P., Pancoast, D.L., & Gordon, R.A. (1992). *The development of the MMPI-A Immaturity Scale: Findings for normal and clinical samples.* Unpublished manuscript.

Archer, R.P., Stolberg, A.L., Gordon, R.A., & Goldman, W.R. (1986). Parent and child MMPI responses: Characteristics among families with adolescents in inpatient and outpatient settings. *Journal of Abnormal Child Psychology, 14,* 181–190.

Arthur, G. (1944). An experience in examining an Indian twelfth-grade group with the Multiphasic Personality Inventory. *Mental Hygiene, 25,* 243–250.

Atkinson, L. (1986). The comparative validities of the Rorschach and MMPI: A meta-analysis. *Canadian Psychology, 27,* 238–247.

Avery, R.D., Mussio, S.J., & Payne, G. (1972). Relationships between MMPI scores and job performance measures of fire fighters. *Psychological Reports, 31,* 199–202.

Baer, R.A., Wetter, M.W., & Berry, D.T.R. (1992). Detection of underreporting of psychopathology on the MMPI: A meta-analysis. *Clinical Psychology Review, 12,* 509–525.

Baillargeon, J., & Danis, C. (1984). Barnum meets the computer: Critical test. *Journal of Personality Assessment, 48,* 415–419.

Ball, J.C. (1960). Comparison of MMPI profile differences among Negro-white adolescents. *Journal of Clinical Psychology, 16,* 304–307.

Barefoot, J.C., Dahlstrom, W.G., & Williams, R.B. (1983). Hostility, CHD incidence, and total mortality: A 25-year follow-up study of 255 physicians. *Psychosomatic Medicine, 45,* 59–63.

Barker, H.R., Fowler, R.D., & Peterson, L.P. (1971). Factor analytic structure of the short form MMPI in a VA hospital population. *Journal of Clinical Psychology, 27,* 228–233.

Barron, F. (1953). An ego strength scale which predicts response to psychotherapy. *Journal of Consulting Psychology, 17,* 327–333.

Barron, F., & Leary, T. (1955). Changes in psychoneurotic patients with and without psychotherapy. *Journal of Consulting Psychology, 19,* 239–245.

Ben-Porath, Y.S., & Butcher, J.N. (1986). Computers in personality assessment: A brief past, an ebullient present, and an expanding future. *Computers in Human Behavior, 2,* 167–182.

Ben-Porath, Y.S., & Butcher, J.N. (1989a). Psychometric stability of rewritten MMPI items. *Journal of Personality Assessment, 53,* 645–653.

Ben-Porath, Y.S., & Butcher, J.N. (1989b). The comparability of MMPI and MMPI-2 scales and profiles. *Psychological Assessment: A Journal of Consulting and Clinical Psychology, 1,* 345–347.

Ben-Porath, Y.S., Butcher, J.N., & Graham, J.R. (1991). Contribution of the MMPI-2 content scales to the differential diagnosis of schizophrenia and major depression. *Psychological Assessment: A Journal of Consulting and Clinical Psychology, 3,* 634–640.

Ben-Porath, Y.S., & Graham, J.R. (1991). Resolutions to interpretive dilemmas created by the Minnesota Multiphasic Personality Inventory-2 (MMPI-2): A reply to Strassberg. *Journal of Psychopathology and Behavioral Assessment, 13,* 173–179.

Ben-Porath, Y.S., Hostetler, K., Butcher, J.N., & Graham, J.R. (1989). New subscales for the MMPI-2 social introversion (Si) scale. *Psychological Assessment: A Journal of Consulting and Clinical Psychology, 1,* 169–174.

Ben-Porath, Y.S., Slutske, W.S., & Butcher, J.N. (1989). A real-data simulation of computerized adaptive administration of the MMPI. *Psychological Assessment: A Journal of Consulting and Clinical Psychology, 1,* 18–22.

Berk, E., Black, J., Locastrok, J., Wickis, J., Simpson, T., Keane, T.M., & Penk, W. (1989). Traumatogenicity: Effects of self-reported noncombat trauma on MMPIs of male Vietnam combat and noncombat veterans treated for substance abuse. *Journal of Clinical Psychology, 45,* 704–717.

Bernstein, I.H. (1980). Security guards' MMPI profiles: Some normative data. *Journal of Personality Assessment, 44,* 377–380.

Berry, D.T.R., Baer, R.A., & Harris, M.J. (1991). Detection of malingering on the MMPI: A meta-analysis. *Clinical Psychology Review, 11,* 585–598.

Beutler, L.E., Storm, A., Kirkish, P., Scogin, F., & Gaines, J.A. (1985). Parameters in the prediction of police officer performance. *Professional Psychology: Research and Practice, 16,* 324–335.

Black, J.D. (1953). *The interpretation of MMPI profiles of college women.* Unpublished doctoral dissertation, University of Minnesota, Minneapolis, MN.

Block, J. (1965). *The challenge of response sets: Unconfounding meaning, acquiescence, and social desirability in the MMPI.* New York: Appleton-Century-Crofts.

Block, J., & Bailey, D.Q. (1955). Q-sort item analyses of a number of MMPI scales. *Officer Education Research Laboratory, Technical Memorandum.* (OERL-TM-55-7).

Boerger, A.R., Graham, J.R., & Lilly, R.S. (1974). Behavioral correlates of single-scale MMPI code types. *Journal of Consulting and Clinical Psychology, 42,* 398–402.

Brayfield, A.H. (Ed.). (1965). Testing and public policy. *American Psychologist, 20,* 857–1005.

Brozek, J. (1955). Personality changes with age: An item analysis of the MMPI. *Journal of Gerontology, 10,* 194–206.

Bubenzer, D.L., Zimpfer, D.G., & Mahrle, C.L. (1990). Standardized individual appraisal in agency and private practice. *Journal of Mental Health Counseling, 12,* 51–66.

Buechly, R., & Ball, H. (1952). A new test of "validity" for the group MMPI. *Journal of Consulting Psychology, 16,* 299–301.

Burkhart, B.R., Christian, W.L., & Gynther, M.D. (1978). Item subtlety and faking on the MMPI: A paradoxical relationship. *Journal of Personality Assessment, 42,* 76–80.

Burkhart, B.R., Gynther, M.D., & Fromuth, M.E. (1980). The relative validity of subtle versus obvious items on the MMPI Depression scale. *Journal of Clinical Psychology, 36,* 748–751.

Butcher, J.N. (1972). *Objective personality assessment: Changing perspectives.* New York: Academic Press.

Butcher, J.N. (1978). Computerized scoring and interpreting services. In O.K. Buros (Ed.), *Eighth mental measurments yearbook* (pp. 942–945). Highland Park, NJ: Gryphon.

Butcher, J.N. (1979). Use of the MMPI in personnel selection. In J.N. Butcher (Ed.), *New developments in the use of the MMPI* (pp. 165–201). Minneapolis: University of Minnesota Press.

Butcher, J.N. (1985). Personality assessment in industry: Theoretical issues and illustrations. In H.J. Bernardin (Ed.), *Personality assessment in organizations* (pp. 277–310). New York: Praeger.

Butcher, J.N. (1987). The use of computers in psychological assessment: An overview of practices and issues. In J.N. Butcher (Ed.), *Computerized psychological assessment* (pp. 3–14). New York: Basic Books.

Butcher, J.N. (1988, March). *Use of the MMPI in personnel screeening.* Paper presented at the 23rd Annual Symposium on Recent Developments in the Use of the MMPI. St. Petersburg Beach, FL.

Butcher, J.N. (1989a). Interpreting defensive profiles. In J.N. Butcher & J.R. Graham (Eds.), *Topics in MMPI-2 interpretation* (No. 7), pp. 28–30. Minneapolis: MMPI-2 Workshops and Symposia, Department of Psychology, University of Minnesota.

Butcher, J.N. (1989b). *Minnesota Multiphasic Personality Inventory-2, user's guide, the Minnesota Report: Adult clinical system.* Minneapolis: National Computer Systems.

Butcher, J.N. (1989c). *User's guide for the Minnesota Personnel Report.* Minneapolis: National Computer Systems.

Butcher, J.N. (1990a). *MMPI-2 in psychological treatment.* New York: Oxford University Press.

Butcher, J.N. (1990b). Education level and MMPI-2 measured pathology: A case of negligible influence. *MMPI-2 News and Profiles: A Newsletter of the MMPI-2 Workshops and Symposia, 1,* 3.

Butcher, J.N., Aldwin, C.M., Levenson, M.R., Ben-Porath, Y.S., Spiro, A., & Bosse, R. (1991). Personality and aging: A study of the MMPI-2 among older men. *Psychology and Aging, 6,* 361–370.

Butcher, J.N., Ball, B., & Ray, E. (1964). Effects of socio-economic level on MMPI differences in Negro-white college students. *Journal of Counseling Psychology, 11,* 183–187.

Butcher, J.N., Braswell, L., & Raney, D. (1983). A cross-cultural comparison of American Indian, Black, and White inpatients on the MMPI and presenting symptoms. *Journal of Consulting and Clinical Psychology, 51,* 587–594.

Butcher, J.N., Dahlstrom, W.G., Graham, J.R., Tellegen, A., & Kaemmer, B. (1989). *Minnesota Multiphasic Personality Inventory-2 (MMPI-2): Manual for administration and scoring.* Minneapolis: University of Minnesota Press.

Butcher, J.N., Graham, J.R., Dahlstrom, W.G., & Bowman, E. (1990). The MMPI-2 with college students. *Journal of Personality Assessment, 54,* 1–15.

Butcher, J.N., Graham, J.R., Williams, C.L., & Ben-Porath, Y.S. (1990). *Development and use of the MMPI-2 content scales.* Minneapolis: University of Minnesota Press.

Butcher, J.N., Keller, L.S., & Bacon, S.F. (1985). Current developments and future directions in computerized personality assessment. *Journal of Consulting and Clinical Psychology, 53,* 803–815.

Butcher, J.N., Kendall, P.C., & Hoffman, N. (1980). MMPI short forms: CAUTION. *Journal of Consulting Psychology, 48,* 275–278.

Butcher, J.N., & Tellegen, A. (1966). Objections to MMPI items. *Journal of Consulting and Clinical Psychology, 30,* 527–534.

Butcher, J.N., & Williams, C.L. (1992). *Essentials of MMPI-2 and MMPI-A interpretation.* Minneapolis: University of Minnesota Press.

Butcher, J.N., Williams, C.L., Graham, J.R., Archer, R.P., Tellegen, A., Ben-Porath, Y.S., & Kaemmer, B. (1992). *Minnesota Multiphasic Personality Inventory—Adolescent (MMPI-A): Manual for administration, scoring, and interpretation.* Minneapolis: University of Minnesota Press.

Butler, R.W., Foy, D.W., Snodgress, L., Hurwicz, M., & Goldfarb, J. (1988). Combat-related posttraumatic stress disorder in a nonpsychiatric population. *Journal of Anxiety Disorders, 2,* 111–120.

Caldwell, A.B. (1969). *MMPI critical items.* Unpublished manuscript. (Available from Caldwell Report, 1545 Sawtelle Bl., No. 14, Los Angeles, CA 90025).

Caldwell, A.B. (1988). *MMPI supplemental scale manual.* Los Angeles: Caldwell Report.

Caldwell, A.B. (1991). Commentary on "The Minnesota Multiphasic Personality Inventory-2: A review." *Journal of Counseling & Development, 69,* 568–569.

Calvin, J. (1974). *Two dimensions or fifty: Factor analytic studies with the MMPI.* Unpublished materials, Kent State University, Kent, OH.

Calvin, J. (1975). *A replicated study of the concurrent validity of the Harris subscales for the MMPI.* Unpublished doctoral dissertation, Kent State University, Kent, OH.

Campos, L.P. (1989). Adverse impact, unfairness, and bias in the psychological screening of Hispanic peace officers. *Hispanic Journal of Behavioral Sciences, 11,* 122–135.

Cannon, D.S., Bell, W.E., Andrews, R.H., & Finkelstein, A.S. (1987). Correspondence between MMPI PTSD measures and clinical diagnosis. *Journal of Personality Assessment, 51,* 517–521.

Carr, A.C., Ghosh, A., & Ancill, R.J. (1983). Can a computer take a psychiatric history? *Psychological Medicine, 13,* 151–158.

Carson, R.C. (1969). Interpretive manual to the MMPI. In J.N. Butcher (Ed.), *Research developments and clinical applications* (pp. 279–296). New York: McGraw-Hill.

Chang, A.F., Caldwell, A.B., & Moss, T. (1973). Stability of personality traits in alcoholics during and after treatment as measured by the MMPI: A one-year follow-up study. *Proceedings of the 81st Annual Convention of the American Psychological Association, 8,* 387–388.

Chang, P.N., Nesbit, M.E., Youngren, N., & Robison, L.L. (1988). Personality characteristics and psychosocial adjustment of long-term survivors of childhood cancer. *Journal of Psychosocial Oncology, 5,* 43–58.

Chojnacki, J.T., & Walsh, W.B. (1992). The consistency of scores and configural patterns between the MMPI and MMPI-2. *Journal of Personality Assessment, 59,* 276–289.

Chu, C. (1966). *Object cluster analysis of the MMPI.* Unpublished doctoral dissertation, University of California, Berkeley, CA.

Clark, C., & Klonoff, H. (1988). Empirically derived MMPI profiles: Coronary bypass surgery. *The Journal of Nervous and Mental Disease, 176,* 101–106.

Clark, C.G., & Miller, H.L. (1971). Validation of Gilberstadt and Duker's 8–6 profile type on a black sample. *Psychological Reports, 29,* 259–264.

Clavelle, P.R. (1992). Clinicians' perceptions of the comparability of the MMPI and MMPI-2. *Psychological Assessment, 4,* 466-472.

Clayton, M.R., & Graham, J.R. (1979). Predictive validity of Barron's Es scale: The role of symptom acknowledgement. *Journal of Consulting and Clinical Psychology, 47,* 424-425.

Clopton, J.R., Shanks, D.A., & Preng, K.W. (1987). Classification accuracy of the MacAndrew scale with and without K corrections. *The International Journal of the Addictions, 22,* 1049–1051.

Cofer, C.N., Chance, J., & Judson, A.J. (1949). A study of malingering on the MMPI. *Journal of Psychology, 27,* 491–499.

Colligan, R.C., & Offord, K.P. (1992). Age, stage, and the MMPI: Changes in response patterns over an 85-year age span. *Journal of Clinical Psychology, 48,* 476–493.

Colligan, R.C., & Offord, K.P. (1988). The risky use of the MMPI hostility scale in assessing risk for coronary heart disease. *Psychosomatics, 29,* 188–196.

Colligan, R.C., & Offord, K.P. (1990). MacAndrew versus MacAndrew: The relative efficacy of the MAC and the SAP scales for the MMPI in screening male adolescents for substance misuse. *Journal of Personality Assessment, 55,* 708–716.

Colligan, R.C., Osborne, D., & Offord, K.P. (1980). Linear transformation and the interpretation of MMPI T scores. *Journal of Clinical Psychology, 36,* 162–165.

Colligan, R.C., Osborne, D., Swenson, W.M., & Offord, K.P. (1983). *The MMPI: A contemporary normative study.* New York: Praeger.

Comrey, A.L. (1957a). A factor analysis of items on the MMPI depression scale. *Educational and Psychological Measurement, 17,* 578–585.

Comrey, A.L. (1957b). A factor analysis of items on the MMPI hypochondriasis scale. *Educational and Psychological Measurement, 17,* 566–577.

Comrey, A.L. (1957c). A factor analysis of items on the MMPI hysteria scale. *Educational and Psychological Measurement, 17,* 586–592.

Comrey, A.L. (1958a). A factor analysis of items on the F scale of the MMPI. *Educational and Psychological Measurement, 18,* 621–632.

Comrey, A.L. (1958b). A factor analysis of items on the MMPI hypomania scale. *Educational and Psychological Measurement, 18,* 313–323.

Comrey, A.L. (1958c). A factor analysis of items on the MMPI paranoia scale. *Educational and Psychological Measurement, 18,* 99–107.

Comrey, A.L. (1958d). A factor analysis of items on the MMPI psychasthenia scale. *Educational and Psychological Measurement, 18,* 293–300.

Comrey, A.L. (1958e). A factor analysis of items on the MMPI psychopathic deviate scale. *Educational and Psychological Measurement, 18,* 91–98.

Comrey, A.L., & Marggraff, W. (1958). A factor analysis of items on the MMPI schizophrenia scale. *Educational and Psychological Measurement, 18,* 301–311.

Cook, W.N., & Medley, D.M. (1954). Proposed hostility and pharisaic-virtue scales for the MMPI. *Journal of Applied Psychology, 38,* 414–418.

Costello, R.M. (1977). Construction and cross-validation of an MMPI black-white scale. *Journal of Personality Assessment, 41,* 514–519.

Costello, R.M., Hulsey, T.L., Schoenfeld, L.S., & Ramamurthy, S. (1987). P-A-I-N: A four cluster MMPI typology for chronic pain. *Pain, 30,* 199–209.

Costello, R.M., Schoenfeld, L.S., & Kobos, J. (1982). Police applicant screening: An analogue study. *Journal of Clinical Psychology, 38,* 216–221.

Costello, R.M., Schoenfeld, L.S., Ramamurthy, S., & Hobbs-Hardee, B. (1989). Sociodemographic and clinical correlates of P-A-I-N. *Journal of Psychosomatic Research, 33,* 315–321.

Costello, R.M., Tiffany, D.W., Gier, R.H. (1972). Methodological issues and racial (black-white) comparisons on the MMPI. *Journal of Consulting and Clinical Psychology, 38,* 161–168.

Cottle, W.C. (1950). Card versus booklet forms of the MMPI. *Journal of Applied Psychology, 34,* 255–259.

Cowan, M.A., Watkins, B.A., & Davis, W.E. (1975). Level of education, diagnosis and race-related differences in MMPI performance. *Journal of Clinical Psychology, 31,* 442–444.

Cox, G.D. (1991, November). *Court enjoins psychological tests.* The National Law Journal.

Crovitz, E., Huse, M.N., & Lewis, D.E. (1973). Selection of physicians' assistants. *Journal of Medical Education, 48,* 551–555.

Dahlstrom, W.G. (1972). Whither the MMPI? In J.N. Butcher (Ed.), *Objective personality assessment: Changing perspectives* (pp. 85–115). New York: Academic Press.

Dahlstrom, W.G. (1980). Altered versions of the MMPI. In W.G. Dahlstrom & L. Dahlstrom (Eds.), *Basic readings on the MMPI: A new selection on personality measurement.* (pp. 386–393). Minneapolis: University of Minnesota Press.

Dahlstrom, W.G. (1992). Comparability of two-point high-point code patterns from original MMPI norms to MMPI-2 norms for the restandardization sample. *Journal of Personality Assessment, 59,* 153–164.

Dahlstrom, W.G., & Dahlstrom, L. (Eds.). (1980). *Basic readings on the MMPI: A new selection on personality measurement.* Minneapolis: University of Minnesota Press.

Dahlstrom, W.G., Lachar, D., & Dahlstrom, L.E. (1986). *MMPI patterns of American minorities.* Minneapolis: University of Minnesota Press.

Dahlstrom, W.G., Welsh, G.S., & Dahlstrom, L.E. (1972). *An MMPI handbook: Vol. I. Clinical interpretation.* Minneapolis: University of Minnesota Press.

Dahlstrom, W.G., Welsh, G.S., & Dahlstrom, L.E. (1975). *An MMPI handbook: Vol. II. Research applications.* Minneapolis: University of Minnesota Press.

Davis, W.E. (1975). Race and the differential "power" of the MMPI. *Journal of Personality Assessment, 39,* 138–140.

Davis, K.R., & Sines, J.O. (1971). An antisocial behavior pattern associated with a specific MMPI profile. *Journal of Consulting and Clinical Psychology, 36,* 229–234.

Deiker, T.E. (1974). A cross-validation of MMPI scales of aggression on male criminal criterion groups. *Journal of Consulting and Clinical Psychology, 42,* 196–202.

Diamond, R., Barth, J.T., & Zillmer, E.A. (1988). Emotional correlates of mild closed head trauma: The role of the MMPI. *The International Journal of Clinical Neuropsychology, 10,* 35–40.

Dikmen, S., Hermann, B.P., Wilensky, A.J., & Rainwater, G. (1983). Validity of the Minnesota Multiphasic Personality Inventory (MMPI) to psychopathology in patients with epilepsy. *Journal of Nervous and Mental Disease, 171,* 114–122.

Distler, L.S., May, P.R., & Tuma, A.H. (1964). Anxiety and ego strength as predictors of response to treatment in schizophrenic patients. *Journal of Consulting Psychology, 28,* 170–177.

Dodrill, C.B. (1986). Psychosocial consequences of epilepsy. In S. Filskov & T. Boll (Eds.), *Handbook of clinical neuropsychology, Vol. 2* (pp. 338–363). New York: Wiley.

Dolan, M.P., Roberts, W.R., Penk, W.E., Robinowitz, R., & Atkins, H.G. (1983). Personality differences among Black, White and Hispanic-American male heroin addicts on MMPI content scales. *Journal of Clinical Psychology, 39,* 807–813.

Drake, L.E. (1946). A social I.E. scale for the MMPI. *Journal of Applied Psychology, 30,* 51–54.

Drake, L.E., & Oetting, E.R. (1959). *An MMPI codebook for counselors.* Minneapolis: University of Minnesota Press.

Dubinsky, S., Gamble, D.J., & Rogers, M.L. (1985). A literature review of subtle-obvious items on the MMPI. *Journal of Personality Assessment, 49,* 62–68.

Duckworth, J.C. (1991a). Response to Caldwell and Graham. *Journal of Counseling & Development, 69,* 572–573.

Duckworth, J.C. (1991b). The Minnesota Multiphasic Personality Inventory-2: A review. *Journal of Counseling & Development, 69,* 564–567.

Duckworth, J., & Anderson, W. (1986). *MMPI interpretation manual for counselors and clinicians.* Muncie, IN: Accelerated Development, Inc.

Duckworth, J.C., & Duckworth, E. (1975). *MMPI interpretation manual for counselors and clinicians.* Muncie, IN: Accelerated Development, Inc.

Duff, F.L. (1965). Item subtlety in the personality inventory scales. *Journal of Consulting Psychology, 29,* 565–570.

Dunnette, M.D., Bownas, D.A., & Bosshardt, M.J. (1981). *Electric power plant study: Prediction of inappropriate, unreliable or aberrant job behavior in nuclear power plant settings.* Minneapolis, MN: Personnel Decisions Research Institute.

Dwyer, S.A., Graham, J.R., & Ott, E.K. (1992). *Psychiatric symptoms associated with the MMPI-2 content scales.* Unpublished manuscript, Kent State University, Kent, OH.

Edwards, A.L. (1957). *The social desirability variable in personality assessment and research.* New York: Dryden.

Edwards, A.L. (1964). Social desirability and performance on the MMPI. *Psychometrika, 29,* 295–308.

Edwards, L.K., & Clark, C.L. (1987). A comparison of the first factor of the MMPI and the first factor of the EMPI: The PSD factor. *Journal of Educational and Psychological Measurement, 47,* 1165–1173.

Edwards, A.L., & Edwards, L.K. (1992). Social desirability and Wiggins's MMPI content scales. *Journal of Personality and Social Psychology, 62,* 147–153.

Ehrenworth, N.V., & Archer, R.P. (1985). A comparison of clinical accuracy ratings of interpretive approaches for adolescent MMPI responses. *Journal of Personality Assessment, 49,* 413–421.

Eichman, W.J. (1961). Replicated factors on the MMPI with female NP patients. *Journal of Consulting Psychology, 25,* 55–60.

Eichman, W.J. (1962). Factored scales for the MMPI: A clinical and statistical manual. *Journal of Clinical Psychology, 18,* 363–395.

Elion, V.H., & Megargee, E.I. (1975). Validity of the MMPI Pd scale among black males. *Journal of Consulting and Clincal Psychology, 43,* 166–172.

Ends, E.J., & Page, C.W. (1957). Functional relationships among measures of anxiety, ego strength, and adjustment. *Journal of Clinical Psychology, 13,* 148–150.

Eschenback, A.E., & Dupree, L. (1959). The influence of stress on MMPI scale scores. *Journal of Clinical Psychology, 15,* 42–45.

Evan, W.M., & Miller, J.R. (1969). Differential effects on response bias of computer vs. conventional administration of a social science questionnaire. *Behavior Science, 14,* 216–227.

Evans, D.R. (1977). Use of the MMPI to predict effective hotline workers. *Journal of Clinical Psychology, 33,* 1113–1114.

Exner, J.E., McDowell, E., Pabst, J., Stackman, W., & Kirk, L. (1963). On the detection of willful falsification in the MMPI. *Journal of Consulting Psychology, 27,* 91–94.

Eyde, L.D., Kowal, D.M., & Fishburne, F.J. (1991). The validity of computer-based test interpretations of the MMPI. In T.B. Gutkin & S.L. Wise (Eds.), *The computer and the decision-making process* (pp. 75–123). Hillsdale, NJ: Lawrence Erlbaum Press.

Fairbank, J., McCaffrey, R., & Keane, T. (1985). Psychometric detection of fabricated symptoms of post-traumatic stress disorder. *American Journal of Psychiatry, 142,* 501–503.

Fashinbauer, T.R. (1974). A 166-item written short form of the group MMPI: The FAM. *Journal of Consulting and Clinical Psychology, 42,* 645–655.

Finn, S.E. (1990, June). *A model for providing feedback with the MMPI and MMPI-2.* Paper presented at the 25th Annual Symposium on Recent Developments in the Use of the MMPI (MMPI-2). Minneapolis, MN.

Finn, S.E., & Tonsager, M.E. (1992). Therapeutic effects of providing MMPI-2 test feedback to college students awaiting therapy. *Psychological Assessment, 4,* 278–287.

Fisher, G. (1970). Discriminating violence eminating from over-controlled vs. under-controlled aggressivity. *British Journal of Social and Clinical Psychology, 9,* 54–59.

Fordyce, W.E. (1979). *Use of the MMPI in the assessment of chronic pain* (Clinical Notes on the MMPI No. 3). Minneapolis, MN: National Computer Systems.

Fowler, R.D. (1969a). Automated interpretation of personality test data. In J.N. Butcher (Ed.), *MMPI: Research developments and clinical applications* (pp. 105–126). New York: McGraw-Hill.

Fowler, R.D. (1969b). The current status of computer interpretation of psychological tests. *American Journal of Psychiatry, 125,* 21–27.

Fowler, R.D. (1975). *A method for the evaluation of the abuse prone patient.* Paper presented at the meeting of the American Academy of Family Physicians, Chicago, IL.

Fowler, R.D., & Butcher, J.N. (1986). Critique of Matarazzo's views on computerized testing: All sigma and no meaning. *American Psychologist, 41,* 94–96.

Fowler, R.D., & Coyle, F.A. (1968). Overlap as a problem in atlas classification of MMPI profiles. *Journal of Clinical Psychology, 24,* 435.

Fowler, R.D., Teel, S.K., & Coyle, F.A. (1967). The measurement of alcoholic response to treatment by Barron's ego strength scale. *Journal of Psychology, 67,* 65–68.

Fredericksen, S.J. (1976, March). *A comparison of selected personality and history variables in highly violent, mildly violent, and nonviolent female offenders.* Paper presented at the 11th Annual MMPI Symposium, Minneapolis, MN.

Gallucci, N.T., Kay, D.C., & Thornby, J.I. (1989). The sensitivity of 11 substance abuse scales from the MMPI to change in clinical status. *Psychology of Addictive Behaviors, 3,* 29–33.

Gantner, A.B., Graham, J.R., & Archer, R.A. (1992). The usefulness of the MAC scale in normal, psychiatric, and substance abuse settings. *Psychological Assessment, 4,* 133–137.

Garb, H.N. (1984). The incremental validity of information used in personality assessment. *Clinical Psychology Review, 4*, 641–655.

Garetz, F.K., & Anderson, R.W. (1973). Patterns of professional activities of psychiatrists: A follow-up of 100 psychiatric residents. *American Journal of Psychiatry, 130*, 981–984.

Gass, C.S. (1992). MMPI-2 interpretation of patients with cerebrovascular disease: A correction factor. *Archives of Clinical Neuropsychology, 7*, 17–27.

Gass, C.S., & Lawhorn, L. (1991). Psychological adjustment following stroke: An MMPI study. *Psychological Assessment: A Journal of Consulting and Clinical Psychology, 3*, 628–633.

Gayton, W.F., Burchstead, G.N., & Matthews, G.R. (1986). An investigation of the utility of an MMPI post-traumatic stress disorder subscale. *Journal of Clinical Psychology, 42*, 916–917.

Getter, H., & Sundland, D.M. (1962). The Barron Ego Strength scale and psychotherapy outcome. *Journal of Consulting Psychology, 26*, 195.

Gilberstadt, H., & Duker, J. (1965). *A handbook for clinical and actuarial MMPI interpretation.* Philadelphia: Saunders.

Gocka, E. (1965). *American Lake norms for 200 MMPI scales.* Unpublished materials, Veterans Administration Hospital, American Lake, WA.

Gocka, E., & Holloway, H. (1963). *Normative and predictive data on the Harris and Lingoes subscales for a neuropsychiatric population* (Report No. 7). American Lake, WA: Veterans Administration Hospital.

Goldberg, L.R. (1965). Diagnosticians vs. diagnostic signs: The diagnosis of psychosis vs. neurosis for the MMPI. *Psychological Monographs, 79* (9, Whole No. 602).

Goldberg, L.R. (1968). Simple models or simple processes. *American Psychologist, 23*, 483–496.

Good, P.K., & Brantner, J.P. (1961). *The physician's guide to the MMPI.* Minneapolis: University of Minnesota Press.

Gottesman, I.I. (1959). More construct validation of the Ego Strength scale. *Journal of Consulting Psychology, 23*, 342–346.

Gottesman, I.I., & Prescott, C.A. (1989). Abuses of the MacAndrew MMPI alcoholism scale: A critical review. *Clinical Psychology Review, 9*, 223–242.

Gough, H.G. (1950). The F minus K dissimulation index for the MMPI. *Journal of Consulting Psychology, 14*, 408–413.

Gough, H.G. (1954). Some common misconceptions about neuroticism. *Journal of Consulting Psychology, 18*, 287–292.

Gough, H.G., McClosky, H., & Meehl, P.E. (1951). A personality scale for dominance. *Journal of Abnormal and Social Psychology, 46*, 360–366.

Gough, H.G., McClosky, H., & Meehl, P.E. (1952). A personality scale for social responsibility. *Journal of Abnormal and Social Psychology, 47*, 73–80.

Gough, H.G., McKee, M.G., & Yandell, R.J. (1955). *Adjective check list analyses of a number of selected psychometric and assessment variables.* Officer Education Research Laboratory. (Technical Memorandom No. OERL-TM-5S-10).

Graham, J.R. (1967). A Q-sort study of the accuracy of clinical descriptions based on the MMPI. *Journal of Psychiatric Research, 5*, 297–305.

Graham, J.R. (1971a). Feedback and accuracy of clinical judgments from the MMPI. *Journal of Consulting and Clinical Psychology, 36*, 286–291.

Graham. J.R. (1971b). Feedback and accuracy of predictions of hospitalization from the MMPI. *Journal of Clinical Psychology, 27*, 243–245.

Graham, J.R. (1977). *Stability of MMPI configurations in a college setting.* Unpublished manuscript, Kent State University, Kent, OH.

Graham, J.R. (1978). A review of some important MMPI special scales. In P. McReynolds (Ed.), *Advances in psychological assessment* (Vol. IV) (pp. 311–331). San Francisco: Jossey-Bass.

Graham, J.R. (1979). Using the MMPI in counseling and psychotherapy. In J. Butcher, W.G. Dahlstrom, M. Gynther, & W. Schofield (Eds.), *Clinical notes on the MMPI (No. 1).* Minneapolis: National Computer Systems.

Graham, J.R. (1977). *The MMPI: A practical guide (1st edition).* New York: Oxford.

Graham, J.R. (1987). *The MMPI: A practical guide (2nd edition).* New York: Oxford.

Graham, J.R. (1988, August). *Establishing validity of the revised form of the MMPI.* Symposium presentation at the 96th Annual Convention of the American Psychological Association, Atlanta, GA.

Graham, J.R. (1990). *MMPI-2: Assessing personality and psychopathology.* New York: Oxford.

Graham, J.R. (1991). Comments on Duckworth's review of the Minnesota Multiphasic Personality Inventory-2. *Journal of Counseling & Development, 69,* 570–571.

Graham, J.R., & Lilly, R.S. (1984). *Psychological testing.* Englewood Cliffs, NJ: Prentice Hall.

Graham, J.R., & Lilly, R.S. (1986, March). *Linear T scores versus normalized T scores: An empirical study.* Paper presented at the 21st Annual Symposium on Recent Developments in the Use of the MMPI, Clearwater Beach, FL.

Graham, J.R., & Mayo, M.A. (1985, March). *A comparison of MMPI strategies for identifying black and white male alcoholics.* Paper presented at the 20th Annual Symposium on Recent Developments in the Use of the MMPI, Honolulu, HI.

Graham, J.R., & McCord, G. (1985). Interpretation of moderately elevated MMPI scores for normal subjects. *Journal of Personality Assessment, 49,* 477–484.

Graham, J.R., Schroeder, H.E., & Lilly, R.S. (1971). Factor analysis of items on the Social Introversion and Masculinity-Femininity scales of the MMPI. *Journal of Clinical Psychology, 27,* 367–370.

Graham, J.R., Smith, R.L., & Schwartz, G.F. (1986). Stability of MMPI configurations for psychiatric inpatients. *Journal of Consulting and Clinical Psychology, 54,* 375–380.

Graham, J.R., & Strenger, V.E. (1988). MMPI characteristics of alcoholics: A review. *Journal of Consulting and Clinical Psychology, 56,* 197–205.

Graham, J.R., Timbrook, R.E., Ben-Porath, Y.S., & Butcher, J.N. (1991). Code-type congruence between MMPI and MMPI-2: Separating fact from artifact. *Journal of Personality Assessment, 57,* 205–215.

Graham, J.R., Watts, D., & Timbrook, R.E. (1991). Detecting fake-good and fake-bad MMPI-2 profiles. *Journal of Personality Assessment, 57,* 264–277.

Grayson, H.M. (1951). *A psychological admissions testing program and manual.* Los Angeles: Veterans Administration Center, Neuropsychiatric Hospital.

Grayson, H.M., & Olinger, L.B. (1957). Simulation of "normalcy" by psychiatric patients on the MMPI. *Journal of Consulting Psychology, 21,* 73–77.

Greene, R.L. (1978). An empirically derived MMPI carelessness scale. *Journal of Clinical Psychology, 34,* 407–410.

Greene, R.L. (1980). *The MMPI: An interpretive manual.* New York: Grune & Stratton.

Greene, R.L. (1987). Ethnicity and MMPI performance: A review. *Journal of Consulting and Clinical Psychology, 55,* 497–512.

Greene, R.L. (1989). Assessing the validity of MMPI profiles in clinical settings. In J. Butcher, W.G. Dahlstrom, M. Gynther, & W. Schofield (Eds.), *Clinical Notes on the MMPI (No. 11).* Minneapolis: National Computer Systems.

Greene, R.L. (1991). *The MMPI-2/MMPI: An interpretive manual.* Boston: Allyn and Bacon.

Greene, R.L., Weed, N.C., Butcher, J.N., Arrendondo, R., & Davis, H.G. (1992). A cross-validation of MMPI-2 substance abuse scales. *Journal of Personality Assessment, 58,* 405–410.

Greist, J.H., & Klein, M.H. (1980). Computer programs for patients, clinicians, and researchers in psychiatry. In J.B. Sidowski, J.H. Johnson, & T.A. Williams (Eds.), *Technology in mental health care delivery systems* (pp. 161–182). Norwood, NJ: Ablex.

Grossman, L.S., Haywood, T.W., Ostrov, E., Wasyliw, O., & Cavanaugh, J.L., Jr. (1990). Sensitivity of MMPI validity scales to motivational factors in psychological evaluations of police officers. *Journal of Personality Assessment, 55,* 549–561.

Grow, R., McVaugh, W., & Eno, T.D. (1980). Faking and the MMPI. *Journal of Clinical Psychology, 36,* 910–911.

Guck, T.P., Meilman, P.W., & Skultety, F.M. (1987). Pain assessment index: Following multidisciplinary pain treatment. *Journal of Pain and Symptom Management, 2,* 23–27.

Guthrie, G.M. (1949). *A study of the personality characteristics associated with the disorders encountered by an internist.* Unpublished doctoral dissertation, University of Minnesota, Minneapolis.

Guthrie, G.M. (1952). Common characteristics associated with frequent MMPI profile types. *Journal of Clinical Psychology, 8*, 141–145.

Gynther, M.D. (1979). Aging in personality. In J.N. Butcher (Ed.), *New developments in the use of the MMPI* (pp. 39–68). Minneapolis: University of Minnesota Press.

Gynther, M.D., Altman, H., & Sletten, I.W. (1973). Replicated correlates of MMPI two-point types: The Missouri Actuarial System. *Journal of Clinical Psychology* (Suppl. 39).

Gynther, M.D., Altman, H., & Warbin, W. (1973). Interpretation of uninterpretable Minnnesota Multiphasic Personality Inventory profiles. *Journal of Consulting and Clinical Psychology, 40*, 78–83.

Gynther, M.D., & Brillant, P.J. (1968). The diagnostic utility of Welsh's A-R categories. *Journal of Projective Techniques and Personality Assessment, 32*, 572–574.

Gynther, M.D., Burkhart, B.R., & Hovanitz, C. (1979). Do face-valid items have more predictive validity than subtle items? The case of the MMPI Pd scale. *Journal of Consulting and Clinical Psychology, 47*, 295–300.

Hanvik, L.J. (1949). *Some psychological dimensions of low back pain.* Unpublished doctoral dissertation, University of Minnesota, Minneapolis, MN.

Hanvik, L.J. (1951). MMPI profiles in patients with low back pain. *Journal of Consulting Psychology, 15*, 250–253.

Harrell, T.W., & Harrell, M.S. (1973). The personality of MBAs who reach general management early. *Personnel Psychology, 26*, 127–134.

Harrell, T.H., Honaker, L.M., & Parnell, T. (1992). Equivalence of the MMPI-2 with the MMPI in psychiatric patients. *Psychological Assessment, 4*, 460–465.

Harris, R., & Christiansen, C. (1946). Prediction of response to brief psychotherapy. *Journal of Psychology, 21*, 269–284.

Harris, R., & Lingoes, J. (1955). *Subscales for the Minnesota Multiphasic Personality Inventory.* Mimeographed materials, The Langley Porter Clinic.

Harris, R., & Lingoes, J. (1968). *Subscales for the Minnesota Multiphasic Personality Inventory.* Mimeographed materials, The Langley Porter Clinic.

Harrison, P.L., Kaufman, A.S., Hickman, J.A., & Kaufman, N.L. (1988). A survey of tests used for adult assessment. *Journal of Psychoeducational Assessment, 6*, 188–198.

Harrison, R.H., & Kass, E.H. (1968). MMPI correlates of Negro acculturation in a northern city. *Journal of Personality and Social Psychology, 10*, 262–270.

Hartman, B.J. (1987). Psychological screeening of law enforcement candidates. *American Journal of Forensic Psychology, 1*, 5–10.

Hathaway, S.R. (1947). A coding system for MMPI profiles. *Journal of Consulting Psychology, 11*, 334–337.

Hathaway, S.R. (1956). Scales 5 (Masculinity-Femininity), 6 (Paranoia), and 8 (Schizophrenia). In G.S. Welsh & W.G. Dahlstrom (Eds.), *Basic readings on the MMPI in psychology and medicine* (pp. 104–111). Minneapolis: University of Minnesota Press.

Hathaway, S.R. (1965). Personality inventories. In B.B. Wolman (Ed.), *Handbook of clinical psychology* (pp. 451–476). New York: McGraw-Hill.

Hathaway, S.R., & Briggs, P.F. (1957). Some normative data on new MMPI scales. *Journal of Clinical Psychology, 13*, 364–368.

Hathaway, S.R., & McKinley, J.C. (1940). A multiphasic personality schedule (Minnesota): I. Construction of the schedule. *Journal of Psychology, 10*, 249–254.

Hathaway, S.R., & McKinley, J.C. (1942). A multiphasic personality schedule (Minnesota): III. The measurement of symptomatic depression. *Journal of Psychology, 14*, 73–84.

Hathaway, S.R., & McKinley, J.C. (1983). The Minnesota Multiphasic Personality Inventory manual. New York: Psychological Corporation.

Hathaway, S.R., & Meehl, P.E. (1952). *Adjective check list correlates of MMPI scores.* Unpublished materials, University of Minnesota.

Hathaway, S.R., & Monachesi, E.D. (1953). *Analyzing and predicting juvenile delinquency with the MMPI.* Minneapolis: University of Minnesota Press.

Hathaway, S.R., & Monachesi, E.D. (1957). The personalities of predelinquent boys. *Journal of Criminal Law, Criminology, and Political Science, 48,* 149–163.

Hathaway, S.R., & Monachesi, E.D. (1961). *An atlas of juvenile MMPI profiles.* Minneapolis: University of Minnesota Press.

Hathaway, S.R., & Monachesi, E.D. (1963). *Adolescent personality and behavior: MMPI patterns of normal, delinquent, dropout, and other outcomes.* Minneapolis: University of Minnesota Press.

Hawkinson, J.R. (1961). *A study of the construct validity of Barron's Ego Strength scale with a state mental hospital population.* Unpublished doctoral dissertation, University of Minnesota, Minneapolis, MN.

Hearn, M.D., Murray, D.M., & Luepker, R.V. (1989). Hostility, coronary heart disease, and total mortality: A 33-year follow-up of university students. *Journal of Behavioral Medicine, 12,* 105–121.

Hedlund, J.L. (1977). MMPI clinical scale correlates. *Journal of Consulting and Clinical Psychology, 45,* 739–750.

Hedlund, J.L., Morgan, D.W., & Master, F.D. (1972). The Mayo Clinic automated MMPI program: Cross-validation with psychiatric patients in an army hospital. *Journal of Clinical Psychology, 28,* 505–510.

Henrichs, T.F. (1964). Objective configural rules for discriminating MMPI profiles in a psychiatric population. *Journal of Clinical Psychology, 20,* 157–159.

Henrichs, T.F. (1981). *Using the MMPI in medical consultation* (Clinical Notes on the MMPI, No. 6). Minneapolis, MN: National Computer Systems.

Henrichs, T.F. (1990). The effect of methods of accurate feedback on clinical judgments based upon the MMPI. *Journal of Clinical Psychology, 46,* 778–781.

Henrichs, T.F., & Waters, W.F. (1972). Psychological adjustment and responses to open-heart surgery: Some methodological considerations. *British Journal of Psychiatry, 120,* 491–496.

Herreid, C.F., & Herreid, J.R. (1966). Differences in MMPI scores in native and non-native Alaskans. *The Journal of Social Psychology, 70,* 191–198.

Himelstein, P. (1964). Further evidence of the Ego Strength scale as a measure of psychological health. *Journal of Consulting Psychology, 28,* 90–91.

Hjemboe, S., Almagor, M., & Butcher, J.N. (in press). Empirical assessment of marital distress: The Marital Distress Scale (MDS) for the MMPI-2. In J.N. Butcher & C.D. Spielberger (Eds.), *Advances in personality assessment: Volume 9.* Hillsdale, NJ: Lawrence Erlbaum.

Hjemboe, S., & Butcher, J.N. (1991). Couples in marital distress: A study of personality factors as measured by the MMPI-2. *Journal of Personality Assessment, 57,* 216–237.

Hoffman, H., Loper, R.G., & Kammeier, M.L. (1974). Identifying future alcoholics with the MMPI alcoholism scales. *Quarterly Journal of Studies on Alcohol, 35,* 490–498.

Holmes, D.S. (1967). Male-female differences in MMPI Ego Strength: An artifact. *Journal of Consulting Psychology, 31,* 408–410.

Honaker, L.M. (1988). The equivalency of computerized and conventional MMPI administration: A critical review. *Clinical Psychology Review, 8,* 561–577.

Honaker, L.M., Harrell, T.H., & Buffaloe, J.D. (1988). Equivalency of microtest computer MMPI administration for standard and special scales. *Computers in Human Behavior, 4,* 323–337.

Hovey, H.B. (1953). MMPI profiles and personality characteristics. *Journal of Consulting Psychology, 17,* 142–146.

Hovey, H.B., & Lewis, E.G. (1967). *Semiautomatic interpretation of the MMPI.* Brandon, VT: Clinical Psychology Publishing Company.

Hsu, L.M. (1984). MMPI T scores: Linear versus normalized. *Journal of Consulting and Clinical Psychology, 52,* 821–823.

Hsu, L.M. (1986). Implications of differences in elevations of K-corrected and non-K-corrected MMPI T scores. *Journal of Consulting and Clinical Psychology, 54,* 552–557.

Huber, N.A., & Danahy, S. (1975). Use of the MMPI in predicting completion and evaluating changes in a long-term alcoholism treatment program. *Journal of Studies on Alcohol, 36,* 1230–1237.

Huff, F.W. (1965). Use of actuarial description of abnormal personality in a mental hospital. *Psychological Reports, 17*, 224.

Hunt, H.F. (1948). The effect of deliberate deception on MMPI performance. *Journal of Consulting Psychology, 12*, 396–402.

Hyer, L., Fallon, J.H., Jr., Harrison, W.R., & Boudewyns, P.A. (1987). MMPI overreporting by Vietnam combat veterans. *Journal of Clinical Psychology, 43*, 79–83.

Hyer, L., O'Leary, W.C., Saucer, R.T., Blount, J., Harrison, W.R., & Boudewyns, P.A. (1986). Inpatient diagnosis of post-traumatic stress disorder. *Journal of Consulting and Clinical Psychology, 54*, 698–702.

Hyer, L., Woods, M.G., Summers, M.N., Boudewyns, P., & Harrison, W.R. (1990) Alexithymia among Vietnam veterans with post-traumatic stress disorder. *Journal of Clinical Psychiatry, 51*, 243–247.

Inwald, R.E. (1988). Five-year follow-up study of departmental terminations as predicted by 16 preemployment psychological indicators. *Journal of Applied Psychology, 4*, 703–710.

Jansen, D.G., & Garvey, F.J. (1973). High-, average-, and low-rated clergymen in a state hospital clinical program. *Journal of Clinical Psychology, 29*, 89–92.

Johnson, J.R., Null, C., Butcher, J.N., & Johnson, K.N. (1984). Replicated item level factor analysis of the full MMPI. *Journal of Personality and Social Psychology, 47*, 105–114.

Johnson, M.E., & Brems, C. (1990). Psychiatric inpatient MMPI profiles: An exploration for potential racial bias. *Journal of Counseling Psychology, 37*, 213–215.

Katz, M.M., & Lyerly, S.B. (1963). Methods for measuring adjustment and social behavior in the community: I. Rationale, description, discriminative validity and scale development. *Psychological Reports, 13*, 503–535.

Keane, T.M., Malloy, P.F., & Fairbank, J.A. (1984). Empirical development of an MMPI subscale for the assessment of combat-related post-traumatic stress disorder. *Journal of Consulting and Clinical Psychology, 52*, 888–891.

Keiller, S.W., & Graham, J.R. (1993). Interpreting low scores on the MMPI-2 clinical scales. *Journal of Personality Assessment*.

Keller, J.W., & Piotrowski, C. (1989, March). *Psychological testing patterns in outpatient mental health facilities: A national study.* Paper presented at the meeting of the Southeastern Psychological Association, Washington, D.C.

Keller, L.S., & Butcher, J.N. (1991). *Assessment of chronic pain with the MMPI-2.* Minneapolis: University of Minnesota Press.

Kelley, C.K., & King, G.D. (1979a). Cross validation of the 2–8/8–2 MMPI code type for young adult psychiartic outpatients. *Journal of Personality Assessment, 43*, 143–149.

Kelley, C.K., & King, G.D. (1979b). Behavioral correlates of the 2–7–8 MMPI profile type in students at a university mental health center. *Journal of Consulting and Clinical Psychology, 47*, 679–685.

Kelley, C.K., & King, G.D. (1979c). Behavioral correlates of infrequent two-point MMPI code types at a univeristy mental health center. *Journal of Clinical Psychology, 35*, 576–585.

Kelly, W.L. (1974). Psychological prediction of leadership in nursing. *Nursing Research, 23*, 38–42.

Kincannon, J.C. (1968). Prediction of the standard MMPI scale score from 71 items: The Mini-Mult. *Journal of Consulting and Clinical Psychology, 32*, 319–325.

King, G.D. (1978). Minnesota Multiphasic Personality Inventory. In O.K. Buros (Ed.), *Eighth mental measurments yearbook* (pp. 935–938). Highland Park, NJ: Gryphon.

Kirkcaldy, B.D., & Kobylinska, E. (1988). Psychological characteristics of breast cancer patients. *Psychotherapy & Psychosomatics, 48*, 32–43.

Kleinmuntz, B. (1960). An extension of the construct validity of the Ego Strength scale. *Journal of Consulting Psychology, 24*, 463–464.

Kleinmuntz, B. (1961). The college maladjustment scale (Mt): Norms and predictive validity. *Educational and Psychological Measurement, 21*, 1029–1033.

Kleinmuntz, B. (1963). MMPI decision rules for the identification of college maladjustment: A digital computer approach. *Psychological Monographs, 77* (14, Whole No. 577).

Klett, W. (1971). The utility of computer interpreted MMPIs at St. Cloud VA Hospital. *Newsletter of Research in Psychology, 13,* 45–47.

Klinge, V., Lachar, D., Grisell, J., & Berman, W. (1978). Effects of scoring norms on adolescent psychiatric drug users' and nonusers' MMPI profiles. *Adolescence, 13,* 1–11.

Klinge, V., & Strauss, M.E. (1976). Effects of scoring norms on adolescent psychiatric patients' MMPI profiles. *Journal of Personality Assessment, 40,* 13–17.

Kline, J.A., Rozynko, V.V., Flint, G., & Roberts, A.C. (1973). Personality characteristics of male Native-American alcoholic patients. *International Journal of the Addictions, 8,* 729–732.

Knapp, R.R. (1960). A reevaluation of the validity of MMPI scales of dominance and responsibility. *Educational and Psychological Measurement, 20,* 381–386.

Koeppl, P.M., Bolla-Wilson, K., & Bleecker, M.L. (1989). The MMPI: Regional difference or normal aging? *Journal of Gerontology: Psychological Sciences, 44,* P95–99.

Koretzky, M.B., & Peck, A.H. (1990). Validation and cross-validation of the PTSD subscale of the MMPI with civilian trauma victims. *Journal of Clinical Psychology, 46,* 296–300.

Koss, M.P. (1979). MMPI item content: Recurring issues. In J.N. Butcher (Ed.), *New developments in the use of the MMPI* (pp. 3–38). Minneapolis: University of Minnesota Press.

Koss, M.P. (1980). Assessing psychological emergencies with the MMPI. In J. Butcher, W. Dahlstrom, M. Gynther, & W. Schofield (Eds.), *Clinical notes on the MMPI* (No. 4). Minneapolis, MN: National Computer Systems.

Koss, M.P., Butcher, J.N., & Hoffman, N. (1976). The MMPI critical items: How well do they work? *Journal of Consulting and Clinical Psychology, 44,* 921–928.

Kostlan, A. (1954). A method for the empirical study of psychodiagnosis. *Journal of Consulting Psychology, 18,* 83–88.

Kramer, J.J., & Conoley, J.C. (Eds.). (1992). *The eleventh mental measurements yearbook.* Lincoln, NE: Buros Institute of Mental Measurements.

Kranitz, L. (1972). Alcoholics, heroine addicts and non-addicts: Comparisons on the MacAndrew Alcoholism scale on the MMPI. *Quarterly Journal of Studies on Alcohol, 33,* 807–809.

Lachar, D. (1974a). Accuracy and generalization of an automated MMPI interpretation system. *Journal of Consulting and Clinical Psychology, 42,* 267–273.

Lachar, D. (1974b). *The MMPI: Clinical assessment and automated interpretation.* Los Angeles: Western Psychological Services.

Lachar, D. (1974c). Prediction of early U.S. Air Force cadet adaptation with the MMPI. *Journal of Counseling Psychology, 21,* 404–408.

Lachar, D., Hays, J.R., & Buckle, K.E. (1991, August). *An exploratory study of MMPI-2 single-scale correlates.* Symposium presentation at the 99th Annual Convention of the American Psychological Association, San Francisco, CA.

Lachar, D., Klinge, V., & Grisell, J.L. (1976). Relative accuracy of automated MMPI narratives generated from adult norm and adolescent norm profiles. *Journal of Consulting and Clinical Psycholgy, 44,* 20–24.

Lachar, D., & Wrobel, T.A. (1979). Validation of clinicians' hunches: Construction of a new MMPI critical item set. *Journal of Consulting and Clinical Psychology, 47,* 277–284.

Lane, P.J. (1976). *Annotated bibliography of the Megargee et al.'s Overcontrolled-Hostility (O-H) scale and the overcontrolled personality literature.* Unpublished materials. Florida State University, Tallahassee, FL.

Lanyon, R.I. (1967). Simulation of normal and psychopathic MMPI personality patterns. *Journal of Consulting Psychology, 31,* 94–97.

Lanyon, R.I. (1968). *A handbook of MMPI group profiles.* Minneapolis: University of Minnesota Press.

Lauber, M., & Dahlstrom, W.G. (1953). MMPI findings in the rehabilitation of delinquent girls. In S.R. Hathaway & E.D. Monachesi (Eds.), *Analyzing and predicting juvenile delinquency with the MMPI* (pp. 61–69). Minneapolis, MN: University of Minnesota Press.

Lees-Haley, P.R., English, L.T., & Glenn, W.J. (1991). A fake-bad scale on the MMPI-2 for personal injury claimants. *Psychologial Reports, 68,* 203–210.

Leon, G.R., Finn, S.E., Murray, D., & Bailey, J.M. (1988). Inability to predict cardiovascular disease from hostility scores or MMPI items related to Type A behavior. *Journal of Consulting and Clinical Psychology, 56,* 597–600.

Leon, G.R., Gillum, B., Gillum, R., & Gouze, M. (1979). Personality stability and change over a 30-year period—middle age to old age. *Journal of Consulting and Clinical Psychology, 47,* 517–524.

Levenson, M.R., Aldwin, C.M., Butcher, J.N., DeLabry, L., Workman-Daniels, K., & Bosse, R. (1990). The MAC scale in a normal population: The meaning of "false positives." *Journal of Studies on Alcohol, 51,* 457–462.

Levitt, E.E. (1990). A structural analysis of the impact of MMPI-2 on MMPI-1. *Journal of Personality Assessment, 55,* 562–577.

Levitt, E.E., Browning, J.M., & Freeland, L.J. (1992). The effect of MMPI-2 on the scoring of special scales derived from MMPI-1. *Journal of Personality Assessment, 59,* 22–31.

Lewandowski, D., & Graham, J.R. (1972). Empirical correlates of frequently occurring two-point code types: A replicated study. *Journal of Consulting and Clinical Psychology, 39,* 467–472.

Lewinsohn, P.M. (1965). Dimensions of MMPI change. *Journal of Clinical Psychology, 21,* 37–43.

Lezak, M.D. (1987). Norms for growing older. *Developmental Neuropsychology, 3,* 1–12.

Lichtenstein, E., & Bryan, J.H. (1966). Short-term stability of MMPI profiles. *Journal of Consulting Psychology, 30,* 172–174.

Lingoes, J. (1960). MMPI factors of the Harris and Wiener subscales. *Journal of Consulting Psychology, 24,* 74–83.

Little, K.B., & Shneidman, E.S. (1959). Congruencies among interpretations of psychological test and anamnestic data. *Psychological Monographs, 73* (6, Whole No. 476).

Litz, B.T., Penk, W.E., Walsh, S., Hyer, L., Blake, D.D., Marx, B., Keane, T.M., & Bitman, D. (1992). Similarities and differences between MMPI and MMPI-2 applications to the assessment of post-traumatic stress disorder. *Journal of Personality Assessment, 57,* 238–253.

Lubin, B., Larsen, R.M., & Matarazzo, J.D. (1984). Patterns of psychological test usage in the United States: 1935–1982. *American Psychologist, 39,* 451–454.

Lubin, B., Larsen, R.M., Matarazzo, J.D., & Seever, M. (1985). Psychological test usage patterns in five professional settings. *American Psychologist, 40,* 857–861.

Lucas, R.W., Mullin, P.J., Luna, C.B., & McInroy, D.C. (1977). Psychiatrists and a computer as interrogators of patients with alcohol-related illnesses: A comparison. *British Journal of Psychiatry, 131,* 160–167.

Lushene, R.E. (1967). *Factor structure of the MMPI item pool.* Unpublished master's thesis, Florida State University, Tallahassee, FL.

Lushene, R.E., & Gilberstadt, H. (1972, March). *Validation of VA MMPI computer-generated reports.* Paper presented at the Veterans Administration Cooperative Studies Conference, St. Louis, MO.

Lyons, J.A., & Keane, T.M. (1992). Keane PTSD scale: MMPI and MMPI-2. *Journal of Traumatic Stress, 5,* 111–117.

MacAndrew, C. (1965). The differentiation of male alcoholic out-patients from nonalcoholic psychiatric patients by means of the MMPI. *Quarterly Journal of the Studies on Alcohol, 26,* 238–246.

MacAndrew, C. (1986). Toward the psychometric detection of substance misuse in young men: The SAP scale. *Journal of Studies on Alcohol, 47,* 161–166.

MacDonald, G.L. (1952). A study of the shortened group and individual forms of the MMPI. *Journal of Clinical Psychology, 8,* 309–311.

Marks, P.A., & Seeman, W. (1963). *Actuarial description of abnormal personality.* Baltimore, MD: Williams & Wilkins.

Marks, P.A., Seeman, W., & Haller, D.L. (1974). *The actuarial use of the MMPI with adolescents and adults.* Baltimore, MD: Williams & Wilkins.

Marsella, A.J., Sanborn, K.O., Kameoka, V., Shizuru, L., & Brennan, J. (1975). Cross-validation of self-report measures of depression among normal populations of Japanese, Chinese, and Caucasian ancestry. *Journal of Clinical Psychology, 31,* 281–287.

Matarazzo, J.D. (1986). Computerized clinical psychological test interpretations: Unvalidated plus all mean and no sigma. *American Psychologist, 41,* 14–24.

McAnulty, D.P., Rappaport, N.B., & McAnulty, R.D. (1985). An aposteriori investigation of standard MMPI validity scales. *Psychological Reports, 57,* 95–98.

McCaffrey, R.J., Hickling, E.J., & Marrazo, M.J. (1989). Civilian-related post-traumatic stress disorder: Assessment-related issues. *Journal of Clinical Psychology, 45,* 72–76.

McCrae, R.R., Costa, P.T., Dahlstrom, W.G., Barefoot, J.C., Siegler, I.C., & Williams, R.B. (1989). A caution on the use of the MMPI K-correction in research on psychosomatic medicine. *Psychosomatic Medicine, 51,* 58–65.

McCranie, E.W., Watkins, L.O., Brandsma, J.M., & Sisson, B.D. (1986). Hostility, coronary heart disease (CHD) incidence, and total mortality: Lack of association in a 25-year follow-up study of 478 physicians. *Journal of Behavioral Medicine, 9,* 119–125.

McCreary, C., & Padilla, E. (1977). MMPI differences among Black, Mexican-American, and White male offenders. *Journal of Clinical Psychology, 33,* 171–177.

McFall, M.E., Smith, D.E., Roszell, D.K., Tarver, D.J., & Malas, K.L. (1990). Convergent validity of measures of PTSD in Vietnam combat veterans. *American Journal of Psychiatry, 147,* 645–648.

McKinley, J.C., & Hathaway, S.R. (1940). A multiphasic personality schedule (Minnesota): II. A differential study of hypochondriasis. *Journal of Psychology, 10,* 255–268.

McKinley, J.C., & Hathaway, S.R. (1944). The MMPI: V. Hysteria, hypomania, and psychopathic deviate. *Journal of Applied Psychology, 28,* 153–174.

McKinley, J.C., Hathaway, S.R., & Meehl, P.E. (1948). The MMPI: VI. The K scale. *Journal of Consulting Psychology, 12,* 20–31.

Meehl, P.E. (1951). *Research results for counselors.* St. Paul, MN: State Department of Education.

Meehl, P.E. (1956). Wanted—a good cookbook. *American Psychologist, 11,* 263–272.

Meehl, P.E., & Dahlstrom, W.G. (1960). Objective configural rules for discriminating psychotic from neurotic MMPI profiles. *Journal of Consulting Psychology, 24,* 375–387.

Meehl, P.E., & Hathaway, S.R. (1946). The K factor as a suppressor variable in the MMPI. *Journal of Applied Psychology, 30,* 525–564.

Megargee, E.I. (1979). Development and validation of an MMPI-based system for classifying criminal offenders. In J.N. Butcher (Ed.), *New developments in the use of the MMPI* (pp. 303–324). Minneapolis: University of Minnesota Press.

Megargee, E.I., Bohn, M.J., Meyer, J.E., Jr., & Sink, F. (1979). *Classifying criminal offenders: A new system based on the MMPI.* Beverly Hills, CA: Sage.

Megargee, E.I., Cook, P.E., & Mendelsohn, G.A. (1967). The development and validation of an MMPI scale of assaultiveness in overcontrolled individuals. *Journal of Abnormal Psychology, 72,* 519–528.

Meikle, S., & Gerritse, R. (1970). MMPI cookbook pattern frequencies in a psychiatric unit. *Journal of Clinical Psychology, 26,* 82–84.

Messick, S., & Jackson, D.N. (1961). Acquiescence and the factorial interpretation of the MMPI. *Psychological Bulletin, 58,* 299–304.

Miller, H.R., & Streiner, D.L. (1990). Using the Millon Clinical Multiaxial Inventory's Scale B and the MacAndrew Alcoholism Scale to identify alcoholics with concurrent psychiatric diagnoses. *Journal of Personality Assessment, 54,* 736–746.

Montgomery, G.T., Arnold, B.R., & Orozco, S. (1990). MMPI supplemental scale performance of Mexican Americans and level of acculturation. *Journal of Personality Assessment, 54,* 328–342.

Moore, J.E., McFall, M.E., Kivlahan, D.R., & Capestany, F. (1988). Risk of misinterpretation of MMPI Schizophrenia scale elevations in chronic pain patients. *Pain, 32,* 207–213.

Moreland, K.L. (1984). Intelligent use of automated psychological reports. *Critical Items, 1,* 4–5. (Distributed by National Computer Systems, Minneapolis, MN).

Moreland, K.L. (1985a). Computer-assisted psychological assessment in 1986: A practical guide. *Computers in Human Behavior, 1,* 221–233.

Moreland, K.L. (1985b). *Test-retest reliability of 80 MMPI scales.* Unpublished materials. (Available from National Computer Systems, Minneapolis, MN).

Moreland, K.L. (1985c). Validation of computer-based test interpretations: Problems and prospects. *Journal of Consulting and Clinical Psychology, 53*, 816–825.

Moreland, K.L., & Onstad, J.A. (1985, March). *Validity of the Minnesota Clinical Report, I: Mental health outpatients.* Paper presented at the 20th Annual Symposium on Recent Developments in the Use of the MMPI, Honolulu, HI.

Moreland, K.L., & Walsh, S. (1991, August). *Comparative concurrent validity of the MMPI-2 using MMPI and MMPI-2 based descriptors.* Symposium presentation at the 99th Annual Convention of the American Psychological Association, San Francisco, CA.

Muller, B.P., & Bruno, L.N. (1988, March). *The MMPI and the Inwald Personality Inventory in the psychological screening of police candidates.* Paper presented at the 23rd Annual Symposium on Recent Developments in the Use of the MMPI. St. Petersburg, FL.

Munley, P.H. (1991). A comparison of MMPI-2 and MMPI T scores for men and women. *Journal of Clinical Psychology, 47*, 87–91.

Munley, P.H., & Zarantonello, M.M. (1990). A comparison of MMPI profile types with corresponding estimated MMPI-2 profiles. *Journal of Clinical Psychology, 46*, 803–810.

Nelson, L.D., & Marks, P.A. (1985). Empirical correlates of infrequently occurring MMPI code types. *Journal of Clinical Psychology, 41*, 477–482.

Nichols, D.S. (1992). Minnesota Multiphasic Personality Inventory-2. In J.J. Kramer & J.C. Conoly (Eds.), *Eleventh mental measurements yearbook* (pp. 562–565). Lincoln, NE: Buros Institute of Mental Measurements.

Nichols, W. (1980). The classification of law offenders with the MMPI: A methodological study. (Doctoral Dissertation, University of Alabama, 1979). *Dissertation Abstracts International, 41*, No. 1, 333B.

Nockleby, D.M., & Deaton, A.V. (1987). Denial versus distress: Coping patterns in post head trauma patients. *The International Journal of Clinical Neuropsychology, 10*, 145–148.

Olmstead, D.W., & Monachesi, E.D. (1956). A validity check on MMPI scales of responsibility and dominance. *Journal of Abnormal and Social Psychology, 53*, 140–141.

Orr, S.P., Claiborn, B.A., Forgue, D.F., DeJong, J.B., Pitman, R.K., & Herz, L.R. (1990). Psychometric profile of post-traumatic stress disorder, anxious, and healthy Vietnam veterans and correlations with psychophysiologic responses. *Journal of Consulting and Clinical Psychology, 58*, 329–335.

Osborne, D. (1979). Use of the MMPI with medical patients. In J.N. Butcher (Ed.), *New developments in the use of the MMPI* (pp. 141–163). Minneapolis: University of Minnesota Press.

Otto, R.K., Lang, A.R., Megargee, E.I., & Rosenblatt, A.I. (1988). Ability of alcoholics to escape detection by the MMPI. *Journal of Consulting and Clinical Psychology, 56*, 452–457.

Page, R.D., & Bozlee, S. (1982). A cross-cultural MMPI comparison of alcoholics. *Psychological Reports, 50*, 639–646.

Panton, J.H. (1958). MMPI profile configurations among crime classification groups. *Journal of Clinical Psychology, 14*, 305–308.

Panton, J. (1959). The response of prison inmates to MMPI subscales. *Journal of Social Therapy, 5*, 233–237.

Parker, C.A. (1961). The predictive use of the MMPI in a college counseling center. *Journal of Counseling Psychology, 8*, 154–158.

Pauker, J.D. (1966). Stability of MMPI profiles of female psychiatric inpatients. *Journal of Clinical Psychology, 22*, 209–212.

Parker, K.C.H., Hanson, R.K., & Hunsley, J. (1988). MMPI, Rorschach, and WAIS: A meta-analytic comparison of reliability, stability, and validity. *Psychological Bulletin, 103*, 367–373.

Penk, W.E., Robinowitz, R., Roberts, W.R., Dolan, M.P., & Atkins, H.G. (1981). MMPI differences of male Hispanic-American, Black and White heroin addicts. *Journal of Consulting and Clinical Psychology, 49*, 488–490.

Pepper, L.J., & Strong, P.N. (1958). *Judgmental subscales for the Mf scale of the MMPI.* Unpublished materials, Hawaii Department of Health, Honolulu, HI.

Persky, V.W., Kempthorne-Rawson, J., & Shekelle, R.B. (1987). Personality and risk of cancer: 20-year follow-up of the Western Electric Study. *Psychosomatic Medicine, 49*, 435–449.

Persons, R.W., & Marks, P.A. (1971). The violent 4–3 MMPI personality type. *Journal of Consulting and Clinical Psychology, 36,* 189–196.

Peterson, C.D., & Dahlstrom, W.G. (1992). The derivation of gender-role scales GM and GF for MMPI-2 and their relationship to scale 5 (Mf). *Journal of Personality Assessment, 59,* 486–499.

Peterson, D.R. (1954). Predicting hospitalization of psychiatric outpatients. *Journal of Abnormal and Social Psychology, 49,* 260–265.

Piotrowski, C., & Keller, J.W. (1984). Attitudes toward clinical assessment by members of AABT. *Psychological Reports, 55,* 831–838.

Piotrowski, C., & Keller, J.W. (1989). Psychological testing in outpatient mental health facilities: A national survey. *Professional Psychology: Research and Practice, 20,* 423–425.

Piotrowski, C., & Lubin, B. (1990). Assessment practices of health psychologists: Survey of APA Division 38 practitioners. *Professional Psychology: Research and Practice, 21,* 99–106.

Piotrowski, C., Shery, D., & Keller, J.W. (1985). Psychodiagnostic test usage: A survey of the Society for Personality Assessment. *Journal of Personality Assessment, 49,* 115–119.

Pollack, D., & Shore, J.H. (1980). Validity of the MMPI with Native Americans. *American Journal of Psychiatry, 137,* 946–950.

Pope, K.S. (1992). Responsibilities in providing psychological test feedback to clients. *Psychological Assessment, 4,* 268–271.

Pritchard, D.A., & Rosenblatt, A. (1980). Racial bias in the MMPI: A methodological review. *Journal of Consulting and Clinical Psychology, 48,* 263–267.

Prokop, C.K. (1986). Hysteria scale elevations in low back pain patients: A risk factor for misdiagnosis. *Journal of Consulting and Clinical Psychology, 54,* 558–562.

Quay, H. (1955). The performance of hospitalized psychiatric patients on the ego-strength scale of the MMPI. *Journal of Clinical Psychology, 11,* 403–405.

Query, W.T., Megran, J., & McDonald, G. (1986). Applying post-traumatic stress disorder MMPI subscale to World War II POW veterans. *Journal of Clinical Psychology, 42,* 315–317.

Rapaport, G.M. (1958). "Ideal self" instructions, MMPI profile changes, and the prediction of clinical improvement. *Journal of Consulting Psychology, 27,* 459–463.

Rhodes, R.J. (1969). The MacAndrew Alcoholism scale: A replication. *Journal of Clinical Psychology, 25,* 189–191.

Rice, M.E., Arnold, L.S., & Tate, D.L. (1983). Faking good and bad adjustment on the MMPI and overcontrolled-hostility in maximum security psychiatric patients. *Canadian Journal of Behavioral Sciences, 15,* 43–51.

Rich, C.C., & Davis, H.G. (1969). Concurrent validity of MMPI alcoholism scales. *Journal of Clinical Psychology, 25,* 425–426.

Richard, L.S., Wakefield, J.A., & Lewak, R. (1990). Similarity of personality variables as predictors of marital satisfaction: A Minnesota Multiphasic Personality Inventory (MMPI) item analysis. *Personality and Individual Differences, 11,* 39–43.

Rohan, W.P. (1972). MMPI changes in hospitalized alcoholics: A second study. *Quarterly Journal of Studies on Alcohol, 33,* 65–76.

Rohan, W.P., Tatro, R.L., & Rotman, S.R. (1969). MMPI changes in alcoholics during hospitalization. *Quarterly Journal of Studies on Alcohol, 30,* 389–400.

Roper, B.L., Ben-Porath, Y.S., & Butcher, J.N. (1991). Comparability of computerized adaptive and conventional testing with the MMPI-2. *Journal of Personality Assessment, 57,* 278–290.

Rosen, A. (1963). Diagnostic differentiation as a construct validity indication for the MMPI Ego Strength scale. *Journal of General Psychology, 69,* 65–68.

Rosenberg, N. (1972). MMPI alcoholism scales. *Journal of Clinical Psychology, 28,* 515–522.

Schaffer, C.E., Pettigrew, C.G., Blouin, D., & Edwards, D.W. (1983). Multivariate classification of female offender MMPI profiles. *Journal of Crime and Justice, 6,* 57–66.

Schill, T., & Wang, S. (1990). Correlates of the MMPI-2 Anger content scale. *Psychological Reports, 67,* 800–802.

Schlenger, W.E., & Kulka, R.A. (1989). *PTSD scale development for the MMPI-2.* Research Triangle Park, NC: Research Triangle Institute.

Schlenger, W.E., Kulka, R.A., Fairbank, J.A., Hough, R.L., Jordan, B.K., Marmar, C.R., & Weiss,

D.S. (1989). *The prevalence of post-traumatic stress disorder in the Vietnam generation: Findings from the National Vietnam Veterans Readjustment study.* Report from Research Triangle Institute, Research Triangle Park, NC.

Schretlen, D.J. (1988). The use of psychological tests to identify malingered symptoms of mental disorder. *Clinical Psychology Review, 8,* 451–476.

Schubert, H.J.P. (1973). *A wide-range MMPI manual.* Unpublished materials.

Schuldberg, D. (1992). Ego Strength revised: A comparison of the MMPI-2 and MMPI-1 versions of the Barron Ego Strength scale. *Journal of Clinical Psychology, 48,* 500–505.

Schwartz, G.F. (1977). *An investigation of the stability of single scale and two-point MMPI code types for psychiatric patients.* Unpublished doctoral dissertation, Kent State University, Kent, OH.

Schwartz, M.F., & Graham, J.R. (1979). Construct validity of the MacAndrew Alcoholism scale. *Journal of Consulting and Clinical Psychology, 47,* 1090–1095.

Serkownek, K. (1975). *Subscales for scales 5 and 0 of the Minnesota Multiphasic Personality Inventory.* Unpublished materials.

Shekelle, R.B., Gale, M., Ostfeld, A.M., & Paul, O. (1983). Hostility, risk of coronary heart disease, and mortality. *Psychosomatic Medicine, 45,* 109–114.

Sherriffs, A.C., & Boomer, D.S. (1954). Who is penalized by the penalty for guessing? *Journal of Educational Psychology, 45,* 81–90.

Shondrick, D.D., Ben-Porath, Y.S., & Stafford, K.P. (1992, May). *Forensic applications of the MMPI-2.* Paper presented at the 27th Annual Symposium on Recent Developments in the Use of the MMPI (MMPI-2 and MMPI-A). Minneapolis, MN.

Sieber, K.O., & Meyers, L.S. (1992). Validation of the MMPI-2 Social Introversion subscales. *Psychological Assessment, 4,* 185–189.

Silver, R.J., & Sines, L.K. (1962). Diagnostic efficiency of the MMPI with and without the K correction. *Journal of Clinical Psychology, 18,* 312–314.

Sines, L.K. (1959). The relative contribution of four kinds of data to accuracy in personality assessment. *Journal of Consulting Psychology, 23,* 483–492.

Sinnett, E.R. (1962). The relationship between the Ego Strength scale and rated in-hospital improvement. *Journal of Clinical Psychology, 18,* 46–47.

Sivanich, G. (1960). *Test-retest changes during the course of hospitalization among some frequently occurring MMPI profiles.* Unpublished doctoral dissertation, University of Minnesota.

Sloan, P. (1988). Post-traumatic stress in survivors of an airplane crash-landing: A clinical and exploratory research intervention. *Journal of Traumatic Stress, 1,* 211–229.

Smith, C.P., & Graham, J.R. (1981). Behavioral correlates for the MMPI standard F scale and for a modified F scale for black and white psychiatric patients. *Journal of Consulting and Clinical Psychology, 49,* 455–459.

Snyder, D.K., Kline, R.B., & Podany, E.C. (1985). Comparison of external correlates of MMPI substance abuse scales across sex and race. *Journal of Consulting and Clinical Psychology, 53,* 520–525.

Snyter, C.M., & Graham, J.R. (1984). The utility of subtle and obvious MMPI subscales. *Journal of Clinical Psychology, 40,* 981–985.

Sobel, H.J., & Worden, W. (1979). The MMPI as a predictor of psychosocial adaptation to cancer. *Journal of Consulting and Clinical Psychology, 47,* 716–724.

Solway, K.S., Hays, J.R., & Zieben, M. (1976). Personality characteristics of juvenile probation officers. *Journal of Community Psychology, 4,* 152–156.

Spanier, G.B. (1976). Measuring dyadic adjustment: New scales for assessing the quality of marriage and similar dyads. *Journal of Marriage and the Family, 38,* 15–28.

Spiegel, D.E. (1969). SPI and MMPI predictors of psychopathology. *Journal of Projective Techniques and Personality Assessment, 33,* 265–273.

Stein, K.B. (1968). The TSC scales: The outcome of a cluster analysis of the 550 MMPI items. In P. McReynolds (Ed.), *Advances in psychological assessment, Vol. I* (pp. 80–104). Palo Alto, CA: Science and Behavior Books.

Stone, L.A., Bassett, G.R., Brousseau, J.D., Demers, J., & Stiening, J.A. (1972). Psychological test scores for a group of MEDEX trainees. *Psychological Reports, 31,* 827–831.

Strassberg, D.S. (1991). Interpretive dilemmas created by the Minnesota Multiphasic Personality Inventory-2 (MMPI-2). *Journal of Psychopathology and Behavioral Assessment, 13,* 53–59.

Strauss, M.E., Gynther, M.D., & Wallhermfechtel, J. (1974). Differential misdiagnosis of blacks and whites by the MMPI. *Journal of Personality Assessment, 38,* 55–60.

Strenger, V.E. (1989). *Content homogeneous subscales for scale 7 of the MMPI.* Unpublished masters thesis, Kent State University, Kent, OH.

Strupp, H.H., & Bloxom, A.L. (1975). An approach to defining a patient population in psychotherapy research. *Journal of Counseling Psychology, 22,* 231–237.

Sue, S., & Sue, D.W. (1974). MMPI comparisons between Asian-American and non-Asian students utilizing a student health psychiatric clinic. *Journal of Counseling Psychology, 21,* 423–427.

Sullivan, D.L., Miller, C., & Smelser, W. (1958). Factors in the length of stay and progress in psychotherapy. *Journal of Consulting Psychology, 22,* 1–9.

Swenson, W.M. (1961). Structured personality testing in the aged: An MMPI study of the gerontic population. *Journal of Clinical Psychology, 17,* 302–304.

Swenson, W.M., Pearson, J.S., & Osborne, D. (1973). *An MMPI source book: Basic item, scale, and pattern data for 50,000 medical patients.* Minneapolis: University of Minnesota Press.

Swenson, W.M., Rome, H.P., Pearson, J.S., & Brannick, T.L. (1965). A totally automated psychological test: Experience in a medical center. *Journal of the American Medical Association, 191,* 925–927.

Taft, R. (1957). The validity of the Barron Ego Strength scale and the Welsh Anxiety index. *Journal of Consulting Psychology, 21,* 247–249.

Tamkin, A.S. (1957). An evaluation of the construct validity of Barron's Ego Strength scale. *Journal of Consulting Psychology, 13,* 156–158.

Tamkin, A.S., & Klett, C.J. (1957). Barron's Ego Strength scale: A replication of an evaluation of its construct validity. *Journal of Consulting Psychology, 21,* 412.

Tanner, B.A. (1990). Composite descriptions associated with rare MMPI two-point code types: Codes that involve scale 5. *Journal of Clinical Psycholgy, 46,* 425–431.

Taulbee, E.S., & Sisson, B.D. (1957). Configural analysis of MMPI profiles of psychiatric groups. *Journal of Consulting Psychology, 21,* 413–417.

Taylor, J.R., Strassberg, D.S., & Turner, C.W. (1989). Utility of the MMPI in a geriatric population. *Journal of Personality Assessment, 53,* 665–676.

Tellegen, A., & Ben-Porath, Y.S. (1992). The new uniform T scores for the MMPI-2: Rationale, derivation, and appraisal. *Psychological Assessment, 4,* 145–155.

Terman, L.M., & Miles, C.C. (1936). *Sex and personality: Studies in masculinity and femininity.* New York: McGraw-Hill.

Timbrook, R.E., & Graham, J.R. (1992). *The meaning of low scores on the MMPI-2 clinical and content scales in a psychiatric setting.* Unpublished manuscript, Kent State University.

Timbrook, R.E., Graham, J.R., Keiller, S.W., & Watts, D. (1993). Comparison of the Wiener-Harmon subtle-obvious scales and the standard validity scales in detecting valid and invalid MMPI-2 profiles. *Psychological Assessment.*

Tryon, R.C. (1966). Unrestricted cluster and factor analysis, with application to the MMPI and Holzinger-Harman problems. *Multivariate Behavioral Research, 1,* 229–244.

Tryon, R.C., & Bailey, D. (Eds.). (1965). *Users' manual of the BC TRY system of cluster and factor analysis* (Taped version). Berkeley, CA: University of California Computer Center.

Tsushima, W.T., & Onorato, V.A. (1982). Comparison of MMPI scores of White and Japanese-American medical patients. *Journal of Consulting and Clinical Psychology, 50,* 150–151.

Uecker, Á.E. (1969). Comparability of two methods of administering the MMPI to brain-damaged geriatric patients. *Journal of Clinical Psychology, 25,* 196–198.

Uecker, A.E. (1970). Differentiating male alcoholics from other psychiatric inpatients: Validity of the MacAndrew scale. *Quarterly Journal of Studies on Alcohol, 31,* 379–383.

Uecker, A.E., Boutilier, L.R., & Richardson, E.H. (1980). "Indianism" and MMPI scores of men alcoholics. *Journal of Studies on Alcohol, 41,* 357–362.

Vanderploeg, R.D., Sisson, G.F.P., Hickling, E.J. (1987). A reevaluation of the use of the MMPI in the assessment of combat-related post-traumatic stress disorder. *Journal of Personality Assessment, 51,* 140–150.

Velasquez, R.J. (1992). Hispanic-American MMPI research (1949–1992): A comprehensive bibliography. *Psychological Reports, 70,* 743–754.

Velasquez, R.J., & Callahan, W.J. (1990a). MMPI comparisons of Hispanic- and White-American veterans seeking treatment for alcoholism. *Psychological Reports, 67,* 95–98.

Velasquez, R.J., & Callahan, W.J. (1990b). MMPIs of Hispanic, Black, and White DSM-III schizophrenics. *Psychological Reports, 66,* 819–822.

Velasquez, R.J., Callahan, W.J., & Carrillo, R. (1989). MMPI profiles of Hispanic-American inpatient and outpatient sex offenders. *Psychological Reports, 65,* 1055–1058.

Walters, G.D. (1988). Assessing dissimulation and denial on the MMPI in a sample of maximum security, male inmates. *Journal of Personality Assessment, 52,* 465–474.

Walters, G.D., Greene, R.L., & Jeffrey, T.B. (1984). Discriminating between alcoholic and nonalcoholic blacks and whites on the MMPI. *Journal of Personality Assessment, 48,* 486–488.

Walters, G.D., Greene, R.L., Jeffrey, T.B., Kruzich, D.J., & Haskin, J.J. (1983). Racial variations on the MacAndrew Alcoholism scale of the MMPI. *Journal of Consulting and Clinical Psychology, 51,* 947–948.

Ward, L.C., & Jackson, D.B. (1990). A comparison of primary alcoholics, secondary alcoholics, and nonalcoholic psychiatric patients on the MacAndrew Alcoholism scale. *Journal of Personality Assessment, 54,* 729–735.

Watkins, C.E., Jr., Campbell, V.L., & McGregor, P. (1988). Counselling psychologists' uses of and opinions about psychological tests: A contemporary perspective. *Counseling Psychologist, 16,* 476–486.

Watson, C.G., Juba, M., Anderson, E.D., & Manifold, V. (1990). What does the Keane et al. PTSD scale for the MMPI measure? *Journal of Clinical Psychology, 46,* 600–606.

Watson, C.G., Kucala, T., & Manifold, V. (1986). A cross-validation of the Keane and Penk MMPI scales as measures of post-traumatic stress disorder. *Journal of Clinical Psychology, 42,* 727–732.

Watson, C.G., Kucala, T., Manifold, V., Vassar, P. & Juba, M. (1988). Differences between post-traumatic stress disorder patients with delayed and undelayed onsets. *Journal of Nervous and Mental Disease, 176,* 568–572.

Webb, J.T. (1970). Validity and utility of computer-produced reports with Veterans Administration psychiatric populations. *Proceedings of the 78th Annual Convention of the American Psychological Association, 5,* 541–542.

Webb, J.T., Levitt, E.E., & Rojdev, R. (1993, March). *After three years: Comparing clinical use of MMPI-1 and MMPI-2.* Paper presented at the Midwinter Meeting of the Society for Personality Assessment. San Francisco, CA.

Webb, J.T., Miller, M.L., & Fowler, R.D. (1969). Validation of a computerized MMPI interpretation system. *Proceedings of the 77th Annual Convention of the American Psychological Association, 4,* 523–524.

Webb, J.T., Miller, M.L., & Fowler, R.D. (1970). Extending professional time: A computerized MMPI interpretation service. *Journal of Clinical Psychology, 26,* 210–214.

Weed, N.C., Ben-Porath, Y.S., & Butcher, J.N. (1990). Failure of Wiener and Harmon Minnesota Multiphasic Personality Inventory (MMPI) subtle scales as personality descriptors and as validity indicators. *Psychological Assessment: A Journal of Consulting and Clinical Psychology, 2,* 281–285.

Weed, N.C., Butcher, J.N., McKenna, T., & Ben-Porath, Y.S. (1992). New measures for assessing alcohol and drug abuse with the MMPI-2: The APS and AAS. *Journal of Personality Assessment, 58,* 389–410.

Weed, N.C., Butcher, J.N., & Williams, C.L. (1992). *Development of the Alcohol and Drug Problem Proneness Scale (PRO) and the Alcohol and Drug Problem Acknowledgment Scale (ACK).* Unpublished manuscript.

Welsh, G.S. (1948). An extension of Hathaway's MMPI profile coding system. *Journal of Consulting Psychology, 12,* 343–344.

Welsh, G.S. (1956). Factor dimensions A and R. In G.S. Welsh & W.G. Dahlstrom (Eds.), *Basic readings on the MMPI in psychology and medicine* (pp. 264–281). Minneapolis: University of Minnesota Press.

Welsh, G.S. (1965). MMPI profiles and factors A and R. *Journal of Clinical Psychology, 21,* 43–47.

Wetter, M.W., Baer, R.A., Berry, D.T.R., Smith, G.T., & Larsen, L.H. (1992). Sensitivity of MMPI-2 validity scales to random responding and malingering. *Psychological Assessment, 4,* 369–374.

Wiener, D.N. (1947). Differences between the individual and group forms of the MMPI. *Journal of Consulting Psychology, 11,* 104–106.

Wiener, D.N. (1948). Subtle and obvious keys for the MMPI. *Journal of Consulting Psychology, 12,* 164–170.

Wiggins, J.S. (1969). Content dimensions in the MMPI. In J.N. Butcher (Ed.), *MMPI: Research developments and clinical applications* (pp. 127–180). New York: McGraw-Hill.

Wiggins, J.S. (1973). *Personality and prediction: Principles of personality assessment.* Reading, MA: Addison-Wesley.

Wilderman, J.E. (1984). *An investigation of the clinical utility of the College Maladjustment scale.* Unpublished master's thesis, Kent State University, Kent, OH.

Williams, A.F., McCourt, W.F., & Schneider, L. (1971). Personality self-descriptions of alcoholics and heavy drinkers. *Quarterly Journal of Studies on Alcohol, 32,* 310–317.

Williams, C.L. (1986). MMPI profiles from adolescents: Interpretive strategies and treatment considerations. *Journal of Child and Adolescent Psychotherapy, 3,* 179–193.

Williams, C.L., Ben-Porath, Y.S., & Hevern, V. (1991, March). *Item-level improvements for the MMPI-A.* Paper presented at the 26th Annual Symposium on Recent Developments in the Use of the MMPI (MMPI-2 and MMPI-A). St. Petersburg Beach, FL.

Williams, C.L., & Butcher, J.N. (1989a). An MMPI study of adolescents: I. Empirical validity of the standard scales. *Psychological Assessment: A Journal of Consulting and Clinical Psychology, 1,* 251–259.

Williams, C.L., & Butcher, J.N. (1989b). An MMPI study of adolescents: II. Verification and limitations of code type classifications. *Psychological Assessment: A Journal of Consulting and Clinical Psychology, 1,* 260–265.

Williams, C.L., Butcher, J.N., Ben-Porath, Y.S., & Graham, J.R. (1992). *MMPI-A content scales: Assessing psychopathology in adolescents.* Minneapolis: University of Minnesota Press.

Williams, H.L. (1952). The development of a caudality scale for the MMPI. *Journal of Clinical Psychology, 8,* 293–297.

Williams, R.B., Haney, T.L., Lee, K.L., Kong, Y.H., Blumenthal, J.A., & Whalen, R.E. (1980). Type A behavior, hostility, and coronary atherosclerosis. *Psychosomatic Medicine, 42,* 539–549.

Wimbish, L.G. (1984). *The importance of appropriate norms for the computerized interpretation of adolescent MMPI profiles.* Unpublished doctoral dissertation, Ohio State University, Columbus, OH.

Wirt, R.D. (1955). Further validation of the Ego Strength scale. *Journal of Consulting Psychology, 19,* 444.

Wirt, R.D. (1956). Actuarial prediction. *Journal of Consulting Psychology, 20,* 123–124.

Wisniewski, N.M., Glenwick, D.S., & Graham, J.R. (1985). MacAndrew scale and socio-demographic correlates of adolescent alcohol and drug use. *Addictive Behaviors, 10,* 55–67.

Wolf, A.W., Schubert, D.S.P., Patterson, M., Grande, T., & Pendleton, L. (1990). The use of the MacAndrew Alcoholism scale in detecting substance abuse and antisocial personality. *Journal of Personality Assessment, 54,* 747–755.

Wolfson, K.P., & Erbaugh, S.E. (1984). Adolescent responses to the MacAndrew Alcoholism scale. *Journal of Consulting and Clinical Psychology, 52,* 625–630.

Wooten, A.J. (1984). Effectiveness of the K correction in the detection of psychopathology and its impact on profile height and configuration among young adult men. *Journal of Consulting and Clinical Psychology, 52,* 468–473.

Wrobel, N.H., & Lachar, D. (1992). Refining adolescent MMPI interpretations: Moderating effects of gender in prediction of descriptions from parents. *Psychological Assessment, 4,* 375–381.

Zager, L.D. (1988). The MMPI-based criminal classification system: A review, current status, and future directions. *Criminal Justice and Behavior, 15,* 39–57.

Zager, L.D., & Megargee, E.I. (1981). Seven MMPI drug abuse scales: An empirical investigation of their interrelationships, convergent, and discriminant validity, and degree of racial bias. *Journal of Personality and Social Psychology, 40,* 532–544.

Zalewski, C.E., & Gottesman, I.I. (1991). (Hu)Man versus mean revisited: MMPI group data and psychiatric diagnosis. *Journal of Abnormal Psychology, 100,* 562–568.

Ziskin, J. (1981). Use of the MMPI in forensic settings (*Clinical Notes on the MMPI, No. 9*). Minneapolis, MN: National Computer Systems.

Appendixes

Appendix A. Composition of the Standard Validity and Clinical Scales

L Scale
  True: None
  False: 16, 29, 41, 51, 77, 93, 102, 107,
    123, 139, 153, 183, 203, 232, 260
F Scale
  True: 18, 24, 30, 36, 42, 48, 54, 60, 66,
    72, 84, 96, 114, 138, 144, 150,
    156, 162, 168, 180, 198, 216,
    228, 234, 240, 246, 252, 258,
    264, 270, 282, 288, 294, 300,
    306, 312, 324, 336, 349, 355, 361
  False: 6, 12, 78, 90, 102, 108, 120, 126,
    132, 174, 186, 192, 204, 210,
    222, 276, 318, 330, 343
K Scale
  True: 83
  False: 29, 37, 58, 76, 110, 116, 122,
    127, 130, 136, 148, 157, 158,
    167, 171, 196, 213, 243, 267,
    284, 290, 330, 338, 339, 341,
    346, 348, 356, 365
Scale 1—Hypochondriasis (Hs)
  True: 18, 28, 39, 53, 59, 97, 101, 111,
    149, 175, 247
  False: 2, 3, 8, 10, 20, 45, 47, 57, 91,
    117, 141, 143, 152, 164, 173,
    176, 179, 208, 224, 249, 255
Scale 2—Depression (D)
  True: 5, 15, 18, 31, 38, 39, 46, 56, 73,
    92, 117, 127, 130, 146, 147, 170,
    175, 181, 215, 233
  False: 2, 9, 10, 20, 29, 33, 37, 43, 45,
    49, 55, 68, 75, 76, 95, 109, 118,
    134, 140, 141, 142, 143, 148,
    165, 178, 188, 189, 212, 221,
    223, 226, 238, 245, 248, 260,
    267, 330
Scale 3—Hysteria (Hy)
  True: 11, 18, 31, 39, 40, 44, 65, 101,
    166, 172, 175, 218, 230
  False: 2, 3, 7, 8, 9, 10, 14, 26, 29, 45,

47, 58, 76, 81, 91, 95, 98, 110,
115, 116, 124, 125, 129, 135,
141, 148, 151, 152, 157, 159,
161, 164, 167, 173, 176, 179,
185, 193, 208, 213, 224, 241,
243, 249, 253, 263, 265
Scale 4—Psychopathic Deviate (Pd)
  True: 17, 21, 22, 31, 32, 35, 42, 52, 54,
    56, 71, 82, 89, 94, 99, 105, 113,
    195, 202, 219, 225, 259, 264, 288
  False: 9, 12, 34, 70, 79, 83, 95, 122,
    125, 129, 143, 157, 158, 160,
    167, 171, 185, 209, 214, 217,
    226, 243, 261, 263, 266, 267
Scale 5—Masculinity-Femininity (Mf)—Male
  True: 4, 25, 62, 64, 67, 74, 80, 112,
    119, 122, 128, 137, 166, 177,
    187, 191, 196, 205, 209, 219,
    236, 251, 256, 268, 271
  False: 1, 19, 26, 27, 63, 68, 69, 76, 86,
    103, 104, 107, 120, 121, 132,
    133, 163, 184, 193, 194, 197,
    199, 201, 207, 231, 235, 237,
    239, 254, 257, 272
Scale 5—Masculinity-Femininity (Mf)—
Female
  True: 4, 25, 62, 64, 67, 74, 80, 112,
    119, 121, 122, 128, 137, 177,
    187, 191, 196, 205, 219, 236,
    251, 256, 271
  False: 1, 19, 26, 27, 63, 68, 69, 76, 86,
    103, 104, 107, 120, 132, 133,
    163, 166, 184, 193, 194, 197,
    199, 201, 207, 209, 231, 235,
    237, 239, 254, 257, 268, 272
Scale 6—Paranoia (Pa)
  True: 16, 17, 22, 23, 24, 42, 99, 113,
    138, 144, 145, 146, 162, 234,
    259, 271, 277, 285, 305, 307,
    333, 334, 336, 355, 361
  False: 81, 95, 98, 100, 104, 110, 244,

Source: Butcher, J.N., Dahlstrom, W.G., Graham, J.R., Tellegen, A., & Kaemmer, B. (1989). *Minnesota Multiphasic Personality Inventory-2 (MMPI-2): Manual for administration and scoring.* Minneapolis: University of Minnesota Press. Copyright © 1989 by the University of Minnesota. Reproduced by permission.

255, 266, 283, 284, 286, 297, 314, 315

Scale 7—Psychasthenia (Pt)

True: 11, 16, 23, 31, 38, 56, 65, 73, 82, 89, 94, 130, 147, 170, 175, 196, 218, 242, 273, 275, 277, 285, 289, 301, 302, 304, 308, 309, 310, 313, 316, 317, 320, 325, 326, 327, 328, 329, 331

False: 3, 9, 33, 109, 140, 165, 174, 293, 321

Scale 8—Schizophrenia (Sc)

True: 16, 17, 21, 22, 23, 31, 32, 35, 38, 42, 44, 46, 48, 65, 85, 92, 138, 145, 147, 166, 168, 170, 180, 182, 190, 218, 221, 229, 233, 234, 242, 247, 252, 256, 268, 273, 274, 277, 279, 281, 287, 291, 292, 296, 298, 299, 303, 307, 311, 316, 319, 320, 322, 323, 325, 329, 332, 333, 355

False: 6, 9, 12, 34, 90, 91, 106, 165, 177, 179, 192, 210, 255, 276, 278, 280, 290, 295, 343

Scale 9—Hypomania (Ma)

True: 13, 15, 21, 23, 50, 55, 61, 85, 87, 98, 113, 122, 131, 145, 155, 168, 169, 182, 190, 200, 205, 206, 211, 212, 218, 220, 227, 229, 238, 242, 244, 248, 250, 253, 269

False: 88, 93, 100, 106, 107, 136, 154, 158, 167, 243, 263

Scale 0—Social Introversion (Si)

True: 31, 56, 70, 100, 104, 110, 127, 135, 158, 161, 167, 185, 215, 243, 251, 265, 275, 284, 289, 296, 302, 308, 326, 337, 338, 347, 348, 351, 352, 357, 364, 367, 368, 369

False: 25, 32, 49, 79, 86, 106, 112, 131, 181, 189, 207, 209, 231, 237, 255, 262, 267, 280, 321, 328, 335, 340, 342, 344, 345, 350, 353, 354, 358, 359, 360, 362, 363, 366, 370

Appendix B. T-Score Conversions for the Standard Validity and Clinical Scales

Table B1. T-Score Conversions with K-corrections*

Raw Score	Men													Women													Raw Score
	L	F	K	Hs +.5K	D	Hy	Pd +.4K	Mf	Pa	Pt +1K	Sc +1K	Ma +.2K	Si	L	F	K	Hs +.5K	D	Hy	Pd +.4K	Mf	Pa	Pt +1K	Sc +1K	Ma +.2K	Si	
73																								120			73
72																								119			72
71																								118			71
70													100											116			70
69													99											115		95	69
68													98											113		94	68
67											120		97											112		93	67
66											118		96										120	111		91	66
65											117		94										119	109		90	65
64											115		93										117	108		89	64
63											113		92										115	106		88	63
62											111		91										114	105		87	62
61											110		90										112	103		86	61
60										120	108		89										110	102		85	60
59										119	106		87										108	100		84	59
58										117	105		86										106	99		83	58
57										115	103		85										105	97		82	57
56								109		113	101		84										103	96		81	56
55								107		111	99		83										101	94		79	55
54					120			105		109	98		82					120					99	93		78	54
53					119			103		106	96		80					118	120				97	91		77	53
52					117		120	101		104	94		79					116	118				95	90		76	52
51					115			99		102	93		78										94	88		75	51
50					112	120	117	97		100	91		77					114	115	120			92	87		74	50
49					110	119	115	95		98	89		76					112	113	118			90	85		73	49
48					108	116	112	93		96	87		75					109	111	115			88	84		72	48
47					106	114	110	91		94	86		73					107	108	113			86	82		71	47
46					104	111	107	89		91	84		72				120	105	106	110			84	81		70	46
45					102	109	105	87		89	82		71				118	103	104	107			83	79		69	45
44				120	100	106	102	85		87	81	120	70				116	101	101	105	30		81	78	120	67	44
43				119	98	104	100	83		85	79	120	69				113	99	99	102	33		79	76	118	66	43
42				116	97	101	97	81		83	77	117	68				111	96	96	100	35		77	75	115	65	42
41				114	95	99	95	79		81	75	114					109	94	94	97	38		75	73	112	64	41

354

Per the source note, Uniform T scores are given for Hs, D, Hy, Pd, Pa, Pt, Sc, Ma and Linear T scores for L, F, K, Mf, Si. The interior columns give the corresponding T‑scores. (The grid is extremely dense; values are transcribed to the best reading of the image.)

Linear T panel (L, F, K, Mf, Si) and Uniform T panel (Hs, D, Hy, Pd, Pa, Pt, Sc, Ma)

Raw	F	Mf	Si	Ma	Pa	D	Hy	Pd	Sc	Hs	Pt	K	L	Raw
40	112	40	63	109	120	107	92	94	72					40
39	110	43	62	106	118	105	90	92	70					39
38	108	45	61	103	114	103	89	89	69					38
37	105	47	60	100	111	101	87	87	67					37
36	103	50	59	97	107	99	84	84	66					36
35	101	52	58	94	103	97	82	82	65					35
34	99	55	57	91	100	95	80	80	63					34
33	97	57	55	88	96	92	77	77	62					33
32	94	60	54	85	92	90	75	75	60	120				32
31	92	62	53	82	89	88	73	73	59	118				31
30	90	65	52	79	86	86	71	71	57	114		81		30
29	88	67	51	76	84	84	69	69	56	111		79		29
28	86	69	50	74	82	82	66	66	54	107	120	77		28
27	84	72	49	71	80	80	64	64	52	104	119	75		27
26	81	74	48	68	78	78	61	61	50	111	116	72		26
25	79	77	47	65	76	76	59	59	48	107	113	70		25
24	77	79	46	62	74	74	57	57	46	103	110	68		24
23	75	82	45	59	71	71	54	54	44	100	107	66		23
22	73	84	43	56	69	69	52	52	42	96	104	64		22
21	70	87	42	53	67	67	50	50	40	90	101	62		21
20	68	89	41	51	65	65	44	47	38	86	98	60		20
19	66	92	40	49	63	63	42	45	36	83	95	58		19
18	64	94	39	47	61	61	40	43	34	79	92	56		18
17	62	96	38	45	59	59	39	42	32	75	89	54		17
16	59	99	37	43	57	57	37	40	30	72	85	51		16
15	57	101	36	41	54	54	36	38		68	82	49		15
14	54	104	35	39	51	51	35	36		64	79	47		14
13	52	106	34	37	48	48	33	35		61	76	45		13
12	49	109	33	35	46	45	31	34		57	73	43		12
11	45	111	32	33	43	42	30	32		53	70	41		11
10	42	114	31	31	40	39		31		49	67	39		10
9	39	116	30	30	38	37		30		46	64	37		9
8	37	118			35	35				42	61	35		8
7	35	120			33	33				39	58	33		7
6	33					30				37	55	30		6
5	30									34	51			5
4										32	48			4
3										31	45			3
2										30	42			2
1											39			1
0											36			0

*Uniform T scores are presented for scales Hs, D, Hy, Pd, Pa, Pt, Sc, Ma. Linear T scores are presented for scales L, F, K, Mf, Si.

Source: Butcher, J.N., Dahlstrom, W.G., Graham, J.R., Tellegen, A., & Kaemmer, B. (1989). *Minnesota Multiphasic Personality Inventory-2 (MMPI-2): Manual for administration and scoring.* Minneapolis: University of Minnesota Press. Copyright © 1989 by the University of Minnesota. Reproduced by permission.

Appendix B (continued)

Table B2. T-Score Conversions without K-corrections*

	Men													Women													
Raw Score	L	F	K	Hs	D	Hy	Pd	Mf	Pa	Pt	Sc	Ma	Si	L	F	K	Hs	D	Hy	Pd	Mf	Pa	Pt	Sc	Ma	Si	Raw Score
72																								120			72
71																								118			71
70													100											117			70
69											120		99											116		95	69
68											119		98											115		94	68
67											117		97											114		93	67
66											116		96											113		91	66
65											115		94											112		90	65
64											114		93											111		89	64
63											113		92											109		88	63
62											111		91											108		87	62
61																								107		86	61
60											110		90											106		85	60
59											109		89											105		84	59
58											108		87											104		83	58
57											107		86											103		82	57
56								109			105		85											102		81	56
55					120			107			104		84											100		79	55
54					119			105			103		83					120						99		78	54
53					117			103			102		82					118	120					98		77	53
52					115			101			100		80					116	118					97		76	52
51					114			99			99		79											96		75	51
50					112	120		97			98		78					114	115					95		74	50
49					110	119		95			97		77					112	113					94		73	49
48					108	116	120	93		104	96		76					109	111				98	93		72	48
47					106	114	117	91		103	94		75					107	108	120			97	92		71	47
46					104	111	115	89		101	93		73					105	106	118			95	90		70	46
45					102	109	113	87		100	92		72					103	104	116			94	89		69	45
44					100	106	111	85		98	91		71					101	101	113	30		93	88		67	44
43					98	104	109	83		97	90	120	70					99	99	111	33		91	87	120	66	43
42					97	101	106	81		95	88	118	69					96	96	109	35		90	86	119	65	42
41					95	99	104	79		94	87		68					94	94	106	38		89	85	116	64	41

356

40					93	96	102	78		92	86	115	66					92	92	104	40		87	84	114	63	**40**
39					91	94	100	76		91	85	112	65					90	89	102	43		86	83	111	62	**39**
38					89	91	97	74		89	84	109	64					88	87	99	45		84	81	108	61	**38**
37					87	89	95	72		88	82	106	63					86	84	97	47		83	80	105	60	**37**
36					85	86	93	70		86	81	103	62					83	82	95	50		82	79	102	59	**36**
35					83	84	91	68		85	80	100	61					81	80	92	52		80	78	100	58	**35**
34					81	81	89	66		83	79	97	59					79	77	90	55		79	77	97	57	**34**
33					80	79	86	64		82	78	94	58					77	75	88	57		78	76	94	55	**33**
32				107	78	76	84	62		80	76	91	57				100	75	73	85	60		76	75	91	54	**32**
31				105	76	74	82	60		79	75	88	56				98	72	70	83	62		75	74	89	53	**31**
30			81	103	74	71	80	58	120	78	74	85	55			83	97	70	68	81	65	120	73	72	86	52	**30**
29			79	101	72	69	77	56	119	76	73	82	54			81	95	68	65	78	67	118	72	71	83	51	**29**
28		120	77	99	70	66	75	54	116	75	72	79	52			78	93	66	63	76	69	114	71	70	80	50	**28**
27		119	75	97	68	64	73	52	112	73	70	76	51			76	91	64	61	74	72	111	69	69	77	49	**27**
26		116	72	95	66	61	71	50	108	72	69	73	50			74	89	62	58	72	74	107	68	68	75	48	**26**
25		113	70	93	64	59	68	48	105	70	68	70	49			72	87	59	56	69	77	103	67	67	72	47	**25**
24		110	68	91	62	57	66	46	101	69	67	67	48		120	70	85	57	54	67	79	100	65	66	69	46	**24**
23		107	66	89	61	54	64	44	97	67	66	64	47		116	67	83	55	51	65	82	96	64	65	66	45	**23**
22		104	64	87	59	52	62	42	94	66	64	61	45		113	65	82	53	49	62	84	92	62	63	64	43	**22**
21		101	62	85	57	50	60	40	90	64	63	58	44		109	63	80	51	47	60	87	89	61	62	61	42	**21**
20		98	60	83	54	47	57	38	86	63	62	56	43		106	61	78	49	45	58	89	85	60	61	58	41	**20**
19		95	58	81	52	45	55	36	83	61	61	53	42		103	59	76	47	43	55	92	81	58	60	55	40	**19**
18		92	56	79	50	43	53	34	79	60	60	51	41		99	56	74	46	41	53	94	78	57	59	53	39	**18**
17		89	54	77	47	42	51	32	75	58	58	49	40		96	54	72	44	39	51	96	74	56	58	51	38	**17**
16		85	51	75	45	40	49	30	72	57	57	47	38		92	52	70	42	38	49	99	70	55	57	49	37	**16**
15	100	82	49	73	42	38	46		68	56	56	45	37	105	89	50	69	40	36	47	101	67	53	56	47	36	**15**
14	96	79	47	71	40	37	44		64	54	55	43	36	100	85	48	67	38	35	45	104	63	52	55	45	35	**14**
13	91	76	45	69	38	35	42		61	53	54	42	35	95	82	46	65	36	34	43	106	59	51	53	43	34	**13**
12	87	73	43	67	36	34	40		57	52	52	40	34	90	79	43	63	35	32	41	109	56	50	52	41	33	**12**
11	83	70	41	65	34	33	38		53	50	51	38	33	86	75	41	61	32	32	39	111	52	48	51	40	32	**11**
10	78	67	39	63	32	32	36		49	49	49	37	31	81	72	39	59	30	31	37	114	49	47	50	38	30	**10**
9	74	64	37	61	30	31	34		46	47	48	35	30	76	68	37	57		30	35	116	45	46	48	36		**9**
8	70	61	35	59		30	33		42	46	46	33		71	65	35	56			33	118	42	44	47	34		**8**
7	65	58	33	57			31		39	44	45	31		66	61	32	54			31	120	39	43	45	32		**7**
6	61	55	30	54			30		37	42	43	30		62	58	30	52			30		37	41	43	30		**6**
5	56	51		52					34	41	41			57	55		49					34	39	42			**5**
4	52	48		49					32	39	39			52	51		47					32	37	39			**4**
3	48	45		46					31	37	37			47	48		44					31	35	37			**3**
2	43	42		42					30	34	35			43	44		41					30	33	35			**2**
1	39	39		38						32	32			38	41		37						31	32			**1**
0	35	36		34						30	30			33	37		33						30	30			**0**

*Uniform T scores are presented for scales Hs, D, Hy, Pd, Pa, Pt, Sc, Ma. Linear T scores are presented for scales L, F, K, Mf, Si.

Source: Butcher, J.N., Dahlstrom, W.G., Graham, J.R., Tellegen, A., & Kaemmer, B. (1989). *Minnesota Multiphasic Personality Inventory-2 (MMPI-2): Manual for administration and scoring.* Minneapolis: University of Minnesota Press. Copyright © 1989 by the University of Minnesota. Reproduced by permission.

Appendix C. Composition of the Harris-Lingoes Subscales

SCALE 2—DEPRESSION

D1—Subjective Depression
True: 31, 38, 39, 46, 56, 73, 92, 127, 130, 146, 147, 170, 175, 215, 233
False: 2, 9, 43, 49, 75, 95, 109, 118, 140, 148, 178, 188, 189, 223, 260, 267, 330

D2—Psychomotor Retardation
True: 38, 46, 170, 233
False: 9, 29, 37, 49, 55, 76, 134, 188, 189, 212

D3—Physical Malfunctioning
True: 18, 117, 175, 181
False: 2, 20, 45, 141, 142, 143, 148

D4—Mental Dullness
True: 15, 31, 38, 73, 92, 147, 170, 233
False: 9, 10, 43, 75, 109, 165, 188

D5—Brooding
True: 38, 56, 92, 127, 130, 146, 170, 215
False: 75, 95

SCALE 3—HYSTERIA

Hy1—Denial of Social Anxiety
True: None
False: 129, 161, 167, 185, 243, 265

Hy2—Need for Affection
True: 230
False: 26, 58, 76, 81, 98, 110 124, 151, 213, 241, 263

Hy3—Lassitude-malaise
True: 31, 39, 65, 175, 218
False: 2, 3, 9, 10, 45, 95, 125, 141, 148, 152

Hy4—Somatic Complaints
True: 11, 18, 40, 44, 101, 172
False: 8, 47, 91, 159, 164, 173, 176, 179, 208, 224, 249

Hy5—Inhibition of Aggression
True: None
False: 7, 14, 29, 115, 116, 135, 157

SCALE 4—PSYCHOPATHIC DEVIATE

Pd1—Familial Discord
True: 21, 54, 195, 202, 288
False: 83, 125, 214, 217

Pd2—Authority Problems
True: 35, 105
False: 34, 70, 129, 160, 263, 266

Pd3—Social Imperturbability
True: None
False: 70, 129, 158, 167, 185, 243

Pd4—Social Alienation
True: 17, 22, 42, 56, 82, 99, 113, 219, 225, 259
False: 12, 129, 157

Pd5—Self-alienation
True: 31, 32, 52, 56, 71, 82, 89, 94, 113, 264
False: 9, 95

SCALE 6—PARANOIA

Pa1—Persecutory Ideas
True: 17, 22, 42, 99, 113, 138, 144, 145, 162, 234, 259, 305, 333, 336, 355, 361
False: 314

Pa2—Poignancy
True: 22, 146, 271, 277, 285, 307, 334
False: 100, 244

Pa3—Naïveté
True: 16
False: 81, 98, 104, 110, 283, 284, 286, 315

SCALE 8—SCHIZOPHRENIA

Sc1—Social Alienation
True: 17, 21, 22, 42, 46, 138, 145, 190, 221, 256, 277, 281, 291, 292, 320, 333
False: 90, 276, 278, 280, 343

Sc2—Emotional Alienation
True: 65, 92, 234, 273, 303, 323, 329, 332

Source: Butcher, J.N., Dahlstrom, W.G., Graham, J.R., Tellegen, A., & Kaemmer, B. (1989). *Minnesota Multiphasic Personality Inventory-2 (MMPI-2): Manual for administration and scoring.* Minneapolis: University of Minnesota Press. Copyright © 1989 by the University of Minnesota. Reproduced by permission.

False: 9, 210, 290

Sc3—Lack of Ego Mastery, Cognitive
   True: 31, 32, 147, 170, 180, 299, 311,
      316, 325
   False: 165

Sc4—Lack of Ego Mastery, Conative
   True: 31, 38, 48, 65, 92, 233, 234, 273,
      299, 303, 325
   False: 9, 210, 290

Sc5—Lack of Ego Mastery, Defective Inhibition
   True: 23, 85, 168, 182, 218, 242, 274,
      320, 322, 329, 355
   False: None

Sc6—Bizarre Sensory Experiences
   True: 23, 32, 44, 168, 182, 229, 247,
      252, 296, 298, 307, 311, 319, 355

False: 91, 106, 177, 179, 255, 295

SCALE 9—HYPOMANIA

Ma1—Amorality
   True: 131, 227, 248, 250, 269
   False: 263

Ma2—Psychomotor Acceleration
   True: 15, 85, 87, 122, 169, 206, 218,
      242, 244
   False: 100, 106

Ma3—Imperturbability
   True: 155, 200, 220
   False: 93, 136, 158, 167, 243

Ma4—Ego Inflation
   True: 13, 50, 55, 61, 98, 145, 190, 211,
      212
   False: None

Appendix D. Linear T-Score Conversions for the Harris-Lingoes Subscales

| | Men | | | | | | | | | | | | | | | | | | | | | | | | | | | | |
Raw Score	D1	D2	D3	D4	D5	Hy1	Hy2	Hy3	Hy4	Hy5	Pd1	Pd2	Pd3	Pd4	Pd5	Pa1	Pa2	Pa3	Sc1	Sc2	Sc3	Sc4	Sc5	Sc6	Ma1	Ma2	Ma3	Ma4	Raw Score
32	116																												32
31	114																												31
30	111																												30
29	108																												29
28	106																												28
27	103																												27
26	100																												26
25	98																												25
24	95																												24
23	93																												23
22	90																												22
21	87																												21
20	85																		120										20
19	82																		117										19
18	79																		113										18
17	77					120													109					120					17

360

Men

Raw Score	D1	D2	D3	D4	D5	Hy1	Hy2	Hy3	Hy4	Hy5	Pd1	Pd2	Pd3	Pd4	Pd5	Pa1	Pa2	Pa3	Sc1	Sc2	Sc3	Sc4	Sc5	Sc6	Ma1	Ma2	Ma3	Ma4	Raw Score
16	74								115										105					119					16
15	71	98		110				106	111										101					114					15
14	69	92		105				102	106							120			97			114		109					14
13	66	87		101				97	101					99		118			92			109		104					13
12	64	81		96			71	93	96					94	91	112			88			103		99					12
11	61	76	116	91			67	88	91					88	87	106			84			98	117	95		78			11
10	58	70	108	86	96		63	84	86					83	82	100	96		80		103	92	110	90		73			10
9	56	65	100	82	91		59	79	82		98			78	77	94	89	70	76	120	96	87	103	85		68		89	9
8	53	59	91	77	85		55	75	77		91	81		73	72	88	82	65	72	117	90	82	96	80		63	77	82	8
7	50	54	83	72	79		51	70	72	78	84	74		67	67	82	76	60	68	107	84	76	89	75	81	58	71	76	7
6	48	48	75	67	74	61	47	66	67	71	78	68	64	62	63	76	69	56	64	98	78	71	82	70	74	53	65	69	6
5	45	43	67	62	68	56	43	61	62	63	71	61	58	57	58	70	62	51	59	88	72	65	75	65	66	49	59	63	5
4	42	37	59	58	62	51	40	57	57	55	65	55	52	51	53	64	55	46	55	78	66	60	68	60	58	44	53	56	4
3	40	32	51	53	56	45	36	52	52	48	58	48	46	46	48	58	48	41	51	69	60	55	61	55	50	39	47	50	3
2	37	30	43	48	51	40	32	48	48	40	51	42	40	41	43	52	41	36	47	59	54	49	54	51	42	34	41	43	2
1	35		35	43	45	34	30	43	43	33	45	35	35	36	38	46	34	32	43	50	48	44	47	46	35	30	35	37	1
0	32		30	38	40	30		38	38	30	38	30	30	30	34	40	30	30	39	40	42	39	40	41	35	30	30	30	0

Appendix D (*continued*)

Women

Raw Score	D1	D2	D3	D4	D5	Hy1	Hy2	Hy3	Hy4	Hy5	Pd1	Pd2	Pd3	Pd4	Pd5	Pa1	Pa2	Pa3	Sc1	Sc2	Sc3	Sc4	Sc5	Sc6	Ma1	Ma2	Ma3	Ma4	Raw Score
32	108																												**32**
31	105																												**31**
30	103																												**30**
29	101																												**29**
28	98																												**28**
27	96																												**27**
26	94																												**26**
25	91																												**25**
24	89																												**24**
23	86																												**23**
22	84																												**22**
21	82																			119										**21**
20	79																			115										**20**
19	77																			111										**19**
18	75									105										108				120						**18**
17	72																			104				118						**17**

362

Women

Raw Score	D1	D2	D3	D4	D5	Hy1	Hy2	Hy3	Hy4	Hy5	Pd1	Pd2	Pd3	Pd4	Pd5	Pa1	Pa2	Pa3	Sc1	Sc2	Sc3	Sc4	Sc5	Sc6	Ma1	Ma2	Ma3	Ma4	Raw Score
16	70								101										100					113					16
15	67			106				99	97										96					109					15
14	65	95		102				95	93						120				92			111		104					14
13	63	90		97				91	89					97	117				88			106		100					13
12	60	84		93			71	87	85					92	111				84			100		95					12
11	58	79	107	88			67	83	81					87	105				81			95	110	91					11
10	56	73	100	84	89		63	79	77		92			81	99	91			77		104	90	104	86		80		86	10
9	53	68	93	79	83		59	75	73		86	92		76	93	84	84	69	73	120	98	85	97	81		75	82	80	9
8	51	62	85	75	78		55	71	69		80	85		71	87	78	78	65	69	113	92	80	91	77		70	75	74	8
7	48	57	78	70	73		50	67	65	77	74	77		66	81	72	73	60	65	104	86	75	85	72		65	69	68	7
6	46	51	70	66	68	61	46	63	61	70	68	70	65	60	75	65	67	55	61	95	80	70	78	68	75	60	62	62	6
5	44	46	63	61	63	56	42	59	57	62	62	62	59	55	69	59	62	50	57	86	74	65	72	63	69	55	56	56	5
4	41	41	56	57	58	51	38	55	53	54	56	55	54	50	63	53	56	45	53	76	67	59	65	59	63	50	50	49	4
3	39	35	48	52	53	45	34	51	49	46	50	47	48	45	57	46	51	41	50	67	61	54	59	54	57	45	43	43	3
2	37	30	41	48	47	40	30	47	45	39	44	40	42	40	51	40	45	36	46	58	55	49	54	50	50	40	37	37	2
1	34		34	43	42	35		43	41	31			36	35	45	34	40	31	42	49	49	44	50	45	44	35		37	1
0	32		30	38	37	30		39	37	30	38	32	31	30	39	30	34	30	38	40	43	39	46	41	38	30	30	31	0

Source: Butcher, J.N., Dahlstrom, W.G., Graham, J.R., Tellegen, A., & Kaemmer, B. (1989). *Minnesota Multiphasic Personality Inventory–2 (MMPI-2): Manual for administration and scoring.* Minneapolis: University of Minnesota Press. Copyright © 1989 by the University of Minnesota. Reproduced by permission.

363

Appendix E. Composition of the Si Subscales

Si1—Shyness/Self-consciousness
  True: 158, 161, 167, 185, 243, 265, 275, 289
  False: 49, 262, 280, 321, 342, 360
Si2 —Social Avoidance
  True: 337, 367
  False: 86, 340, 353, 359, 363, 370
Si3—Self/Other Alienation
  True: 31, 56, 104, 110, 135, 284, 302, 308, 326, 328, 338, 347, 348, 358, 364, 368, 369
  False: None

Source: Butcher, J.N., Dahlstrom, W.G., Graham, J.R., Tellegen, A., & Kaemmer, B. (1989). *Minnesota Multiphasic Personality Inventory-2 (MMPI-2): Manual for administration and scoring.* Minneapolis: University of Minnesota Press. Copyright © 1989 by the University of Minnesota. Reproduced by permission.

Appendix F. Linear T-Score Conversions for the Si Subscales

Raw Score	Men			Women		
	Si1	Si2	Si3	Si1	Si2	Si3
17			86			83
16			83			80
15			80			77
14	77		77	73		74
13	74		74	71		72
12	71		71	68		69
11	68		68	65		66
10	65		65	63		63
9	62		62	60		60
8	59	71	59	57	74	58
7	56	67	56	55	69	55
6	53	62	53	52	65	52
5	51	58	50	49	60	49
4	48	54	47	46	56	47
3	45	49	44	44	51	44
2	42	45	41	41	47	41
1	39	41	38	38	42	38
0	36	37		36	37	

Source: Butcher, J.N., Dahlstrom, W.G., Graham, J.R., Tellegen, A., & Kaemmer, B. (1989). *Minnesota Multiphasic Personality Inventory-2 (MMPI-2): Manual for administration and scoring.* Minneapolis: University of Minnesota Press. Copyright © 1989 by the University of Minnesota. Reproduced by permission.

Appendix G. Composition of the Content Scales

ANX—Anxiety
True: 15, 30, 31, 39, 170, 196, 273,
290, 299, 301, 305, 339, 408,
415, 463, 469, 509, 556
False: 140, 208, 223, 405, 496
FRS—Fears
True: 154, 317, 322, 329, 334, 392,
395, 397, 435, 438, 441, 447,
458, 468, 471, 555
False: 115, 163, 186, 385, 401, 453,
462
OBS—Obsessiveness
True: 55, 87, 135, 196, 309, 313, 327,
328, 394, 442, 482, 491, 497,
509, 547, 553
False: None
DEP—Depression
True: 38, 52, 56, 65, 71, 82, 92, 130,
146, 215, 234, 246, 277, 303,
306, 331, 377, 399, 400, 411,
454, 506, 512, 516, 520, 539,
546, 554
False: 3, 9, 75, 95, 388
HEA—Health Concerns
True: 11, 18, 28, 36, 40, 44, 53, 59, 97,
101, 111, 149, 175, 247
False: 20, 33, 45, 47, 57, 91, 117, 118,
141, 142, 159, 164, 176, 179,
181, 194, 204, 224, 249, 255,
295, 404
BIZ—Bizarre Mentation
True: 24, 32, 60, 96, 138, 162, 198,
228, 259, 298, 311, 316, 319,
333, 336, 355, 361, 466, 490,
508, 543, 551
False: 427
ANG—Anger
True: 29, 37, 116, 134, 302, 389, 410,
414, 430, 461, 486, 513, 540,
542, 548

False: 564
CYN—Cynicism
True: 50, 58, 76, 81, 104, 110, 124,
225, 241, 254, 283, 284, 286,
315, 346, 352, 358, 374, 399,
403, 445, 470, 538
False: None
ASP—Antisocial Practices
True: 26, 35, 66, 81, 84, 104, 105, 110,
123, 227, 240, 248, 250, 254,
269, 283, 284, 374, 412, 418,
419
False: 266
TPA—Type A
True: 27, 136, 151, 212, 302, 358,
414, 419, 420, 423, 430, 437,
507, 510, 523, 531, 535, 541, 545
False: None
LSE—Low Self-esteem
True: 70, 73, 130, 235, 326, 369, 376,
380, 411, 421, 450, 457, 475,
476, 483, 485, 503, 504, 519,
526, 562
False: 61, 78, 109
SOD—Social Discomfort
True: 46, 158, 167, 185, 265, 275,
281, 337, 349, 367, 479, 480,
515
False: 49, 86, 262, 280, 321, 340, 353,
359, 360, 363, 370
FAM—Family Problems
True: 21, 54, 145, 190, 195, 205, 256,
292, 300, 323, 378, 379, 382,
413, 449, 478, 543, 550, 563,
567
False: 83, 125, 217, 383, 455
WRK—Work Interference
True: 15, 17, 31, 54, 73, 98, 135, 233,
243, 299, 302, 339, 364, 368,
394, 409, 428, 445, 464, 491,

        505, 509, 517, 525, 545, 554,
        559, 566
False: 10, 108, 318, 521, 561
TRT—Negative Treatment Indicators
    True: 22, 92, 274, 306, 364, 368, 373,

        375, 376, 377, 391, 399, 482,
        488, 491, 495, 497, 499, 500,
        504, 528, 539, 554
False: 493, 494, 501

Appendix H. Uniform T-Score Conversions for the Content Scales

	Men														
Raw Score	ANX	FRS	OBS	DEP	HEA	BIZ	ANG	CYN	ASP	TPA	LSE	SOD	FAM	WRK	TRT
36					112										
35					110										
34					108										
33				100	106									98	
32				99	105									96	
31				97	103									94	
30				95	101									92	
29				94	99									90	
28				92	97									89	
27				90	95									87	
26				88	93									85	104
25				87	91								105	83	101
24				85	89						101	89	102	81	99
23	92	113		83	87	120		83			98	86	99	79	96
22	90	110		82	85	119		80	94		96	84	97	78	94
21	87	107		80	83	115		77	90		93	81	94	76	91
20	85	103		78	81	112		74	87		91	78	91	74	89
19	82	100		77	80	108		71	83	89	88	76	88	72	86
18	80	97		75	78	105		68	79	85	85	73	85	70	84
17	77	93		73	76	101		65	76	81	83	71	82	68	81
16	75	90	87	71	74	98	86	62	72	77	80	68	80	67	79
15	72	87	84	70	72	94	82	59	69	72	77	65	77	65	76
14	70	84	80	68	70	91	78	56	65	68	75	63	74	63	74
13	67	80	77	66	68	88	74	54	62	64	72	60	71	61	71
12	65	77	73	65	66	84	70	52	58	60	70	58	68	59	69
11	62	74	70	63	64	81	67	51	55	56	67	55	66	57	66
10	60	70	66	61	62	77	63	49	53	53	64	54	63	56	64
9	57	67	63	59	60	74	59	48	51	50	62	52	60	54	61
8	55	64	59	58	58	70	56	47	49	48	59	50	57	52	59
7	53	60	56	56	56	67	53	46	47	46	57	49	55	50	56
6	52	57	53	55	53	63	50	44	46	44	55	47	52	48	54
5	50	54	50	53	51	60	48	43	44	43	53	45	50	46	52
4	47	51	47	51	48	57	46	41	42	41	51	43	47	44	49
3	45	48	44	48	44	54	43	40	40	38	48	41	44	41	47
2	42	45	41	45	41	51	40	38	37	36	45	39	41	39	43
1	39	41	37	41	37	46	36	35	34	32	41	35	37	36	39
0	35	35	33	36	33	39	32	32	30	30	35	32	33	33	35

							Women									Raw Score
ANX	FRS	OBS	DEP	HEA	BIZ	ANG	CYN	ASP	TPA	LSE	SOD	FAM	WRK	TRT		
				107												**36**
				105												**35**
				103												**34**
			97	101									99			**33**
			95	100									97			**32**
			93	98									95			**31**
			92	96									92			**30**
			90	94									90			**29**
			88	92									88			**28**
			87	90									86			**27**
			85	89									84	102		**26**
			83	87								99	82	100		**25**
			82	85						97	87	96	80	97		**24**
89	101		80	83	113		83			94	84	94	78	95		**23**
86	98		78	81	110		80	98		92	82	91	76	92		**22**
84	94		77	79	108		77	94		89	80	89	73	89		**21**
81	91		75	77	105		75	91		86	77	86	71	87		**20**
79	88		73	76	102		72	88	94	84	75	83	69	84		**19**
76	85		72	74	99		69	85	90	81	72	81	67	82		**18**
74	81		70	72	96		67	82	86	78	70	78	65	79		**17**
71	78	87	68	70	93	88	64	79	81	76	68	75	63	77		**16**
69	75	83	67	68	90	84	61	75	77	73	65	73	61	74		**15**
66	72	79	65	66	87	80	58	72	73	70	63	70	59	72		**14**
64	68	75	63	64	84	76	56	69	69	68	60	68	57	69		**13**
61	65	71	62	63	81	72	54	66	64	65	58	65	55	67		**12**
59	62	67	60	61	79	68	53	63	60	62	56	62	54	64		**11**
56	59	63	58	59	76	64	51	59	56	60	54	60	52	61		**10**
55	56	59	57	57	73	60	50	56	53	57	52	57	51	59		**9**
53	53	56	55	55	70	56	48	54	50	55	51	55	50	57		**8**
51	51	53	54	53	67	53	47	52	48	54	49	52	48	55		**7**
49	48	50	52	51	64	50	46	49	45	52	48	50	46	53		**6**
47	46	48	50	49	61	47	44	47	43	51	46	47	45	51		**5**
45	43	46	48	46	58	45	42	45	41	49	44	45	43	49		**4**
43	41	44	45	43	56	42	40	42	38	47	41	42	40	46		**3**
40	38	41	42	40	52	39	38	39	36	44	39	39	37	43		**2**
37	35	37	39	36	47	36	35	36	33	40	35	36	34	39		**1**
34	31	32	34	32	39	31	32	33	30	35	32	32	31	35		**0**

Source: Butcher, J.N., Dahlstrom, W.G., Graham, J.R., Tellegen, A., & Kaemmer, B. (1989). *Minnesota Multiphasic Personality Inventory-2 (MMPI-2): Manual for administration and scoring.* Minneapolis: University of Minnesota Press. Copyright © 1989 by the University of Minnesota Press. Reproduced by permission.

Appendix I. Critical-Item Lists

KOSS-BUTCHER CRITICAL ITEMS

Acute Anxiety State

2. I have a good appetite. (F)
3. I wake up fresh and rested most mornings. (F)
5. I am easily awakened by noise. (T)
10. I am about as able to work as I ever was. (F)
15. I work under a great deal of tension. (T)
28. I am bothered by an upset stomach several times a week. (T)
39. My sleep is fitful and disturbed. (T)
59. I am troubled by discomfort in the pit of my stomach every few days or oftener. (T)
140. Most nights I go to sleep without thoughts or ideas bothering me. (F)
172. I frequently notice my hand shakes when I try to do something. (T)
208. I hardly ever notice my heart pounding and I am seldom short of breath. (F)
218. I have periods of such great restlessness that I cannot sit long in a chair. (T)
223. I believe I am no more nervous than most others. (F)
301. I feel anxiety about something or someone almost all the time. (T)
444. I am a high-strung person. (T)
463. Several times a week I feel as if something dreadful is about to happen. (T)
469. I sometimes feel that I am about to go to pieces. (T)

Depressed Suicidal Ideation

9. My daily life is full of things that keep me interested. (F)
38. I have had periods of days, weeks, or months when I couldn't take care of things because I couldn't "get going." (T)
65. Most of the time I feel blue. (T)
71. These days I find it hard not to give up hope of amounting to something. (T)
75. I usually feel that life is worthwhile. (F)
92. I don't seem to care what happens to me. (T)
95. I am happy most of the time. (F)
130. I certainly feel useless at times. (T)
146. I cry easily. (T)
215. I brood a great deal. (T)
233. I have difficulty in starting to do things. (T)
273. Life is a strain for me much of the time. (T)
303. Most of the time I wish I were dead. (T)
306. No one cares much what happens to you. (T)
388. I very seldom have spells of the blues. (F)
411. At times I think I am no good at all. (T)
454. The future seems hopeless to me. (T)
485. I often feel that I'm not as good as other people. (T)
506. I have recently considered killing myself. (T)
518. I have made lots of bad mistakes in my life. (T)
520. Lately I have thought a lot about killing myself. (T)
524. No one knows it but I have tried to kill myself. (T)

Threatened Assault

37. At times I feel like smashing things. (T)
85. At times I have a strong urge to do

something harmful or shocking. (T)

134. At times I feel like picking a fist fight with someone. (T)
213. I get mad easily and then get over it soon. (T)
389. I am often said to be hotheaded. (T)

Situational Stress Due to Alcoholism

125. I believe that my home life is as pleasant as that of most people I know. (F)
264. I have used alcohol excessively. (T)
487. I have enjoyed using marijuana. (T)
489. I have a drug or alcohol problem. (T)
502. I have some habits that are really harmful. (T)
511. Once a week or more I get high or drunk. (T)
518. I have made lots of bad mistakes in my life. (T)

Mental Confusion

24. Evil spirits possess me at times. (T)
31. I find it hard to keep my mind on a task or job. (T)
32. I have had very peculiar and strange experiences. (T)
72. My soul sometimes leaves my body. (T)
96. I see things or animals or people around me that others do not see. (T)
180. There is something wrong with my mind. (T)
198. I often hear voices without knowing where they come from. (T)
299. I cannot keep my mind on one thing. (T)
311. I often feel as if things are not real. (T)
316. I have strange and peculiar thoughts. (T)
325. I have more trouble concentrating than others seem to have. (T)

Persecutory Ideas

17. I am sure I get a raw deal from life. (T)
42. If people had not had it in for me, I would have been much more successful. (T)
99. Someone has it in for me. (T)
124. I often wonder what hidden reason another person may have for doing something nice for me. (T)
138. I believe I am being plotted against. (T)
144. I believe I am being followed. (T)
145. I feel that I have often been punished without cause. (T)
162. Someone has been trying to poison me. (T)
216. Someone has been trying to rob me. (T)
228. There are persons who are trying to steal my thoughts and ideas. (T)
241. It is safer to trust nobody. (T)
251. I have often felt that strangers were looking at me critically. (T)
259. I am sure I am being talked about. (T)
314. I have no enemies who really wish to harm me. (F)
333. People say insulting and vulgar things about me. (T)
361. Someone has been trying to influence my mind. (T)

LACHAR-WROBEL CRITICAL ITEMS

Anxiety and Tension

15. I work under a great deal of tension. (T)
17. I am sure I get a raw deal from life. (T)
172. I frequently notice my hand shakes when I try to do something. (T)
218. I have periods of such great restlessness that I cannot sit long in a chair. (T)

Appendix I *(continued)*

223. I believe I am no more nervous than most others. (F)
261. I have very few fears compared to my friends. (F)
299. I cannot keep my mind on one thing. (T)
301. I feel anxiety about something or someone almost all the time. (T)
320. I have been afraid of things or people that I knew could not hurt me. (T)
405. I am usually calm and not easily upset. (F)
463. Several times a week I feel as if something dreadful is about to happen. (T)

Depression and Worry

2. I have a good appetite. (F)
3. I wake up fresh and rested most mornings. (F)
10. I am about as able to work as I ever was. (F)
65. Most of the time I feel blue. (T)
73. I am certainly lacking in self-confidence. (T)
75. I usually feel that life is worthwhile. (F)
130. I certainly feel useless at times. (T)
150. Sometimes I feel as if I must injure either myself or someone else. (T)
165. My memory seems to be all right. (F)
180. There is something wrong with my mind. (T)
273. Life is a strain for me much of the time. (T)
303. Most of the time I wish I were dead. (T)
339. I have sometimes felt that difficulties were piling up so high that I could not overcome them. (T)
411. At times I think I am no good at all. (T)

415. I worry quite a bit over possible misfortunes. (T)
454. The future seems hopeless to me. (T)

Sleep Disturbance

5. I am easily awakened by noise. (T)
30. I have nightmares every few nights. (T)
39. My sleep is fitful and disturbed. (T)
140. Most nights I go to sleep without thoughts or ideas bothering me. (F)
328. Sometimes some unimportant thought will run through my mind and bother me for days. (T)
471. I have often been frightened in the middle of the night. (T)

Deviant Beliefs

42. If people had not had it in for me I would have been much more successful. (T)
99. Someone has it in for me. (T)
106. My speech is the same as always (not faster or slower, no slurring or hoarseness). (F)
138. I believe I am being plotted against. (T)
144. I believe I am being followed. (T)
162. Someone has been trying to poison me. (T)
216. Someone has been trying to rob me. (T)
228. There are persons who are trying to steal my thoughts and ideas. (T)
259. I am sure I am being talked about. (T)
314. I have no enemies who really wish to harm me. (F)
333. People say insulting and vulgar things about me. (T)
336. Someone has control over my mind. (T)

355. At one or more times in my life I felt that someone was making me do things by hypnotizing me. (T)
361. Someone has been trying to influence my mind. (T)
466. Sometimes I am sure that other people can tell what I am thinking. (T)

Deviant Thinking and Experiences

32. I have had very peculiar and strange experiences. (T)
60. When I am with people, I am bothered by hearing very strange things. (T)
96. I see things or animals or people around me that others do not see. (T)
122. At times my thoughts have raced ahead faster than I could speak them. (T)
198. I often hear voices without knowing where they come from. (T)
298. Peculiar odors come to me at times. (T)
307. At times I hear so well it bothers me. (T)
316. I have strange and peculiar thoughts. (T)
319. I hear strange things when I am alone. (T)
427. I have never seen a vision. (F)

Substance Abuse

168. I have had periods in which I carried on activities without knowing later what I had been doing. (T)
264. I have used alcohol excessively. (T)
429. Except by doctor's orders I never take drugs or sleeping pills. (F)

Antisocial Attitude

27. When people do me a wrong, I feel I should pay them back if I can, just for the principle of the thing. (T)
35. Sometimes when I was young I stole things. (T)

84. I was suspended from school one or more times for bad behavior. (T)
105. In school I was sometimes sent to the principal for bad behavior. (T)
227. I don't blame people for trying to grab everything they can get in this world. (T)
240. At times it has been impossible for me to keep from stealing or shoplifting something. (T)
254. Most people make friends because friends are likely to be useful to them. (T)
266. I have never been in trouble with the law. (F)
324. I can easily make other people afraid of me, and sometimes do for the fun of it. (T)

Family Conflict

21. At times I have very much wanted to leave home. (T)
83. I have very few quarrels with members of my family. (F)
125. I believe that my home life is as pleasant as that of most people I know. (F)
288. My parents and family find more fault with me than they should. (T)

Problematic Anger

85. At times I have a strong urge to do something harmful or shocking. (T)
134. At times I feel like picking a fist fight with someone. (T)
213. I get mad easily and then get over it soon. (T)
389. I am often said to be hotheaded. (T)

Sexual Concern and Deviation

12. My sex life is satisfactory. (F)
34. I have never been in trouble because of my sex behavior. (F)
62. I have often wished I were a girl. (or if you are a girl) I have never been sorry that I am a girl. (T/F)

Appendix I *(continued)*

121. I have never indulged in any unusual sex practices. (F)
166. I am worried about sex. (T)
268. I wish I were not bothered by thoughts about sex. (T)

Somatic Symptoms

18. I am troubled by attacks of nausea and vomiting. (T)
28. I am bothered by an upset stomach several times a week. (T)
33. I seldom worry about my health. (F)
40. Much of the time my head seems to hurt all over. (T)
44. Once a week or oftener I suddenly feel hot all over, for no real reason. (T)
47. I am almost never bothered by pains over my heart or in my chest. (F)
53. Parts of my body often have feelings like burning, tingling, crawling, or like "going to sleep." (T)
57. I hardly ever feel pain in the back of my neck. (F)
59. I am troubled by discomfort in the pit of my stomach every few days or oftener. (T)
101. Often I feel as if there is a tight band around my head. (T)

111. I have a great deal of stomach trouble. (T)
142. I have never had a fit or convulsion. (F)
159. I have never had a fainting spell. (F)
164. I seldom or never have dizzy spells. (F)
175. I feel weak all over much of the time. (T)
176. I have very few headaches. (F)
182. I have had attacks in which I could not control my movements or speech but in which I knew what was going on around me. (T)
224. I have few or no pains. (F)
229. I have had blank spells in which my activities were interrupted and I did not know what was going on around me. (T)
247. I have numbness in one or more places on my skin. (T)
255. I do not often notice my ears ringing or buzzing. (F)
295. I have never been paralyzed or had any unusual weakness of any of my muscles. (F)
464. I feel tired a good deal of the time. (T)

Appendix J. Composition of the Supplementary Scales

A Scale—Anxiety

True: 31, 38, 56, 65, 82, 127, 135, 215, 233,
243, 251, 273, 277, 289, 301, 309,
310, 311, 325, 328, 338, 339, 341,
347, 390, 391, 394, 400, 408, 411,
415, 421, 428, 442, 448, 451, 464,
469

False: 388

R Scale—Repression

True: None

False: 1, 7, 10, 14, 37, 45, 69, 112, 118, 120,
128, 134, 142, 168, 178, 189, 197,
199, 248, 255, 256, 297, 330, 346,
350, 353, 354, 359, 363, 365, 422,
423, 430, 432, 449, 456, 465

Es Scale—Ego Strength

True: 2, 33, 45, 98, 141, 159, 169, 177, 179,
189, 199, 209, 213, 230, 245, 323,
385, 406, 413, 425

False: 23, 31, 32, 36, 39, 53, 60, 70, 82, 87,
119, 128, 175, 196, 215, 221, 225,
229, 236, 246, 307, 310, 316, 328,
391, 394, 441, 447, 458, 464, 469,
471

MAC-R Scale—MacAndrew Alcoholism Scale-Revised

True: 7, 24, 36, 49, 52, 69, 72, 82, 84, 103,
105, 113, 115, 128, 168, 172, 202,
214, 224, 229, 238, 257, 280, 342,
344, 407, 412, 414, 422, 434, 439,
445, 456, 473, 502, 506, 549

False: 73, 107, 117, 137, 160, 166, 251, 266,
287, 299, 325, 387

Fb Scale—Back F

True: 281, 291, 303, 311, 317, 319, 322,
323, 329, 332, 333, 334, 387, 395,
407, 431, 450, 454, 463, 468, 476,
478, 484, 489, 506, 516, 517, 520,
524, 525, 526, 528, 530, 539, 540,
544, 555

False: 383, 404, 501

TRIN—True Response Inconsistency

3 T– 39 T	99 T– 314 T	125 F– 195 F
12 T– 166 T	125 T– 195 T	140 F– 196 F
40 T– 176 T	209 T– 351 T	152 F– 464 F
48 T– 184 T	359 T– 367 T	165 F– 565 F
63 T– 127 T	377 T– 534 T	262 F– 275 F
65 T– 95 T	556 T– 560 T	265 F– 360 F
73 T– 239 T	9 F– 56 F	359 F– 367 F
83 T– 288 T	65 F– 95 F	

VRIN—Variable Response Inconsistency

3 T– 39 T	125 T– 195 T	349 T– 515 F
6 T– 90 F	125 F– 195 F	349 F– 515 T
6 F– 90 T	135 F– 482 T	350 F– 521 T
9 F– 56 F	136 T– 507 F	353 T– 370 F
28 T– 59 F	136 F– 507 T	353 F– 370 T
31 T– 299 F	152 F– 464 F	364 F– 554 T
32 F– 316 T	161 T– 185 F	369 F– 421 T
40 T– 176 T	161 F– 185 T	372 T– 405 F
46 T– 265 F	165 F– 565 F	372 F– 405 T
48 T– 184 T	166 T– 268 F	380 T– 562 F
49 T– 280 F	166 F– 268 T	395 T– 435 F
73 T– 377 F	167 T– 243 F	395 F– 435 T
81 T– 284 F	167 F– 243 T	396 F– 403 F
81 F– 284 T	196 F– 415 T	396 T– 403 T
83 T– 288 T	199 T– 467 F	411 T– 485 F
84 T– 105 F	199 F– 467 T	411 F– 485 T
86 T– 359 F	226 T– 267 F	472 T– 533 F
95 F– 388 T	259 F– 333 T	472 F– 533 T
99 F– 138 T	262 F– 275 F	491 F– 509 F
103 T– 344 F	290 T– 556 F	506 T– 520 F
110 T– 374 F	290 F– 556 T	506 F– 520 T
110 F– 374 T	339 F– 394 T	513 T– 542 F
116 T– 430 F		

O-H Scale—Overcontrolled Hostility

True: 67, 79, 207, 286, 305, 398, 471

False: 1, 15, 29, 69, 77, 89, 98, 116, 117,
129, 153, 169, 171, 293, 344, 390,
400, 420, 433, 440, 460

Do Scale—Dominance

True: 55, 207, 232, 245, 386, 416

Source: Butcher, J.N., Dahlstrom, W.G., Graham, J.R., Tellegen, A., & Kaemmer, B. (1989). *Minnesota Multiphasic Personality Inventory-2 (MMPI-2): Manual for administration and scoring.* Minneapolis: University of Minnesota Press. Copyright © 1989 by the University of Minnesota. Reproduced by permission.

Appendix J *(continued)*

False: 31, 52, 70, 73, 82, 172, 201, 202,
         220, 227, 243, 244, 275, 309, 325,
         399, 412, 470, 473

Re Scale—Social Responsibility

True: 100, 160, 199, 266, 440, 467

False: 7, 27, 29, 32, 84, 103, 105, 145, 164,
         169, 201, 202, 235, 275, 358, 412,
         417, 418, 430, 431, 432, 456, 468,
         470

Mt Scale—College Maladjustment

True: 15, 16, 28, 31, 38, 71, 73, 81, 82,
         110, 130, 215, 218, 233, 269, 273,
         299, 302, 325, 331, 339, 357, 408,
         411, 449, 464, 469, 472

False: 2, 3, 9, 10, 20, 43, 95, 131, 140, 148,
         152, 223, 405

GM Scale—Masculine Gender Role

True: 8, 20, 143, 152, 159, 163, 176, 199,
         214, 237, 321, 331, 350, 385, 388,
         401, 440, 462, 467, 474

False: 4, 23, 44, 64, 70, 73, 74, 80, 100,
         137, 146, 187, 289, 351, 364, 392,
         395, 435, 438, 441, 469, 471, 498,
         509, 519, 532, 536

GF Scale—Feminine Gender Role

True: 62, 67, 119, 121, 128, 263, 266, 353,

         384, 426, 449, 456, 473, 552

False: 1, 27, 63, 68, 79, 84, 105, 123, 133,
         155, 197, 201, 203, 220, 231, 238,
         239, 250, 257, 264, 272, 287, 406,
         417, 465, 477, 487, 510, 511, 537,
         548, 550

PK Scale—Post-traumatic Stress Disorder-Keane

True: 16, 17, 22, 23, 30, 31, 32, 37, 39, 48,
         52, 56, 59, 65, 82, 85, 92, 94, 101,
         135, 150, 168, 170, 196, 221, 274,
         277, 302, 303, 305, 316, 319, 327,
         328, 339, 347, 349, 367

False: 2, 3, 9, 49, 75, 95, 125, 140

PS Scale—Post-traumatic Stress Disorder-Schlenger

True: 17, 21, 22, 31, 32, 37, 38, 44, 48, 56,
         59, 65, 85, 94, 116, 135, 145, 150,
         168, 170, 180, 218, 221, 273, 274,
         277, 299, 301, 304, 305, 311, 316,
         319, 325, 328, 377, 386, 400, 463,
         464, 469, 471, 475, 479, 515, 516,
         565

False: 3, 9, 45, 75, 95, 141, 165, 208, 223,
         280, 372, 405, 564

Appendix K. Linear T-Score Conversions for the Supplementary Scales

Men

Raw Score	A	R	Es	MAC-R	Fb	TRIN	VRIN	O-H	Do	Re	Mt	GM	GF	PK	PS
60															112
59															111
58															110
57															108
56															107
55															106
54															104
53															103
52			83												102
51			81												101
50			78												99
49			76	113											98
48			74	111											97
47			72	109								71			96
46			69	106								69	90	113	94
45			67	104								66	88	112	93
44			65	102								64	85	110	92
43			63	99								62	83	108	91
42			60	97								60	81	107	89
41			58	95							96	58	79	105	88
40			56	92							95	56	77	103	87
39	91		54	90							93	53	75	102	86
38	89		51	88							91	51	73	100	84
37	88	98	49	85							90	49	71	98	83
36	87	96	47	83							88	47	68	97	82
35	85	94	45	81							87	45	66	95	81
34	84	92	42	78							85	42	64	93	79
33	82	89	40	76							84	40	62	92	78
32	81	87	38	74							82	38	60	90	77
31	80	85	36	72							81	36	58	88	76

Women

Raw Score	A	R	Es	MAC-R	Fb	TRIN	VRIN	O-H	Do	Re	Mt	GM	GF	PK	PS
60															104
59															103
58															102
57															100
56															99
55															98
54															97
53															96
52			86												95
51			84												94
50			82												93
49			80												92
48			78												90
47			76												89
46			74	120								78	74	107	88
45			72	119								76	71	106	87
44			70	116								75	69	104	86
43			68	114								73	66	103	85
42			66	111								71	63	101	84
41			64	108							91	70	61	100	83
40			61	105							90	68	58	98	81
39	85		59	103							88	67	56	96	80
38	83		57	100							87	65	53	95	79
37	82	104	55	97							85	63	50	93	78
36	81	102	53	94							84	62	48	92	77
35	80	99	51	92							82	60	45	90	76
34	78	96	49	89							81	58	43	89	75
33	77	94	47	86							80	57	40	87	74
32	76	91	45	84							78	55	37	86	73
31	75	88	43	81							77	54	35	84	71

Appendix K (continued)

Men

Raw Score	A	R	Es	MAC-R	Fb	TRIN	VRIN	O-H	Do	Re	Mt	GM	GF	PK	PS
30	78	83	34	69						76	79	34	56	87	74
29	77	81	31	67						73	77	31	54	85	73
28	75	78	30	65					103	70	76	30	51	83	72
27	74	76		62					99	68	74		49	82	71
26	73	74		60					96	65	73		47	80	69
25	71	72		58					93	78	63	71		45	78	68
24	70	69		55				120	89	75	60	70		43	77	67
23	68	67		53				118	86	72	57	68		41	75	66
22	67	65		51				115	82	68	55	67		39	73	64
21	65	63		48				111	79	65	52	65		37	72	63
20	64	61		46				107	76	61	50	64		34	70	62
19	63	58		44	120	120T	103	72	58	47	62		32	68	61
18	61	56		41	116	114T	99	69	55	45	60		30	67	59
17	60	54		39	112	107T	96	65	51	42	59			65	58
16	58	52		37	108	100T	92	62	48	39	57			63	57

Women

Raw Score	A	R	Es	MAC-R	Fb	TRIN	VRIN	O-H	Do	Re	Mt	GM	GF	PK	PS
30	73	86	41	78						77	75	52	32	83	70
29	72	83	39	75						74	74	50	30	81	69
28	71	81	37	73					103	71	72	49		80	68
27	69	78	35	70					99	68	71	47		78	67
26	68	75	33	67					96	65	70	45		77	66
25	67	73	31	64					92	80	62	68	44		75	63
24	66	70	30	62					88	77	59	67	42		74	64
23	64	67		59				120	85	73	56	65	41		72	62
22	63	65		56				118	81	70	53	64	39		71	61
21	62	62		53				114	77	66	50	62	37		69	60
20	61	60		51	120		110	74	63	47	61	36		68	59
19	59	57		48	116	120T	106	70	59	44	60	34		66	58
18	58	54		45	112	118T	102	66	56	41	58	32		64	57
17	57	52		42	108	111T	98	63	53	38	57	31		63	56
16	56	49		40	105	103T	94	59	49	35	55			61	55

Men

Raw Score	A	R	Es	MAC-R	Fb	TRIN	VRIN	O-H	Do	Re	Mt	GM	GF	PK	PS
15	57	50		34	104	93T	88	59	45	37	56			62	56
14	56	47		32	100	86T	84	55	41	34	54			60	54
13	54	45		30	96	79T	80	52	38	32	53			58	53
12	53	43			92	72T	76	48	34	30	51			57	52
11	51	41			87	65T	73	45	31		50			55	51
10	50	39			83	57T	69	41	30		48			53	49
9	49	36			79	50	65	38			46			52	48
8	47	34			75	57F	61	35			45			50	47
7	46	32			71	64F	57	31			43			48	46
6	44	30			67	71F	54	30			42			47	44
5	43				63	78F	50				40			45	43
4	42				59	85F	46				39			43	42
3	40				55	92F	42				37			42	42
2	39				51	99F	38				36			40	39
1	37				46	107F	34				34			38	38
0	36				42	114F	31				32			37	37

Women

Raw Score	A	R	Es	MAC-R	Fb	TRIN	VRIN	O-H	Do	Re	Mt	GM	GF	PK	PS
15	54	46		37	101	95T	90	55	46	32	54			60	54
14	53	44		34	97	88T	86	52	42	30	52			58	52
13	52	41		31	93	80T	82	48	39		51			57	51
12	50	39		30	89	73T	78	44	35		50			55	50
11	49	36			85	65T	74	41	32		48			54	49
10	48	33			81	58T	70	37	30		47			52	48
9	47	31			77	50	66	33			45			51	47
8	45	30			74	58F	62	30			44			49	46
7	44				70	65F	58				42			48	45
6	43				66	73F	54				41			46	43
5	42				62	80F	50				40			45	42
4	40				58	88F	46				38			43	41
3	39				54	95F	42				37			42	40
2	38				50	103F	38				35			40	39
1	37				46	111F	34				34			39	38
0	35				42	118F	30				32			37	37

Source: Butcher, J.N., Dahlstrom, W.G., Graham, J.R., Tellegen, A., & Kaemmer, B. (1989). *Minnesota Multiphasic Personality Inventory-2 (MMPI-2): Manual for administration and scoring.* Minneapolis: University of Minnesota Press. Copyright © 1989 by the University of Minnesota. Reproduced by permission.

379

Appendix L. Intercorrelations of the MMPI-2 Scales for the Normative Sample

Scale							Scale						
	L	F	K	Hs	D	Hy	Pd	Mf	Pa	Pt	Sc	Ma	Si
L		-04	37	-05	09	15	-19	-18	-04	-31	-28	-19	-08
F	-09		-36	46	36	07	52	13	29	55	69	32	35
K	28	-40		-33	-10	44	-21	-02	-01	-68	-60	-35	-43
Hs	-06	41	-45		53	41	33	04	23	53	56	15	34
D	00	34	-29	56		35	34	18	26	47	39	-21	51
Hy	10	10	24	53	35		25	22	32	-05	01	-09	-20
Pd	-19	52	-28	36	37	26		23	41	46	55	36	11
Mf	-11	13	-03	01	12	10	01		13	22	24	07	10
Pa	-07	36	-15	24	31	22	41	29		34	39	15	05
Pt	-28	55	-71	59	61	09	51	09	43		84	33	54
Sc	-25	71	-62	60	48	15	64	-02	47	84		46	44
Ma	-17	38	-36	25	-07	01	42	-06	21	37	51		-22
Si	-03	32	-51	36	59	-13	14	09	15	58	44	-17	

Note: Correlations for men are above diagonal. Correlations for women are below diagonal. Decimal points are omitted.

Source: Butcher, J.N., Dahlstrom, W.G., Graham, J.R., Tellegen, A., & Kaemmer, B. (1989). *Minnesota Multiphasic Personality Inventory-2 (MMPI-2): Manual for administration and scoring.* Minneapolis: University of Minnesota Press. Copyright © 1989 by the University of Minnesota. Reproduced by permission.

Appendix M. Percentile Equivalents for Uniform T Scores

Uniform T Score	Percentile Equivalent
30	<1
35	4
40	15
45	34
50	55
55	73
60	85
65	92
70	96
75	98
80	>99

Source: Tellegen, A., & Ben-Porath, Y.S. (1992). The new uniform T scores for the MMPI-2: Rationale, derivation and appraisal. *Psychological Assessment, 4,* 145–155. Copyright © 1992 by the American Psychological Association. Reproduced by permission.

Author Index

Adler, T., 168
Aldwin, C.M., 144, 193, 194, 195
Alker, H.A., 168
Almagor, M., 150, 151
Altman, H., 37, 80, 83, 183, 197
Ancil, R.J., 268
Anderson, B.N., 278
Anderson, E.D., 162
Anderson, G.L., 295
Anderson, R.W., 215
Anderson, W., 83
Andrews, R.H., 162
Anthony, N., 166
Apfeldorf, M., 142, 143
Archer, R.A., 143
Archer, R.P., 8, 12, 15, 168, 182, 193, 291, 292, 293, 294, 295, 296, 297, 299, 300, 301, 302, 303, 313, 317
Arnold, B.R., 200
Arnold, L.S., 49
Arrendondo, R., 146, 147, 148, 149
Arthur, G., 201
Atkins, H.G., 196, 200
Atkinson, L., 182
Avery, R.D., 215

Bacon, S.F., 267, 272
Baer, R.A., 38, 46, 49, 53, 166
Bailey, D.Q., 135, 136, 180, 182
Bailey, J.M., 210
Baillargeon, J., 278
Ball, B., 196
Ball, H., 53
Ball, J.C., 196
Barefoot, J.C., 31, 209
Barker, H.R., 180
Barron, F., 137, 138, 139
Barth, J.T., 207
Bassett, G.R., 215
Bauer, B., 83

Bell, W.E., 162
Ben-Porath, Y.S., 11, 12, 15, 82, 118, 119, 120, 122, 123, 124, 145, 146, 147, 148, 149, 166, 169, 170, 173, 176, 177, 179, 184, 186, 187, 188, 193, 194, 195, 198, 205, 215, 267, 268, 271, 291, 292, 293, 294, 296, 297, 298, 299, 300, 301, 302, 303, 307, 313, 381
Berk, E., 162
Berman, W., 292
Bernstein, I.H., 214
Berry, D.T.R., 38, 46, 49, 53, 166
Beutler, L.E., 214
Bitman, D., 163
Black, J., 162
Black, J.D., 80, 182, 183
Blake, D.D., 163
Bleecker, M.L., 195
Block, J., 134, 135, 136, 179, 181, 182
Blouin, D., 212
Blount, J., 162
Bloxom, A.L., 214
Blumenthal, J.A., 209
Boerger, A.R., 182, 183
Bohn, M.J., 211
Bolla-Wilson, K., 195
Boomer, D.S., 135
Bosse, R., 144, 193, 194, 195
Bosshardt, M.J., 215
Boudewyns, P.A., 162, 163
Boutilier, L.R., 201
Bowman, E., 30, 174
Bownas, D.A., 215
Bozlee, S., 200, 201
Brandsma, J.M., 210
Brannick, T.L., 203
Brantner, J.P., 83
Braswell, L., 197, 202
Brayfield, A.H., 216
Brems, C., 197

Subject Index